The Irvine Ranch

The Irvine Ranch: Fulfilling the Vision, 1977–2025 continues the story of the planning and development of the Irvine Ranch under a partnership of high-powered businessmen to the single ownership of Donald Bren.

The Irvine Ranch transformed from an agricultural empire into the most successful New Town in the United States in less than two generations. Since 1977, some of the most innovative and valuable master-planned residential, commercial, and retail developments have occurred on the Irvine Ranch in the cities of Irvine, Newport Beach, Tustin, and Orange. This book details the Irvine Company's success in planning and its strategic handling of major challenges, including infrastructure needs and responses to anti-growth sentiment and litigation. From the start of comprehensive planning of the Irvine Ranch in 1960 to today, the development is regarded by planning and real estate experts as the most significant and valuable master-planned community in the nation.

Successive generations of urban planners and real estate developers will look to the Irvine Ranch—shaped by Bren's long-term leadership—as a model of successful large-scale community development.

H. Pike Oliver has been working on real estate development strategies and master-planned communities since the 1970s, including nearly 8 years at the Irvine Company, and lived in the City of Irvine for 15 years. As the founder and sole owner of URBANEXUS, he focuses on equitable and sustainable real estate development and natural lands management. Earlier in his career, Oliver worked for public agencies, including the California Governor's Office of Planning and Research, where he was a key contributor to *An Urban Strategy for California*. He taught urban studies and real estate development at Cornell University and the University of Washington. Oliver is a member of the Urban Land Institute and the American Planning Association, and he is a founder of the California Planning Roundtable. He graduated from the urban studies and planning program at San Francisco State University and earned a master's degree in urban planning at UCLA.

C. Michael Stockstill earned a journalism degree from Humboldt State University in 1971. After spending a year as a legislative staff member in the California Assembly, he worked as a reporter and editor in Orange County before joining the Irvine Company in 1978. During his 13-year tenure there, he helped develop and implement strategies for major planning and policy concerns, including a multi-year campaign to pass a half-cent sales tax for local transportation improvements. Stockstill worked for the Transportation Corridor Agencies, which built the first toll roads in modern California history, as well as CalOptima and Southern California Edison, and he served as a public affairs consultant. He moved to Irvine in 1975 and was actively involved in local government, youth sports, and education while raising a family. Now retired, he resides in Irvine. He has written for national magazines, including *Planning* and *Parks and Recreation.*

The American Real Estate Society (ARES) Real Estate Thought Leadership Collection Series

The American Real Estate Society promotes the global dissemination of research representing the forefront of critical thought in real estate, with findings applicable to all areas of industry, policy, education, and practice. ARES publishes a broad range of real estate topics within an international scope, including finance, economics, investment, business, development, housing, sustainability, and environmental and urban studies. ARES publications provide insider access to the latest high-level real estate research, case studies, reports, analyses, pedagogies, and reviews. These constitute an up-to-the-minute resource of interest and relevance to business decisionmakers, academic researchers and educators, industry professionals, and emerging real estate leaders.

ARES publishes six, peer-reviewed, academic journals in partnership with Routledge, Taylor & Francis that provide relevant and timely research of interest to real estate academics and practitioners worldwide: The *Journal of Real Estate Research* (JRER), *Journal of Housing Research* (JHR), *Journal of Real Estate Literature* (JREL), *Journal of Real Estate Portfolio Management* (JREPM), *Journal of Real Estate Practice and Education* (JREPE), and *Journal of Sustainable Real Estate* (JOSRE). For more information visit: https://www.aresnet.org/page/ARES-Journal-Info#rjer

Conference

ARES hosts an academic and professional conference each spring in a unique travel destination that attracts real estate researchers from around the globe. The event features formal paper presentations, panels and roundtables, exhibitions, and social and networking events.

Membership

Membership in ARES includes the subscription to the journals, discounted conference registration, and many other benefits. For more information visit: https://www.aresnet.org/page/Benefits.

ARES
American Real Estate Society

Commercial Real Estate Analysis for Investment, Finance, and Development
David M. Geltner, Norman G. Miller, Alex Van De Minne, Piet Eichholtz, Thies Lindenthal, Lily Shen

The Irvine Ranch
Fulfilling the Vision, 1977–2025
H. Pike Oliver and C. Michael Stockstill

For more information: https://www.routledge.com/The-American-Real-Estate-Society-ARES-Real-Estate-Thought-Leadership/book-series/ARES

The Irvine Ranch

Fulfilling the Vision, 1977–2025

H. Pike Oliver and C. Michael Stockstill

Routledge
Taylor & Francis Group
NEW YORK AND LONDON

Designed cover image: Top: Courtesy of Orange County Parks Dept.; Bottom: H. Pike Oliver and C. Michael Stockstill

First published 2026
by Routledge
605 Third Avenue, New York, NY 10158

and by Routledge
4 Park Square, Milton Park, Abingdon, Oxon, OX14 4RN

Routledge is an imprint of the Taylor & Francis Group, an informa business

ISBN: 978-1-041-16161-5 (hbk)
ISBN: 978-1-041-15358-0 (pbk)
ISBN: 978-1-003-68308-7 (ebk)

DOI: 10.1201/9781003683087

Typeset in Times New Roman
by Apex CoVantage, LLC

Pike Oliver dedicates this book to his former students who are planning and developing communities worldwide.

Michael Stockstill dedicates this book in memory of his parents, Mary and Charles Stockstill, who started him on this journey.

Contents

The Irvine Ranch: Fulfilling the Vision

This book is the companion and sequel to *Transforming the Irvine Ranch: Joan Irvine, William Pereira, Ray Watson, and THE BIG PLAN*, published in June 2022.

That book drew from the writings and oral history of Ray Watson, who worked at the Irvine Company for 17 years, rising to become its president. Watson touched on many of the major elements of business, finance, education, the environment, planning, and history that impacted the Irvine Ranch. The book also chronicled the long and contentious conflict between Joan Irvine Smith, granddaughter of James Irvine II, and the James Irvine Foundation. The conclusion described the public bidding contest that led to the sale of the Irvine Company to a group of individuals who took control of the company in September 1977.

This book, *The Irvine Ranch: Fulfilling the Vision, 1977–2025* chronicles the years between the sale of the company to the group that included Donald Bren, through his purchase of the entire company, and the subsequent decades where—with control of planning, development, and open space donations—he shaped the landscape of the Irvine Ranch almost to completion.

The book also completes the story of Joan Irvine Smith—whose quest to take control of the ranch she considered her birthright finally came to an end—and how she lived the final chapter of her life as an art collector, horse breeder, and philanthropist.

The years from 1977 to the present were socially and politically contentious in Orange County. Enthusiasm for the growth and development planned and implemented by the Irvine Company and other developers in the 1960s gave way to unease about the problems that accompanied continued growth, particularly traffic congestion and the loss of open space. Unease grew into hostility in some cases, forcing builders and developers to contend with sophisticated opposition to specific projects as well as efforts to halt them through elections.

Throughout all this time, Bren created and pursued a vision for his properties with the flair of an architect and artist. He looked upon the Irvine Ranch over a period of decades. Confronted with opposition, Bren deftly compromised, negotiated, and sometimes simply withdrew from the arena, willing to wait for months or years if necessary to achieve his goals.

Bren's legacy comprises three key elements: the design and execution of residential and commercial developments with sophisticated and exacting attention to quality, immense wealth, and extraordinary generosity. The Irvine Company is conservatively valued at over $18 billion. Bren has contributed hundreds of millions of dollars primarily to the University of California, as well as to local school districts and educational enrichment programs. Equally important, more than 50,000 acres of land previously owned by the Irvine Company have been donated to a land conservancy and to public agencies.

The first chapter of this book begins in late 1977. The bidding war for control of the Irvine Company has ended with Bren and a consortium of buyers in control.

Location of the Irvine Ranch in Southern California

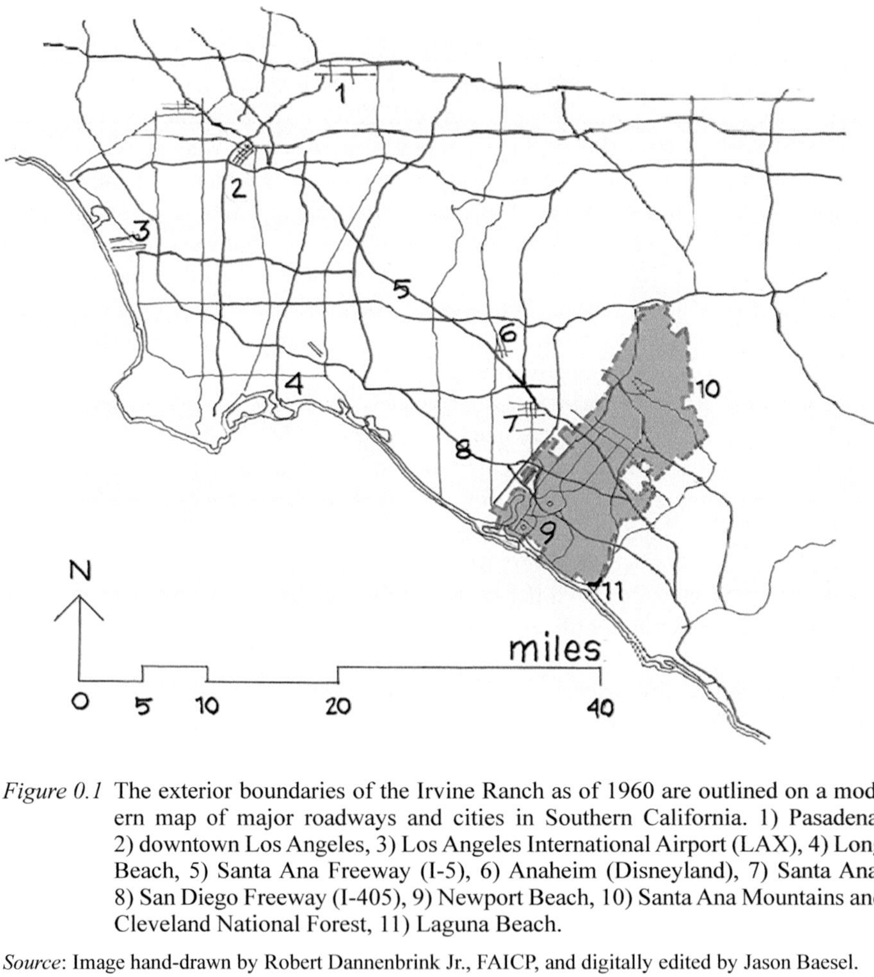

Figure 0.1 The exterior boundaries of the Irvine Ranch as of 1960 are outlined on a modern map of major roadways and cities in Southern California. 1) Pasadena, 2) downtown Los Angeles, 3) Los Angeles International Airport (LAX), 4) Long Beach, 5) Santa Ana Freeway (I-5), 6) Anaheim (Disneyland), 7) Santa Ana, 8) San Diego Freeway (I-405), 9) Newport Beach, 10) Santa Ana Mountains and Cleveland National Forest, 11) Laguna Beach.

Source: Image hand-drawn by Robert Dannenbrink Jr., FAICP, and digitally edited by Jason Baesel.

Acknowledgments

Indigenous peoples

The authors acknowledge the Indigenous peoples whose ancestral homeland includes the Irvine Ranch. These lands were never ceded or deeded during or following the Spanish occupation that began in 1769.

Upon gaining independence from Spain in 1822, Mexico claimed what was then known as Alta California but offered no compensation to Indigenous peoples. And in 1848, the United States formally annexed and paid $15 million for the Mexican Cession that included the land that became the State of California in 1850. No financial or other compensation was offered to Indigenous tribes or individuals.

Between 1864 and 1868, James Irvine and his partners acquired two Mexican land grants and a portion of a Spanish land grant. These men later acquired several adjacent tracts. In 1876, James Irvine acquired his partners' interests, gaining sole control of the land that would become the Irvine Ranch. Indigenous peoples received no compensation from these transactions.

Libraries and librarians

We are thankful for the assistance of the following libraries and librarians.

- Langson Special Library Collections and Archives at the University of California at Irvine
- Lawrence B. DeGraff Center for Oral and Public History
    - Located at California State University, Fullerton
    - Special thanks to Natalie Garcia
- Orange County Archives
    - Special thanks to Chris Jepsen
- Regional Oral History Office of the Bancroft Library
    - Located at the University of California, Berkeley

Individuals who shared memories

The following individuals generously shared their personal memories of the Irvine Ranch, the Irvine Company, UC Irvine, the City of Irvine, and the City of Newport Beach with the authors.

Richard Anderson
Jim Bailey
Leslie Berkman
Michael Bissell
Robert Break
Martin Brower
Robert Dannenbrink Jr., FAICAP, AIA
Joe Davis
Bill Dennis
Pamela Engebretson
William Ferguson
David Fields
Warren Fix
Monica Florian
Paul Freeman
Michael Goode
Larry Heglar
Jeanette Justus
William Kenny
David Kerrigan
Scott Kraft
Michael LeBlanc
Geoff LePlastrier
James Lorman
Roger McErlane
Roberta Marshall
Carol McDermott
William McFarland
Elise Millington
David Mudgett
Thomas H. Nielsen
Brad Olson
Robert Osbrink
Petros Panagopoulos
Mark Pinsky
Mark Rogers
Tom Rogers
Deni Ruggieri
Tom Sakai
Greg Sanders
Sally Anne Sheridan
Robert Schout
Richard Sim
Tom Stephenson
Hardy Strozier
Rudy Svrchek
Robert Thornton
Terry Watt
William Watt
Gary Weber
John Withers
Norm Witt

Special research and editing assistance

Yulan Chung
Warren Fix
Steve Kellenberg
Sherry Main
Craig Reem
Rick Reiff
Teena Spindler
Mary Susa
Robert Thornton
Gary Weber

Maps and illustrations

Robert Dannenbrink Jr., FAICP, AIA

Notes to readers

The story of the Irvine Ranch spans decades and multiple generations of family members, organizations, and individuals who played a role in the transformation of this landmark property. We offer these references to assist and inform the reader.

The Irvine family

James Irvine II, the patriarch, known as **JI**. He was the son of the ranch founder, his father, **James Irvine I**. James Irvine II had three children, **James Jr.** (also known as Jase), **Kathryn** and **Myford**. His first wife was **Frances**, his second **Katherine** (also known as Big Kate).

After divorcing his first wife, James Jr. married his second wife, **Athalie Richardson**. Together they had a daughter, **Athalie Anita Irvine**, who, at age 4, declared that her name would be Joan. She was known for the remainder of her life as Joan.

James Jr. died of tuberculosis at age 42. Kathryn died shortly after childbirth at age 26. Myford died of his own hand at age 60. James Irvine II died at 79 in 1947.

Joan Irvine married four times and had three sons: **James Swinden**, **Russell Penniman IV**, and **Morton Smith**. Her mother married twice after James Jr.'s death; her last husband was **Judge Thurmond Clarke**. **Athalie Clarke** died in 1993. **Joan Irvine Smith** died in 2019.

Key organizations and associated individuals

The James Irvine Foundation

Formed by **James Irvine II** in 1937, its two most influential members were **N. Loyall McLaren**, a San Francisco accountant and business advisor to James Irvine II, and **Arthur J. McFadden**, an Orange County farmer and attorney. By terms of James Irvine II's will, the foundation received a majority of Irvine Company stock and controlled and directed the management of the company until 1977.

Irvine Company

James Irvine II incorporated the Irvine Company under West Virginia law in 1894 and named himself president. After his death in 1947, his son Myford Irvine became president, followed by **AJ McFadden**. **Charles Thomas** became company president in 1960, followed by **William Mason** and **Raymond L. Watson**.

The Tax Reform Act of 1969 changed federal law and forced the foundation to sell its Irvine Company stock, which took place in 1977. The company was sold to a partnership led by **A. Alfred Taubman**, **Max Fisher**, **Milton Petrie**, **Henry Ford II**, **Joan Irvine Smith**, **Athalie Irvine Clarke**, **Howard Margueleas**, and **Donald Bren**. **Peter Kremer** was named president of the Irvine Company in July 1977, followed by **Thomas H. Nielsen** in October 1982. **Donald Bren** bought out his partners in 1983 and named himself chairman of the board of the Irvine Company, a position he holds to this day.

University of California at Irvine

Clark Kerr was the president of the University of California when the 1,000-acre campus of UC Irvine was sold by the Irvine Company to the university for $1 dollar. **Daniel Aldrich** was the founding chancellor, followed by several other chancellors, notably **Jack Peltason.**

City of Irvine

The city was incorporated in 1971. It became the principal site of Irvine Company residential, commercial, and retail development from that point forward. Notable residential villages included the initial villages of University Park and Turtle Rock, followed by Woodbridge, Northwood, Westpark, and Woodbury. The Irvine Spectrum, including the Spectrum Entertainment Center, is located on the eastern edge of the city and, along with the Irvine Business Complex (IBC) surrounding John Wayne Airport, is the city's principal employment center.

City of Newport Beach

This coastal community is home to Newport Center and Fashion Island, the Irvine Company's principal commercial/retail center in the city. Following decades of planning and litigation, development of the Newport Coast began in the late 1980s with annexation to the City of Newport Beach taking place in 2002.

Persons referenced in the book

Frank Abeling	Irvine Company executive
Ken Agid	Irvine Company marketing executive
Larry Agran	City of Irvine city council member and mayor
Steven Albert	CEO, Irvine Apartment Communities
Richard Alden	Attorney
Daniel Aldrich	First chancellor at UC Irvine
Charles Allen Jr.	Investment banker and member of the consortium that acquired the Irvine Company in 1977
Herbert Allen	Investment banker

Peter Alexander	Artist
June Allyson	Hollywood actress
Richard Anderson	U.S. Navy Department planner
Michael Antonovich	Los Angeles County supervisor
George Argyros	Developer and entrepreneur
Ricardo Asch	Physician, UC Irvine Medical School
Al Auer	Irvine Company executive
Frances Ayala	UC Irvine professor
Irving Azoff	Entertainment executive
Bruce Babbitt	Secretary of the Interior
Lauren Bacall	Hollywood actress
Robert Badham	Member of the California Assembly and Member of Congress
David Baker	City of Irvine city council member
Harry Baker	Property owner in Newport Beach
Mark Baldassare	UC Irvine professor
Jose Balmaceda	Physician, UC Irvine Medical School
John Balzer	Reporter for *Los Angeles Times*s
William Barber	Irvine resident who was awarded the Congressional Medal of Honor
Richard Barclay	Donor to UC Irvine
Marjorie Barclay	Donor to UC Irvine
Arnold O. Beckman	Founder of Beckman Instruments
Allan Beek	Newport Beach environmental activist
David Belardes	Native American tribal chief and cultural consultant
Jack Benny	Actor and radio personality
Marian Bergeson	Member of the California Assembly and State Senate
Laura Berman	Reporter for the *Detroit News*
Leslie Berkman	Reporter for the *Los Angeles Times*
Arnold Binder	UC Irvine professor
Stephen Birmingham	Author
Shirley Temple Black	Actor and member board of directors Walt Disney Company
Brigette Bren	Wife of Donald Bren
Charles L. Bren	Step-brother of Donald Bren
Claire Trevor Bren	Second wife of Milton Bren
David L Bren	Son of Donald Bren and Jennifer McKay Gold
Donald L. Bren	Owner of the Irvine Company
Milton Bren	Father of Donald Bren
Peter Bren	Brother of Donald Bren
Marion Newbert Bren	Mother of Donald and Peter Bren, AKA Marion Jorgensen
Ivan Boesky	Financier

Humphrey Bogart	Hollywood actor
David Booher	Irvine Company lobbyist in Sacramento
RJ Brandes	Equestrian business owner
Phil Brigandi	Historian
John Briggs	California state senator and assemblyman
Jonathan Brinsden	Irvine Company executive
Marlys Brooks	Irvine Company executive
Martin Brower	Irvine Company executive, book author
Donald K. BrOwn	Irvine Company lobbyist in Sacramento
Edmund G. "Pat" Brown	Governor of California (1959–1967)
Edmund G. "Jerry" Brown Jr.	Governor of California (1975–1983 and 2011–2019)
Willie L. Brown Jr.	Speaker of the Assembly, mayor of San Francisco
Anita Bryant	Entertainer
Kobe Bryant	Professional basketball player
Jerry Burchfield	Laguna Beach activist
Audrey Steele Burnard	Philanthropist
James M. Burns	Author of books on political science and history
Bill Burt	Former owner Deer Creek Ranch
Phillip Burton	Member House of Representatives
George H. W. Bush	President of the United States
William Butcher	Political consultant
William Campbell	Attorney
Donald Cameron	Irvine Company executive
Ray Catalano	City of Irvine city council member
Ronald Caspers	Orange County supervisor
Mark Chamberlain	Laguna Beach activist
Yong Chen	UCI professor
Chao Family	Donors to UC Irvine
Steven Choi	Irvine mayor and councilman
Robert Citron	Orange County treasurer
Ralph Cicerone	Chancellor UC Irvine
Athalie Irvine Clarke	Wife of James Irvine Jr., mother of Joan Irvine Smith
Thurmond Clarke	Federal judge, third husband of Athalie Richard Irvine
Harry Cohn	Hollywood executive
Jerry Collins	Irvine Company public relations executive
Richard M. Cooke	General, United States Marine Corps
Ron Cordova	California assemblyman
Ross Cortese	Developer
Ken Cory	Member of California Assembly and state controller
Edwin Corwin	Author of books on the U.S. Constitution
Chris Cox	Member of the House of Representatives
John Cox	Newport Beach City Council member
Justin Dart	California business executive

Laura Davick	Crystal Cove resident and advocate
Joseph Davis	Irvine Company executive
Michael Danzi	Newport Beach investment banker
Edward de Bartolo	Developer
Peter de Simone	Environmental activist
Bill Dennis	Real estate executive
Mr. DeVault	Land leasing consultant from Hawaii
James Dickason	President, Newhall Land and Farming
James Dilley	Founder of the Laguna Greenbelt
Roy Disney	Member board of directors Walt Disney Company
Walt Disney	Founder, Walt Disney Company
Don Dorchester	Real estate consultant
M.A. Douglass	Donor to the UC Irvine Medical School
Charlie Dunlap	Donor to UC Irvine
Lanny Eberling	Irvine Company finance executive
James Edwards Sr.	Owner, Edwards Theaters
Michael Eisner	Executive, Walt Disney Company
Tom Ellick	Fluor Corporation executive
Rob Elliott	Irvine Company executive
Dave Ellis	Political consultant
Michael Ellzey	Former CEO, Orange County Great Park
Harvey	Englander Political consultant
Robert Englund	Actor
Arthur Erickson	Architect
Rick Evans	Irvine Company executive
Peter Fairbanks	Art dealer
Tom Fazio	Golf course designer
Charles "Chip" Fedalen	Irvine Company executive
William Ferguson	Irvine Company lobbyist in Washington, D.C.
Max Fisher	Industrialist and member of the consortium that acquired the Irvine Company in 1977
John C. "Jack" Flanigan	Irvine Company executive
Monica Florian	Irvine Company executive
Robert Fluor	President, Fluor Corporation
James Flood	Orange County rancho owner
Henry Ford II	President of the Ford Motor Company and member of the consortium that acquired the Irvine Company in 1977
Arnold Forde	Political consultant
Michael Freed	Developer
Paul Freeman	Political consultant and Laguna Beach mayor and council member
Howard Friedman	Attorney for Joan Irvine Smith
Yehudi Gaffen	Project management company owner
Robert Geddes	Entertainment executive

Diane Ghirado	Planning consultant
Gary Granville	Newspaper reporter and Orange County clerk
Keith Gaede	Husband of Linda Irvine
Linda Irvine Gaede	Daughter of Myford Irvine (later Linda Irvine Smith)
Frank Gehry	Architect
Mary Ann Gaido	City of Irvine city council member
William Geyer	Irvine Company lobbyist in Sacramento
Mossimo Giannulli	Business owner
Lawrence Gilson	American High Speed Rail executive
Doug Gfeller	Irvine Company executive
Rebekah Gladson	Campus architect, UC Irvine
Hector Godinez	Orange County postmaster
Jennifer McKay Gold	Mother of two of Donald Bren's children
Keith Greer	Irvine Company executive
Bill Gross	UC Irvine donor
Emile Haddad	Executive with Lennar and Five Point Communities
Jane Glen Haas	Reporter for *Orange County Register*
Patrick Hanratty	Donor to UC Irvine
Fred Harber	Orange County political consultant
Gary Hart	U.S. senator
Richard Harris	Laguna Beach activist
Tom Hayden	California assemblyman and state senator
Dan Hedigan	Irvine Company and Five Point Communities executive
Brad Hellis	Irvine Ranch general manager
Rogue Hemley	Attorney
Ron Hendricks	Real estate consultant
Peter Herman	Aide to supervisor Thomas Riley
Andrew Hinshaw	Orange County assessor and member of the House of Representatives
Charles Hitch	President of the University of California
Meg Howard	Wife of Jimmy Kohn
Gary Hunt	Irvine Company executive
William Huston	Tustin city manager
Athalie A (Joan) Irvine	Daughter of James Irvine Jr. AKA Joan Irvine Smith
James Irvine I	First owner of the Irvine Ranch
James Irvine II (JI)	Son of James Irvine I, inherited the Irvine Ranch
James Irvine Jr.	Son of James Irvine II
Kathryn Irvine (Lillard)	Daughter of James Irvine II
Myford (Mike) Irvine	Son of James Irvine II
Brad Jacobs	Orange County assessor
Howard Jarvis	Author of Proposition 13
Michael Johnson	Art consultant
Phillip Johnson	Architect
Gordon Jones	Irvine Company executive

Earle Jorgensen	Stepfather of Donald Bren
Jeanette Justus	Irvine Company executive and school consultant
Moundir Kamil	Bank robber
Sue Kappas	Irvine Company executive
Jeanne Keevil	Editor, *Irvine World News*
Bill Kenney	Real estate developer
Ursula Kennedy	City of Tustin city council member
David Kerrigan	Irvine Company attorney
Billie Jean King	Professional tennis player
Frank Knox	Secretary of the Navy
Florence Koch	Landscape architect
Don Koll	Commercial/industrial developer
Len Kranser	Anti-airport activist
Bonnie Kremer	Wife of Peter Kremer
Peter Kremer	President of the Irvine Company, 1977–1982
William Kogerman	Member, Orange County Great Park Board of Directors
Jimmy Kohn	Horse trainer
Walter Knott	Owner of Knott's Berry Farm
Dean Koontz	Author
Tristan Kroius	El Toro Airport opponent
Beth Krom	City of Irvine city council member and mayor
Benjamin Lambert	Investment advisor
Jeff Lalloway	City of Irvine city council member
Dan Lamkin	Irvine Company executive
Jonathan Lansner	*Orange County Register* reporter
Fred Lang	Landscape architect
W.S. Lee	Lieutenant commander, United States Navy
Le Corbusier	Architect, AKA Charles-Édouard Jeanneret
Ricardo Legorreta	Architect
Mia Lehrer	Landscape architect
Lida Lenney	City of Laguna Beach city council member and mayor
Mervyn LeRoy	Hollywood producer
Geoff Le Plastrier	Developer
Craig Leupold	Real estate analyst
Mary Lou Loper	Reporter for *Los Angeles Times*
Hal Lynch	Homebuilder
William Lyon	Homebuilder and general, United States Air Force Reserve
James Lorman	Irvine Company executive
John Lusk	Homebuilder
Lou Malone	Homebuilder
Howard Marguleas	Member of the consortium that acquired the Irvine Company in 1977

Eddie Martin	Aviator
Freddie Martin	Bandleader
John Martin	Irvine Company executive and consultant
Roberta Marshall	Irvine Company executive
William Mason	President of the Irvine Company, 1966–1974
Bernard Maniscalco	Irvine Company executive
Paul Merage	Donor to UC Irvine
Carol McDermott	Irvine Company executive
Marianne McDonald	Donor to UC Irvine
Roger McErlane	Irvine Company executive
William McFarland	Irvine Company executive
Michael McKee	Irvine Company executive
Jeff Melching	Irvine city attorney
Henry Mello	California state senator
Edison Miller	Orange County supervisor
Kim Miller	Irvine Company executive
Ron Miller	Executive at Walt Disney Company
Michael Milken	Financier
Mary Miss	Landscape architect
John Miltner	UC Irvine dean
W.A. Moffett	Rear admiral, United States Navy
Stu Mollrich	Political consultant
John Moorlach	Orange County treasurer, supervisor, California senator
Richard Moran	Irvine Company executive
Anthony “Tony” Moiso	Master-planned community developer
Charles Moore	Architect
Don Morrow	Attorney
Louis Moulton	Owner of the Moulton Ranch
Daniel P. Moynihan	Sociologist and U.S. senator
Brent Muchow	City of Irvine, director of public works
David Mudgett	Irvine Company retail executive
Martina Navratilova	Professional tennis player
David Neish	Irvine Company executive
Robert Nelson	Political consultant
David Neuman	Campus architect, UC Irvine
Bruce Nestande	Orange County supervisor and California Assemblyman
Enrique Norton	Architect
Marilyn Nielsen	Wife of Thomas H. Nielsen
Thomas H. Nielsen	President and vice chairman of the Irvine Company
Richard Nixon	President of the United States
Ken Noll	Irvine Company executive
Enrique Norton	Architect
Michael O’Connell	Irvine Conservancy executive

Richard O'Neill	Owner of Mission Viejo ranchlands
Brad Olson	Irvine Company executive
Richard Ortwein	Koll Company executive
Claude Owens	Superior Court judge, Orange County
Eileen Padberg	Political consultant
Andrea Palladio	Architect
Louella Parsons	Columnist
Wright Patman	Member of the U.S. House of Representatives
Jerry Patterson	Member of the U.S. House of Representatives
Luciano Pavarotti	Singer
William Pereira	Architect and planner
Robert Perkins	Irvine Company executive
Phillip Petty	Superior Court judge, Orange County
Jack Peltason	Chancellor, UC Irvine
Jeffrey Perlman	Reporter for *Los Angeles Times*
Larry Peterson	Reporter for *Orange County Register*
Milton Petrie	Member of the consortium that acquired the Irvine Company in 1977
Michael Pinto	Member, Orange County Great Park Board of Directors
Fern Pirkle	President, Friends of the Irvine Coast
Dick Powell	Hollywood actor
Eddie Power	Irvine Company executive
Miguel Pulido	Mayor of Santa Ana
James "Walkie" Ray	Member of the Great Park Board of Directors
Maureen Reagan	First wife of David Sills, daughter of Ronald Reagan
Ronald Reagan	Governor of California, president of the United States
Richard Reese	Irvine Company executive
Christopher Reeve	Actor
Phillip Reilly	Founder of the Mission Viejo Company
William Rice	Developer
General Thomas Riley	Orange County supervisor
Hal Roach Jr.	Movie producer
Tony Robbins	Author and lecturer
Frances Robinson	Environmental activist, wife of Frank Robinson
Frank Robinson	Environmental activist, husband of Frances Robinson
Ted Robinson	Golf course architect
Thomas Rogers	Irvine Company executive
Tom Rogers	Environmental activist
C.E. Rosendahl	Captain, United States Navy
Michael Roos	California assemblyman
Stan Ross	Accounting executive and real estate analyst
Stephen Roulac	Real estate expert at Deloitte Touche
Sherwood Rowland	UC Irvine professor
Henry Samueli	Donor to UC Irvine

Susan Samueli	Donor to UC Irvine
Norberto Santana	Journalist
Greg Sanders	Congressional aide to Jerry Patterson
Robert Santos	Irvine Company and Five Point Communities executive
Robert Schout	Marketing executive for the Taubman Company
Chuck Schreiber	Real estate developer
E.F. Schumacher	Author
John Schmitz	California State Senator and Member of Congress
Louis Scott	Advertising executive
Henry Segerstrom	Developer and philanthropist
John Seymour	U.S. senator
Roger Seitz	Irvine Company executive
Christina Shea	City of Irvine city council member and mayor
Robert H. Shelton	Irvine Company executive
Clem Shute	Attorney
David Sills	City of Irvine city council member and mayor, Superior Court judge, and California Appellate Court justice
Dan Silver	Environmental activist
Richard Sim	Irvine Company executive
Gene Simmons	Musician
James Slemmons	California assemblyman
Morton “Cappy” Smith	Fourth husband of Joan Irvine
Ken Smith	Landscape architect
Patrick Smith	Irvine Company executive
Sally Smith	Irvine Company attorney
William Spurgeon	Irvine Company executive
Rudy Svrcek	Irvine Company executive
Robert Stern	Architect
Saul Steinberg	Financier
James Stirling	Architect
David Sun	Technology entrepreneur and philanthropist
Diana Sun	Philanthropist
Tom Sutton	Pacific Mutual executive
James Swinden	First son of Joan Irvine
Alfred Taubman	Retail magnate and investor in the consortium that acquired the Irvine Company in 1977
James Taylor	Irvine Company executive
John Taylor	Author
Irving Thalberg	Hollywood producer
Robert Thornton	Attorney
Charles Thomas	President of the Irvine Company, 1960 to 1966
Larry Thomas	Irvine Company executive
Norman Thomas	Candidate for U. S. president

James Toepfer	Founding executive, Mission Viejo Company
Susan Trager	Attorney
Alberto "Al" Treviño	Irvine company planner
Chuck Trevisian	Irvine Company planner
Ted Turner	Owner of Flying D Ranch
Holmes Tuttle	Automobile dealer
Peter Ueberroth	Member of the Irvine Company Board of Directors
Stanley van den Noort	Dean of the UC Irvine medical school
Basil "Bill" Vardoulis	City of Irvine city council member
John Vasconcellos	Member of the California Assembly
Jim Velasques	Native American consultant
Robert Venturi	Architect
Ernie Vasquez	Architect
Jack Wagner	Colonel, United States Marine Corps
Michael Ward	City of Irvine city council member and mayor
Amen Wardy	Fashion designer
Raymond L. Watson	President and vice chairman of the Irvine Company
Jean Watt	City of Newport Beach city council member and environmental advocate
Terry Watt	Planner, daughter of Jean Watt
William Watt	Irvine Company executive
John Wayne	Actor
Gary Weber	Irvine Company executive
Robert B. Webster	Retired judge
Barbara Weiner	City of Irvine city council
Frank Wells	Walt Disney Company executive
Joe C. Wen	UCI donor
Doug Wheeler	California Resources Agency Secretary
Robert "Bob" White	Chief of staff to Pete Wilson
Thomas Wilck	Irvine Company executive
Robin Williams	Actor
Pete Wilson	California governor and U.S. senator
Ray Wirta	Irvine Company/Koll Company executive
Laurel Wilkening	Chancellor, UC Irvine
Janet Wilson	Reporter for *Los Angeles Times*
Franz Wisner	Irvine Company executive and book author
Norm Witt	Irvine Company executive
Herman Wong	Reporter for *Los Angeles Times*
Daniel Yankelovich	Sociologist
William Woolett Jr.	City of Irvine city manager
Tom Wolff	Irvine Company executive
Tiger Woods	Professional golfer
Jon Worden	Planning consultant
Bruce Young	Member, California assemblyman
Barbara Young	Leader of the Committee of 4000

Notes on sources

As was the case with the volume that preceded this book, the records and oral history of Raymond L. Watson form a foundation of factual data and reference materials that give the book repeated insights into people and events that took place from 1960, when he arrived at the Irvine Company as a young planner, until 2008, when he resigned his position as vice chairman. Watson's papers are available for research in Special Collections at the Langson Library at UC Irvine, along with a variety of oral histories, news clippings, and other records the authors regularly accessed to research this book. Similar papers from Alfred Taubman are available at the University of Michigan.

Memoirs from Bruce Babbitt, Bill Dennis, and Peter Kremer supplied valuable information. As was the case in the first book, *California Rich* by Stephen Birmingham, *The Irvine Ranch: A Time for People* by Martin Brower, *The Irvine Ranch* by Robert Glass Cleland, and *A California Woman's Story* by Joan Irvine Smith were helpful. New research was supplied by Dr. Mark Baldassare's three books: *Trouble in Paradise*, *A California State of Mind*, and *When Government Fails*. Also helpful was *Chinese Americans in Irvine* published by the South Coast Chinese Cultural Center, *UCI: Bright Past, Brilliant Future* edited by Sharon Roan and John Westcott, and *The Laguna Canyon Project: Refining Artivism* by Mark Chamberlain. Interviews with dozens of individuals were conducted in person, by phone, and by Zoom. Many of them provided personal records and memorabilia as well. The richness and detail of their recollection of their experiences working on the Irvine Ranch added immeasurably to the book's detail and credibility.

Largely through access to Newspapers.com, dozens of references were obtained, especially from the *Los Angeles Times* and *Orange County Register*, which gave substantial editorial attention to the Irvine Company and related events from 1977 to the early 2000s. A personal visit to the archives of the *Orange County Business Journal* was also helpful.

Finally, the authors drew on their combined 22 years of work experience at the Irvine Company and decades of personal and professional life as residents of Irvine and Orange County. At various points in the book, observations and data were cited from the personal experience of the authors, from individuals the authors worked with or knew in the past who were not interviewed, or from those who have passed away in the intervening years. In those cases, the reference is noted as "author's recollection."

Preface

During the six years of research and writing that produced this book and the one that preceded it, we were often asked about our motivation for doing so.

The answer lies in many ways with the words of Raymond L. Watson, the former president and vice chairman of the Irvine Company.

During and after the years he was employed at the Irvine Company, Watson gathered data for a book, one that would tell the story of how a giant, empty farm in the middle of Orange County was transformed into a New Town, a university, and commercial and residential centers in Newport Beach and beyond.

He wrote six chapters of that book, as well as numerous other critiques, policy papers, and musings on the subject that in many ways defined his professional life. He donated all those papers to the Langson Library Special Collections at the University of California, Irvine, making them available for study and reproduction without restriction.

Among those papers, we found a note he wrote to himself reflecting on his effort to write the book:

> *Where does that put me in reconciling the effort of producing such a book and its value? My conclusion is that without the integrations of the behind the scenes what's and why's, it would be nothing more than another brochure type coffee table book . . . I have dozens of such books . . . never read. Barely scanned . . . The book that would interest me is one that provides insight and understanding behind the obvious . . . no amount of traveling through or living in those (Irvine Company planned and built) communities could ever provide anyone with the joys, trials, aspirations, disappointments, and just plain hard work and dedication that actually produced them. That is the story that would interest me. One that I would read. And one that I, almost uniquely, have the hands on and extended experience and exposure to write.*

We believe, like Ray, this story deserved to be told, and told in detail, including as many of the human, community, institutional, and psychic elements that led to the transformation of the Irvine Ranch, explained in the context of the time wherein it took place. We hope you will enjoy and appreciate this book and find that it reflects the goal Ray set for himself when he began writing the book he wanted to finish.

C. Michael Stockstill
Irvine, CA
H. Pike Oliver
Scottsdale, AZ

1 Kremer takes charge

The new owners of the Irvine Company began searching for a new chief executive officer within days of winning control of the company in a public auction.

Donald Bren, who had joined the consortium of millionaires led by Alfred Taubman at the last minute, had one candidate in mind and set out to bring him on board.

Peter Kremer was just 37 years old and had enjoyed both personal and professional fulfillment at a company that was in many ways like the Irvine Company.[1] Newhall Land and Farming was founded by the Newhall family in 1885, having purchased rancho land in what would become southern Ventura and northern Los Angeles counties. Like James Irvine II, the Newhalls planted citrus orchards and grew lima beans. By the 1950s, the potential for land development presented the company with an opportunity to create a new master-planned community dubbed Valencia. Like Irvine, it sought to blend residential zoning with a commercial/industrial center to create jobs.

Kremer had graduated from Stanford Business School with an MBA when Newhall president James Dickason recruited him to join the company. With his wife and college sweetheart Bonnie Scott, who had recently given birth to a daughter, Kremer found the opportunity appealing, and he agreed to start at the company in the summer of 1964 in an entry-level position.

For the next 13 years, Kremer held various positions at Newhall, assuming increasing responsibility and achieving success. By 1977, he was the number two executive there and widely expected to become CEO when Dickason moved up to chairman of the company board of directors.

Then came the call, as he described in his memoir, "that began a chain of events that would radically change my life, as well as the lives of those closest to me." The caller was Donald Bren, who had met Kremer when Bren built a tract of homes a few years earlier in Valencia.[2]

"Bren's call wasn't exactly a surprise," Kremer wrote:

> Bren confirmed that he was anxious to discuss the possibility of me . . . heading up the new company. My head was reeling with possibilities . . . why would I want to leave Newhall Land? . . . But this was the Irvine Company, the greatest planning and development project in the United States, nearly

DOI: 10.1201/9781003683087-1

> 80,000 acres of pristine, undeveloped land on the Pacific Coast, and I was about to be offered the chance of a lifetime.[3]

In early June 1977, Kremer and his wife went to dinner at Bren's home. "(Al) Taubman was there, and I could tell right away he regarded me with suspicion. He was nervous about the prospect of a 37-year-old managing the Irvine Company, but he seemed to relax as the three of us talked well into the evening about what it would take to succeed in this venture. We talked about the role I might play as president of the Irvine Company, and it became apparent what an incredible opportunity awaited us."

On June 28, 1977, Kremer was introduced at a Newport Beach press conference as the new Irvine Company president. Kremer's corporate uniform—dark suit, white shirt, and rep tie—didn't do much to hide his youthful appearance. Just over six feet tall, Kremer had natural blond hair and striking blue eyes. His fresh complexion made him look even younger than his 37 years (his birthday was just three months and a few days earlier). One reporter told a colleague he thought Kremer looked more like the captain of the Swedish ski team than the president of a major corporation.

Kremer was quick on his feet and spoke easily but with authority. "It was an intensely public time for me, with reporters and journalists wanting to talk endlessly about corporate strategies and development plans and speculation about what we'd do next. It was a far cry from my relatively 'quiet' corporate life with Newhall Land and Farming. I wasn't in Valencia anymore."[4]

***

As Kremer entered the front door of the Irvine Company, half a dozen of the company's highest-ranking executives were passing him on their way out.

Bill Watt had overseen the design and construction of apartments. Al Auer had overseen the creation of Fashion Island and retail centers. Robert Perkins had headed up the personnel division. All departed.

They followed Ray Watson, Lanny Eberling, and Tom Wolff, the most senior managers of the company.

Further down the organization chart, individuals whose personal finances and professional stature prevented an immediate departure bided their time; a few would give the new owners and Kremer some time to determine if the new corporate culture was a fit.

Frank Hughes was promoted to vice president of community development, but he left after a few months. Vice president of planning Richard Reese lasted more than a year before departing for Rancho Mission Viejo and a chance to design the new town just 20 miles south of Irvine. James Taylor and David Neish, who had managed the entitlement of several key company projects, teamed up to form their own firm. Doug Gfeller, who had directed the successful launch of Woodbridge, hung on for a few months and then left, as did marketing chief Ken Agid. These departures caused several openings in the respective departments, and while some of the senior people left, many of their underlings moved into jobs with greater responsibilities.

Two men whose careers benefited from the changes were Donald Cameron and Gordon Jones. Both were Stanford graduates and had been working at the company for several years.

Cameron had been a member of William Pereira's team that created the Irvine Company and UC Irvine master plans, then joined the company in the planning department. Cameron had the intelligence of an academic combined with a light-hearted, engaging personality. He wore a neatly trimmed brown beard and spoke with the diction of a professor, giving him the air of a host of a PBS show (Cameron was decidedly a Scot. He told the story of a visit to Scotland, where he entered a pub and proudly announced himself with his full name, Donald Campbell Cameron. A patron there turned, looked him over, and said, "Laddie, you know you are named after two of the most hated clans in all of Scotland").

Cameron was a director in the Irvine Company corporate ranking just below vice president. He kept that title when Reese left as vice president of the planning department, but with Reese's departure, the planning department had been downsized, leaving Cameron with a small staff but less internal prestige.

Title aside, because of both his knowledge and ability to engender trust, Cameron was often the face of the company in presentations to planning commissions and city councils, traveling from meeting to meeting, lugging a slide carousel and rolls of maps.

Gordon Jones became director of the company's engineering department with the departure of Frank Hughes. Jones had toiled for years in the mire of sewer, water, and transportation planning and, like Cameron, had a winning smile and outlook. He was also one of the company's best-known bachelors and was often seen wearing the most colorful and fashionable tie available on the market that year.

One part of the company that remained relatively stable was public relations and public affairs. Tom Wilck was running the public relations function, as well as the weekly *Irvine World News* and the semi-monthly *New Worlds* magazine, plus the cable television station. Wilck also had the most substantial ties to local political circles, such as the Lincoln Club. Martin Brower and Jerry Collins had significant credibility with the local media—their value increased with the change in ownership. Similarly, Wilck's staff members who interacted with local homeowner associations and the community in general decided to stay on.

Robert Shelton's stature became evident to the new owners and managers. As a result, he was made vice president of the government relations function, overseeing the company's advocacy both locally and in Sacramento. One of Shelton's first hires was Gary Hunt, who had been working for Donald Bren—Bren now held the largest portion of Irvine Company stock, but control remained with Al Taubman and the allies he had recruited into the ownership group. Taubman became chairman of the company's board of directors.

Kremer inherited a company that had been increasingly profitable but carried \$129 million in long-term debt—and that was *before* adding the more than \$240 million the new owners were on the hook for in terms of the loans they had taken out to make the purchase.

Figure 1.1 Irvine Company board meeting in 1977—STANDING (L to R) Warren Fix, Richard Alden, Benjamin Lambert, Howard Marguleas, Keith Gaede, Joan Irvine Smith, Max Fisher, Herbert Allen, Donald Bren, Rogue Hemley—SEATED (L to R) A. Alfred Taubman, Peter Kremer

Source: Image courtesy of Warren Fix

According to the 1976 annual report, the Irvine Company holdings included:

• Apartments	2,636
• Industrial buildings	885,000 square feet
• Village shopping centers	605,000 square feet
• Regional shopping center	937,000 square feet
• Office buildings	1.054 million square feet
• Marinas	23,805 linear feet (Over 400 slips)

The company held ground leases that had been mortgaged to Prudential Insurance for $120 million. They included:

• Apartments	107 acres
• Industrial	293 acres
• Commercial	2,631 acres
• Residential lots	4,974

The company also owned thousands of acres of land in Orange County that was still in crops, primarily citrus, as well as fields where strawberries, asparagus, and other specialty crops were grown. And there were about 11,000 acres of farmland in the Imperial Valley at the southern end of California as well as the Flying D Ranch in Montana.[5]

***

The Flying D sat in the heart of Gallatin County, a few miles north of Yellowstone National Park. It was, and is, breathtakingly beautiful: against a backdrop of rugged, snowy mountains, green meadows and hillsides lush with prairie grasses roll on endlessly toward the horizon. The Madison and Gallatin Rivers run parallel east and west, and large creeks like Cherry Creek and Spanish Creek flow through the land.

James Irvine II purchased the Flying D in 1944, acquiring 82,000 acres of land for a reported price of $1.3 million. Irvine bought the ranch using proceeds from the federal government's condemnation of some of his best lima bean fields for what became Marine Corps Air Station El Toro and for a second condemnation that resulted in the construction of the Lighter Than Air Base in Tustin on the western edge of the ranch (see Chapter 23).

Irvine had established cattle grazing and hay crops on the Flying D. He was fishing on Spanish Creek in August 1947 when he succumbed to a heart attack just two months before his 80th birthday.

The Flying D would remain an Irvine Company asset until 1978, when the new owners decided to sell it to raise funds to lower debt. According to Kremer's memoir, the price was $15 million.

No one was sadder to see the Montana ranch sold than a small group of agriculture division employees who made an annual fishing trip there. The trout were so plentiful in Cherry Creek and Spanish Creek, catching them "was almost too easy," one man recalled fondly.

A few years after the Irvine Company parted with the Flying D, it was purchased by Ted Turner, who sold off the cattle and began raising American Bison. He owns it to this day.[6]

***

Irvine Company's acquisition of farmland in the Imperial Valley came about in the same manner as the purchase of the Flying D—as a way to avoid taxes.

Writing in the 1962 edition of his book *The Irvine Ranch*, author Robert Glass Cleland noted that

> the ranch, constant victim of the needs of burgeoning cities, school systems and public utilities, has for many years been following the policy of placing monies received from condemnations in a replacement fund which is used to buy land elsewhere . . . in Imperial Valley the original acquisition of 2,400 acres in 1953 was augmented . . . until by the end of 1958 the Irvine Company owned 8,700 acres and by 1961 it owned 11,500 acres.[7]

The Imperial Valley lay at the far southern border of California. It was hot and dry, and it looked like a desert. But just beneath the arid crust, the soil was rich, the result of eons of the Colorado River overflowing its banks and depositing layer upon layer of dirt onto the narrow receptacle of land that would eventually produce abundant crops of all kinds.

Through the 1960s and 1970s, the Imperial Valley land owned by the Irvine Company likely produced profits. Water was plentiful, and labor costs were low. Nevertheless, a year after the new owners took control around 5,000 acres had been sold, the proceeds used as with the Flying D to reduce debt, the remaining acreage put in the hands of Howard Marguleas and his company SunWorld to manage.[8]

Hindsight is, of course, 20/20, but one wonders if Peter Kremer, Donald Bren, or Al Taubman ever considered what value the Flying D and Imperial Valley holdings might have created 20, 30, or 40 years in the future. In 2020, an ad appeared in the Wall Street Journal touting the sale of 160-acre home sites in the Gallatin Preserve, a part of the exclusive Spanish Mountain Club, with prices listed from $6 million

Figure 1.2 Peter Kremer in 1979

to $25 million.[9] *And in the Imperial Valley, some farmers are making more money by fallowing their land and receiving payments for the Colorado River water they would have used for irrigation—the water flowing instead to thirsty southern California communities, including Irvine. But according to Robert Schout, who worked for Taubman on the purchase, the Flying D was so remote and its value so uncertain that selling it off was hardly debated.*[10]

More important than selling off assets were two key course redirections Kremer had initiated shortly after evaluating the company management structure and financial outlook.

While Kremer was only one member of the Executive Committee established to guide decisions in between board meetings (Bren and Taubman were the others), the day-to-day operations of the company were largely determined and executed by him. In his memoir, Kremer said that the Irvine Company

> was quite differently structured than most real estate development firms. For starters, the management staff we inherited was large . . . the company had a reputation for being overly bureaucratic and institutional in its approach to the business of real estate.[11]

He went on: "from the outset, our position was that the Irvine Company should be more entrepreneurial in its approach as a way to optimize its assets and return on those assets. In short, there was a desperate need for a change in the corporate culture."[12]

Kremer set out to streamline management, reducing the number of vice presidents from 12 to 7 and encouraging the remaining departments to reduce headcount. He reported that as reductions were made—in all about 20 percent of total employees were given notice—most of the 800 who remained adapted and became more productive.

Kremer's other major initiative was financial—after reviewing past company sales, he determined that money was being left on the table—lots of it: "In multifamily, for example, the company's practice had been to price houses and condominiums based on formal appraisals rather than what the market would bear. The same conservative approach was used in negotiating land prices in sales to developers."[13]

Kremer soon had the opportunity to test his theory:

> the beautifully restored Mai Kai Condominiums in Corona del Mar . . . for its location, the prices (set by the company before his arrival) seemed quite low . . . I didn't want to be overly assertive to the management team, but I suggested we raise the price $50,000 on each unit . . . everyone thought I was crazy. But the following weekend, when sales opened, the reception was tremendous. It took just one week for every unit to sell at the higher price.[14]

Turning his attention to the giant Village of Woodbridge that was continuing to enjoy huge market success, Kremer instituted a new policy targeted at homebuilders.

The company had spent millions of dollars creating the finished lots in the giant village; the money would be recouped when builders purchased the lots. And while the company sought the best price possible for each lot, if demand for housing increased, the builders were free to set their own selling price, reaping increasing profits with every higher dollar they could justify.

Kremer wanted the company to benefit from higher prices, so under his leadership, sales of finished lots in Woodbridge contained language giving the company a percentage of any price for future home sales the builder made above the "pro forma" price agreed upon at the original sale. In the first year of the new policy, Kremer noted in his memoir, the company collected $8 million over and above the land sales revenues from a group of builders.

Kremer also moved to save money by halting the company's policy of constructing industrial buildings for sale and pausing construction of new apartments. Both moves reduced cash outlays that could be saved for debt payoff.

Kremer continued to emphasize the more aggressive and entrepreneurial philosophy, recalling that in meetings he often said, "What would we do if those of us around this table owned this piece of property? I also de-emphasized the importance of outside appraisals . . . and suggested our own observations of market trends and opportunities were more important." Shifting the corporate mindset was, Kremer recalled, "the most defining accomplishment of our first half year of ownership."[15]

Another key decision was to create a division to manage the various properties the company owned. Previously, each operating group handled relations with tenants in their area—the apartment division, headed by Bill Watt, did the leasing of apartments; the commercial division handled tenants in office buildings and retail centers.

Kremer recruited Dan Lamkin to head the new property management division. Younger than Kremer by three years, Lamkin assembled a combination of holdovers and recruits to centralize leasing and management of the array of investment properties the company owned. (Fashion Island would receive a new manager personally selected by Al Taubman, reflecting his retail expertise).

At the marinas, a colorful manager named Eddie Power ruled the company's hundreds of docks, treating multimillionaires with their huge yachts the same as weekend boat owners who struggled to keep up with the ever-rising slip fees. Power was from Chicago and delighted in reminding everyone of the fact with his raspy voice and friendly, "I tell it like it is" demeanor.

Short, a bit pudgy, peering out from behind thick lenses and a stunning set of false teeth that lit up his ever-present smile, Power was a delightful contrast to the army of company corporate staff up the hill at 550 Newport Center Drive, attired in Brooks Brothers (or for the more affluent, At-Ease men's store) uniforms of dark suits and white or blue shirts.

Another individual whose career took a turn for the better after the sale of the company was David Kerrigan. An attorney, he had been hired as a special assistant to William Mason, but shortly after he arrived, Mason died. Ray Watson became president and already had an assistant, Ed Lee. Kerrigan was shifted to become the company's first in-house corporate counsel.

Kerrigan recalled that at the time, each division of the company had a separate attorney from Gibson, Dunn, and Crutcher doing transactional work and, if necessary, litigation. Kerrigan began to bring on other firms and attorneys with specialized expertise, both to meet new legal areas like environmental and political law as well as to keep costs down. When the company changed hands in 1977, new management parted company with Gibson, Dunn, and Crutcher and brought in Latham and Watkins, allowing Kerrigan to accelerate his efforts to build a larger in-house team of attorneys.[16]

***

The combination of Kremer's actions, helped along by a strong economy, resulted in the Irvine Company's surprising declaration just a week short of a year after the 1977 sale that the $240 million loan from the consortium of nine banks had been repaid in full.

It was a stunning achievement. In his memoir, Kremer acknowledged that

> external factors like the market and the rapid inflation of real estate values and prices were significant to our success. However, much of the credit went to me and Taubman. In addition to crediting us to the meeting, Joan Irvine Smith also stood up at the annual meeting and congratulated me personally for a job well done as president.[17]

As for Joan, it would be a brief moment of harmony. Soon thereafter, Kremer's relationship with her would become adversarial as it would with the other owners.

Bren and Taubman had clashed briefly shortly after the company was purchased. Martin Brower, head of public relations at the time, recalled that he was instructed to write a press release announcing the new executive lineup with Bren named as chairman. As he completed it, new instructions arrived: Taubman was chairman. It reflected the fact that Taubman had the votes of the other owners—the friends he had recruited into the deal—even though Bren owned the largest chunk of the company.

Notes

1 Peter Kremer, *Isn't That Nice? Reflections of a Blessed Life* (Self-Published, 2013), 7. Few people outside his family knew that Kremer's name on his birth certificate was Charles Ernst Kremer III. His mother, father, and sister called him Peter, the name they favored from the day he was born, and as he wrote in his memoir, "I would never use the name on my birth certificate."
2 Kremer, *Isn't That Nice?*, 132.
3 Kremer, *Isn't That Nice?*, 132.
4 Kremer, *Isn't That Nice?*, 135.
5 Irvine Company, Annual Report, 1976, 8.
6 "Flying D Ranch Sold to Ted Turner," *Bozeman Daily Chronicle*, July 1989.
7 Robert Glass Cleland, *The Irvine Ranch: A Time and Place Remembered* (Los Angeles: The Huntington Library, 1962), 145.
8 Irvine Company internal memo, "Imperial Valley Holdings Overview," October 1982. Strategic rationale summarized from an internal document.
9 "Full-Page Ad: Irvine Company's Legacy," *Wall Street Journal*, February 2020. Branding language quoted directly from a published advertisement.

10 Kremer, *Isn't That Nice?*, 146.
11 Kremer, *Isn't That Nice?*, 138.
12 Kremer, *Isn't That Nice?*, 138.
13 Kremer, *Isn't That Nice?*, 140.
14 Kremer, *Isn't That Nice?*, 140.
15 Kremer, *Isn't That Nice?*, 141.
16 Irvine Company Legal Affairs Bulletin, "Gibson Dunn to Latham: Legal Counsel Transition," August 1980. Legal counsel transition cited from internal bulletin.
17 Joan Irvine Smith's remarks at Irvine Company annual meeting, May 1978, as noted in her book *A California Woman's Story* (Irvine, CA: The Irvine Museum, 2006).

2 The times, they are a-changing

From the song by Bob Dylan

The dawn of the 1980s was the best of times and the worst of times on the Irvine Ranch.

The national economy was expanding, Orange County was growing, and demand for new homes, apartments, offices, and industrial and retail space was riding a surging wave of prosperity.

But while most residents of the communities on or around the ranch were pleased with rising wages and even faster growth in the value of their homes, change was underway that would turn the county and many of its cities into battlegrounds for the next two decades.

Social scientist Daniel Yankelovich theorized in his 1981 book *New Rules* that public opinion exists on two planes. The first is found at the top of the ground, where everyday activities spin and turn, public opinion being mainly formed in response to political issues and personalities of the moment.[1]

But further below, like the Earth's tectonic plates, there is a more fundamental plane of opinion. It moves invisibly but inexorably, dragging the upper reaches along toward new, more fundamental and lasting positions in the social and political geoplasma. When it reaches a critical point, it shifts both planes, resulting in a dramatic change in the visible world.

In Orange County, the movement of the invisible plane reflected growing unease about the everyday impacts of growth and development.

***

It had taken only four years in the new city of Irvine for the first public vote on development to be held. The issue was the new village of Northwood.

Northwood was an amalgamation of properties owned by individuals, families, and business interests that had acquired them over time from the Irvine Company. James Irvine II had a reputation as a sharp landlord. Still, as Brad Hellis, the general manager of the Irvine Ranch, noted in his oral history, Irvine also had "a sentimental side."[2]

Irvine valued loyalty and rewarded it. There are reports from oral histories and written chronicles of Irvine either deeding small properties to his tenants or allowing them to purchase them. Over time, in the area above the Santa Ana Freeway

DOI: 10.1201/9781003683087-2

between Culver Drive and Jeffrey Boulevard, reaching into the low-lying foothills of the Lomas de Santiago, there was an assemblage of a dozen or so of these properties.

As the potential for development dawned, some owners were eager to give up a few acres of citrus orchards or a field of vegetables and cash in on the same opportunities the Irvine Company was enjoying with its villages of homes and businesses.

The owners formed a loose affiliation and approached the new city council to seek development rights for their properties. However, it soon became apparent that considerable effort was being expended by some owners jockeying for zoning designations that would prove more valuable for their specific properties.

The council instructed the owners' group to hire a planning firm and create a village plan similar to the one the Irvine Company had prepared, complete with open space, parks, and community facilities. The company had considerable acreage above Irvine Boulevard that it kept out of the Northwood zoning, figuring it had a better chance to maximize value and win city approval in a separate entitlement effort. Additionally, the company's development focus was primarily on Woodbridge, where sales were thriving and almost every new tract of homes was achieving higher prices than those built a few months earlier.

Not everyone who owned land in Northwood wanted to sell. The Lambert family's 51 acres of orchards and fields were purchased from the Irvine Company in 1949. It would be 2010 before the two sisters who inherited the property finally decided to sell it for development.

The city approved the Northwood Village plan in early 1975, but opposition was brewing in the small enclave called the Racquet Club, the community most impacted by the new development. Specifically, residents said they did not want apartments in the new village and were prepared to seek a referendum to overturn the zoning if the council allowed apartments. The council wanted apartments as a way to meet demand for lower-cost housing.

Led by attorney David Sills, a former Marine Corps officer who had once been married to Maureen Reagan (daughter of the soon-to-be president), residents rounded up the needed signatures; the council placed the measure on the November 1975 ballot, and the voters overturned the zoning.[3] Sills went on to be elected to the city council in 1976.

The Village of Northwood became much like other Irvine Company villages but lacked a critical component: a master community association like the one in Woodbridge. Also, many neighborhoods did not have mandatory homeowner associations (HOAs). This proved to be a strong marketing point for buyers concerned about costs (HOA fees were relatively low, but could make the difference on a costly monthly loan payment) and who wanted to avoid what many viewed as the stifling and annoying HOA rules in Irvine Company villages.

***

In 1978, there were only 24,000 registered voters in Irvine. The city was small enough that a candidate for city council could, with a handful of motivated volunteers, knock on the door of many, if not most, of the homes in the city where voters lived.

One candidate would do so. His name was Larry Agran.

Agran moved to the city in 1973 with his wife, who was attending UC Irvine Medical School. Settling in University Park, the Harvard-trained attorney eschewed work at a law firm, instead finishing his book, *The Cancer Connection*.[4] He also wrote a weekly column for *Home* magazine in the *Los Angeles Times.*

It likely occurred to Agran that citizens in a city dominated by a near-monopoly landowner might welcome someone on the city council who would directly challenge Irvine Company policies, if only on principle. As one resident observed, "you always need someone to bell the cat."

Agran created a new twist on door-to-door campaigning, handing out or attaching a two-foot-long flyer on stiff cardboard to front doors and garage doors. The flyer included a return card that asked for voters' views on issues, providing Agran with dozens of names of likely voters to follow up with. Agran's relentless campaigning culminated in the twilight of election day as he stood on the shoulder of Culver Drive holding a sign that caught the passing headlights: "I'm Larry Agran, Please Vote for Me."

Agran was the top vote getter, beating Art Anthony, who had served an earlier term on the council. Within a few months, Agran was racking up a series of objections to Irvine Company policies, including rental rates and what became repeated generic questions about traffic and the pace of growth.

His election prompted a long analysis by the Orange County edition of the *Los Angeles Times*. Reporter Herman Wong got right to the point in the first paragraph of his story: "As usual, there was one—and only one—real issue in this year's Irvine City Council election: growth." Wong quoted Agran at length, expounding on housing, integration, and potential new ideas for long-lingering issues. "All forms of concepts should be considered, whether conservative, liberal, or radical, if they make sense."[5]

The article continued:

> Agran said he favored a slower growth land use model the council had debated but rejected last summer (the model set a top population of 175,000). Agran also warned that present city policies would lead to massive, traffic-choked throughways, or what he called the 'Los Angeles-ization of Irvine.[6]

Agran found a kindred spirit on the council in Mary Ann Gaido, a Turtle Rock housewife with two young children. Gaido was blond, blue-eyed, with a demeanor that alternated between girlish charm and a serious passion for progressive, humanitarian politics, nurtured by her Catholic faith and education at Dominion University. Gaido served two terms on the council but, thereafter, was regularly appointed to the city planning commission by Agran. She served on several other advocacy groups and ran unsuccessfully for council or mayor five more times.

Agran proudly assumed the role of lightning rod and contrarian on the council, as well as regularly appearing at other liberal, anti-growth, and anti-establishment events in the county and elsewhere in Southern California. His membership—while pretty much in name only—in Tom Hayden's Campaign for Economic Democracy sent local Republicans into a frenzy. Still, Agran continued to win re-election to the city council for the next 12 years. Every time he won, he would increase the volume on the subject of growth, traffic, and his views on how to control them.

(An Agran initiative resulted in the creation of one of the county's most memorable political nicknames when he invited a baseball team from Nicaragua to visit Irvine. From that point forward, Agran supporters were known to his opponents as Agranistas, a reference to the left-wing Sandinistas that controlled Nicaragua at the time.)

***

The study of attitudes regarding growth and development became Dr. Mark Baldassare's academic specialty at UC Irvine.

As noted in Chapter 8, Baldassare created a survey of Orange County voter attitudes to chart how the rapid rise in the county's population impacted the opinions of residents about their quality of life. He also asked how they felt the government was or was not managing growth, development, and infrastructure.

Baldassare's timing was prescient. In 1970, five years after UCI opened and a year before the City of Irvine incorporated, there were 1.4 million people in the entire county. Twenty years later, the number had risen to 2.4 million—in just two decades, a million people had come to the county to buy or rent a home, many of them in new communities south of the 55 Freeway, the midpoint of the county's landmass.

The annual survey gave Baldassare a platform that enhanced his credibility as an academic, as well as UCI's. It also provided him with a wealth of data on what were becoming the most talked-about issues county leaders and residents were attempting to understand.

Baldassare used that data as a key element in one of a series of books he wrote about growth and development, *Trouble in Paradise: The Suburban Transformation in America.*[7]

The book was published in 1986 but was based in large part on data from the 1982 Orange County Annual Survey (1,000 people were polled). Those findings led Baldassare to write that

> there has been a broad awakening of sentiment against growth, and increasing skepticism that growth offers the average person a better life . . . the subject of future growth is not unfamiliar to the suburbs. In fact, growth controversies have developed into social movements there in the last two decades.[8]

Noting that resistance to growth in areas of the United States can be documented as far back as the 1920s and 1930s, Baldassare wrote that

> suburban people are often ex-city people who left for specific reasons (like) improved housing and crime-free neighborhoods. These feelings heighten

> a sensitivity to community change . . . so there is support for the suburban government's exerting some control over development.[9]

Baldassare believed Orange County attitudes reflected opinions about growth throughout the state. The county had experienced rapid, substantial growth in a relatively short time, and it was affluent, and most of the voters worked in white-collar jobs. And "there had been disagreements about growth, both countywide and in specific locations," further creating an environment where growth was top of mind for many people.

The first question in the survey was straightforward: "Do you think that growth and development in your city should be limited?" The answer: 65 percent said 'yes,' 35 percent said 'no.' Those who answered 'yes' were asked to explain why (multiple answers were allowed). The top reason is traffic congestion, followed by environmental deterioration, maintaining property values, and, finally, avoiding government spending and taxes. Asked for specifics on how to limit growth, those who answered said limiting the construction of high-rise buildings or apartments, but no one wanted to stop building roads.

Views about growth and development throughout the county were useful for Irvine Company executives trying to find a path to allow development to continue. However, in Newport Beach, it became clear that local attitudes on the subject had hardened into a solid wall of opposition, one that would prove nearly immoveable for years to come.

***

The Newport Beach of the 1960s had long left behind its roots as a maritime community, although a few remnants remained. The modern Newport Beach was unashamedly affluent, vibrant, glitzy, sexy, and throbbing with life. Proudly conservative, Newport's residents tolerated the throngs of tourists that descended on the city's beaches and harbor for four months of the year, counting the days until the calm season from Labor Day to July 4 would once again bring peace.

The heart of the city was the harbor, an arm of water about three miles long, emptying into the Pacific Ocean at a rock jetty that created the famous body surfing mecca The Wedge. Near the other end of the bay, actor John Wayne awoke each morning and pulled back the curtains on his bedroom windows to check his boat, the *Wild Goose*, a converted Navy minesweeper 160 feet long. Wayne died in 1979, but the *Wild Goose* lives on, available for charter.

Islands in the bay define individual neighborhoods. Balboa, Bay, Lido, Linda, and Harbor, the latter the most exclusive, Balboa the most democratic, its inner streets filled with seasonal rentals and tiny cottages, while automobiles are banned on Bay Island. Guarding the bay from the relentless Pacific is the long, narrow Newport Peninsula, formed in 1825 when the Santa Ana River leaped its banks during one of the county's massive floods. The detritus and mud from the deluge accumulated over time, eventually allowing homes and businesses to be built there.

The Peninsula neighborhoods are all expensive, but at the northern end, as on Balboa Island, the housing stock is less grand, attracting concentrations of younger, party-hardy men and women. The eastern edge of the bay is the most commercial part of the old city, with the Balboa Bay Club marking a transition point as the land rises above the waterline on its journey south to Corona del Mar, a neighborhood of homes and cottages created not long after incorporation that grew more substantially in the years following World War II. Today, it is rapidly gentrifying as the cottages are torn down and replaced by two-story, narrow but modern homes that take up nearly every square foot of the tiny lots.

Finally, western Newport Beach, from the bluffs overlooking Upper Newport Bay to Westcliff and all the way over to Irvine Boulevard, had filled in with modest homes on land the Irvine Company had sold off in the 1950s to homebuilders like George Holstein. A thriving commercial district had taken hold along 17th Street, extending to Costa Mesa. This was inland Newport Beach, solidly middle class, oriented, not to the bay, but to inland Orange County. (There was another section of Newport Beach where almost no one lived until the mid-2000s; it was a massive square of land bordering John Wayne Airport and three major streets in the Irvine Business Complex. It had been annexed largely at the behest of the Koll Company, which proceeded to build large commercial properties on much of it.)

In 1960, Newport Beach property values were going up. Homes on Lido Island, directly on the waterfront, commanded prices in the $80,000 range, while those on inner streets commanded prices around $40,000. You could get a small cottage on Balboa Island for $30,000, a similar home in Corona del Mar for slightly more. By contrast, a three- to four-bedroom home with a big yard in Anaheim or Fullerton topped out at just under $30,000.

It was not unusual to find professional people, especially from Pasadena, who owned a home in Newport Beach, used on the weekend and for a summer vacation, when Mom and the kids would come down during the week, Dad joining them for the weekend. (Engagement notices in the *Times* often noted the bride or groom's family as hailing from "Pasadena and Newport Beach").

In 1960, the city's growth took on a new complexion as the Irvine Company began to transform its property, almost all of it above Pacific Coast Highway. Newport Center and Fashion Island created an instant locus of commerce previously unknown in Newport Beach, with medical offices, banks, and other financial institutions, as well as lawyers and accountants, flocking to offices with views of the Pacific Ocean. A completely new residential village was underway at Eastbluff. The city had dashed state plans for a freeway paralleling Pacific Coast Highway, freeing up property above Corona del Mar for new homes.

With the opening of the San Diego Freeway in 1969, men who lived in Pasadena and worked in downtown Los Angeles soon discovered that the drive from Newport Beach to downtown was tolerable enough to relocate from elegant—but hot and smoggy—Pasadena to the cool shores of Newport. UC Irvine was now open as well, and some professors and administrators who could afford it chose Newport as a new home (the chancellor's home was located in Newport Beach). All of a sudden, it seemed to many Newport residents, what was once a quiet village was on its way to turning into something very, very different.

The numbers didn't lie; the city's population doubled from 1950 to 1960, from 12,000 to 25,000. Over the next 10 years, it doubled again, reaching 50,000 by 1970. Thus, in not much more time than the journey from elementary school to college, Newport Beach had quadrupled in size. And while for the newcomers, the refugees from Pasadena, the UCI kids four to a cottage on Balboa Island, it seemed like a little slice of heaven.

But not to everyone.

Allan Beek had lived through the changes in Newport Beach, and the more that took place, the less he liked them. Son of the founder and operator of the Balboa ferry, Beek was a graduate of the California Institute of Technology (Cal Tech) and made his living as an aerospace engineer. But his real passion revolved around public issues, first universal healthcare, then the issue that would consume him—growth, traffic, and development in his beloved beachside town.

Tall and lanky, Beek was often described by friend and foe alike as eccentric. He was as outspoken as the color of his bright yellow Volkswagen Beetle. He put more than 500,000 miles on it (his daughter described it as a "deathtrap.")[10] Beek began his crusade by joining hundreds of other Newporters in defeating plans for the Pacific Coast Freeway. That accomplished, in 1972, Beek founded Newport Residents United, seeking a 35-foot limit on new buildings and a cap of ten units to the acre for residential development.

Beek soon had company in his quest to confront environmental and growth issues. Garbage in the Upper Newport Bay was on the minds of several locals, a subset of the years-long campaign being led by Frank and Frances Robinson to preserve the entire area. Jean Watt used the garbage issue as an organizing focus and, combined with her opposition to noisy airplanes flying over the city, founded what would become the premier Newport Beach pro-environment, anti-growth group: Stop Polluting Our Newport, with the acronym SPON.*

A nurse by training, married to a physician, Watt had grown up in Pasadena but spent summers at a vacation home in Newport. As newlyweds, she and her husband moved full-time to Newport Beach and began to raise four children. Building on a foundation of skills obtained as a Girl Scout leader, Watt expanded her efforts into a variety of community service interests, ultimately committing herself to SPON for the remainder of her life.

SPON, Beek, Watt, and dozens, then hundreds of other Newport residents began a steady chorus of comments, criticisms, and attacks on anything directly or remotely connected to the city's quality of life as it might be affected by new development. And what would become a perennial target was Newport Center.

As city general plans were adopted, Newport Center fared well. In the mid-1970s,

> the city council established a ceiling on the center's development, allowing 5.7 million square feet of office and retail space, a hotel, and more than 500 residential units. In December 1978, the council cut almost 600,000 feet from the allowance. One year later, another 600,000 square feet were eliminated

* SPON later changed its name to Still Protecting Our Newport.

> from the center by the council's then anti-development majority, leaving the center with about 4.3 million square feet.
>
> But in 1980, two members of the council's anti-development block were defeated . . . and a few months later the Irvine Company filed a request to expand the center.[11]

The application was a gamble—just two years earlier, the giant Prudential Insurance Company floated a plan to build a 16-story office building and a hotel on land it planned to purchase from the Irvine Company. When the city council rejected the request, Prudential decamped inland to the South Coast Plaza neighborhood, where owner Henry Segerstrom welcomed them with land and enthusiasm. The result was the construction of the 16-story Imperial Bank building.

The company managed to get permission for the Newport Center expansion first from the city planning commission, and finally, after slimming down the project a bit, from the city council in early fall 1981. SPON and its allies launched a signature drive to require a referendum and had no trouble obtaining enough names to force an election. For the company, the timing was terrible—the controversy over dramatically increased lease payments on homes was gaining uncomfortably strong momentum in the city. The company let the city council off the hook and withdrew its application for the expansion.

Four years later, Newport Center was undergoing a dramatic remodeling, led by what the Irvine Company called the Fashion Island Renaissance. The venerable JCPenney building was closed down and hollowed out to become Atrium Court, with specialty shops looking down onto the ground floor, and Irvine Ranch Market, specializing in fresh produce and other expensive foodie delights to serve discerning Newport Beach housewives.

But the most striking makeover for the area was undertaken at the personal direction of Donald Bren, who two years earlier had assumed majority ownership of the company. Former Irvine Company vice president Bill Watt vividly remembers the day he drove onto Newport Center Drive and saw the deciduous street trees being ripped out by their rootballs and tossed into dump trucks.

Watt called Ken Noll, a friend who had remained at the company after Watt departed, to find out what was going on. Noll explained that Bren was bringing in fully grown palm trees that would be planted in the median of Newport Center Drive, which encircled the entire business complex. Date palms would be bunched in the interior of Fashion Island, along with bougainvillea, all part of a new palette of landscaping to give the outdoor retail center a more Mediterranean feel.

Irvine Company vice president of planning Roger Seitz explained that "when Newport Center opened . . . its landscape design was formal and urban. The concept then was to give the center an eastern look, with sycamores selected as the principal street trees. It didn't work out well. The climate wasn't right, the trees didn't flourish . . . our new Mediterranean theme we feel is far more appropriate."[12]

The remodel and relandscaping, which cost $60 million, set the stage for the next great Newport Center battle, which kicked off in mid-1986 with the company filing plans for a $300 million expansion of Newport Center, asking to build "several new office buildings, hundreds of new shops, town houses, cultural amenities and road improvements on 518 acres in and around the Fashion Island complex."[13,14]

SPON and other allies formed up under the new label of Gridlock, using the playbook that backed off the company in 1981. They gathered signatures days after the city council approved the plan, forcing a referendum. But this time, the Irvine Company decided to fight.

The company turned to Nelson and Associates and its brilliant but mercurial owner, Bob Nelson, to organize its campaign.

Nelson had been working in a mid-level position in the county healthcare agency when he caught the eye of supervisor Ronald Caspers. Nelson never graduated from high school but was an autodidact who consumed hundreds of books and newspapers. Nelson could write clear and compelling prose and displayed a creative, sophisticated grasp of issues. He helped Caspers win re-election to a second term, but a boating accident claimed Caspers' life just days later. Nelson worked for his successor, Thomas Riley, then left after a few months to learn the political consulting trade with Butcher and Forde, the county's most successful and notorious campaign firm.

Nelson was well known to Gary Hunt and Jack Flanigan, Bren's top political advisors. Both were impressed by Nelson's winning record as a campaigner but had also experienced Nelson's sometimes erratic and even childish reaction when anyone dared to challenge his strategic advice.

Nelson's business partner, Eileen Padberg, was a veteran of local and national Republican campaigns. In addition to helping craft and implement strategy, she sometimes had to intercede with clients to patch up relations after Nelson had stormed out of a meeting or sulked because his advice was questioned.

Nelson convinced Hunt, Flanigan, and Bren that to secure the attention of busy, affluent Newport Beach voters, a comprehensive campaign using direct mail, cable television, phone banks, and precinct walking was needed. Nelson created multipage, full-color mailers extolling the benefits of an even larger and more sophisticated "Town Center for Newport Beach," along with millions of dollars for traffic improvements. Irvine Company executives signed up to go door-to-door in search of "yes" votes.

Nelson even prevailed upon the famously private Bren to send out a letter directly from him, signed personally, to a select set of voters Nelson believed would be flattered to receive a missive from Donald Bren in their mailbox.

The voters didn't buy it. The measure lost 11,390 "no" to 8,260 "yes."[15]

Campaign reports showed the company spent $500,000 (about $1.4 million in 2025 dollars) for the persuasion effort, a stunning $25 per vote cast.

Flanigan theorized that the election offered every voter in town who felt they had ever been mistreated by the Irvine Company a chance to settle the score by voting "no," regardless of the merits of Newport Center.

The 1986 vote proved to SPON and their allies that passions in Newport Beach were hot when it came to development, and that channeling those passions into action was relatively easy if it involved the ballot box. Four years later, not waiting for a council action to challenge, SPON took to the streets to place an initiative on the ballot for an anti-growth measure it called Greenlight.

Greenlight amended the city charter by requiring a public vote on any project—residential, commercial, or otherwise—that generated more than 100 peak-hour automobile trips. It also limited residential limits to 100 units without a vote. Commercial or retail developments exceeding 40,000 square feet would also require a public vote.[16]

In addition to spending half a million dollars to oppose Greenlight, the Irvine Company also published an ad signed by Gary Hunt promising that the company would never build another high-rise building in Newport Beach. Apparently, Hunt either had his fingers crossed or assumed that the promise would only be binding if Greenlight failed. It passed, and a few years later, the Irvine Company built a 20-story building between 500 and 550 Newport Center Drive.[17]

The city processed the building, and no public vote was required. The reason, most people knowledgeable about zoning law concluded, was that the Irvine Company had, over time, been quietly emptying office buildings in the lower reaches of Newport Center. After the building was devoid of humanity, the traffic numbers those tenants had generated were banked and applied to the new building . . . presto, no Greenlight.

As the years passed, anti-growth initiatives appeared regularly on the ballot in Newport Beach. Most of them failed (Measure K in 1988, Measure X in 2006), but anti-growthers believed that restrictions like Greenlight probably kept millions of square feet of development from occurring. One prominent example occurred in 2017, when plans for a 24-story residential condominium were first approved, only to be withdrawn in the face of a possible voter referendum.

Another location slated for development was the Banning Ranch on the west side of the city. A consortium of developers worked for years crafting a plan to clean up what had been a working oilfield and build homes on a sliver of it, leaving the vast majority of the property in open space. But for SPON and others, it was an all-or-nothing equation, and it wasn't until a $50 million gift from a local multimillionaire placed the property in public ownership that the Banning Ranch battle was over.[18]

What may become the final chapter in Newport's anti-growth history will unfold in 2030, when the Settlement Agreement governing operations at John Wayne Airport is set to expire. The fight, if there is one, will not be joined by Jean Watt or Allan Beek, who both died in their mid-90s, Watt in late 2023, Beek in early 2024.[19, 20]

***

Momentum for growth limitations reached an apex in Orange County in 1988, when a coalition of activists led by Thomas Rogers announced that they would write and qualify an anti-growth initiative to place before county voters in June 1988.

Rogers owned a home on Spotted Bull Lane in San Juan Capistrano, on a bluff overlooking Interstate 5. He described himself as a rancher, with the appearance and countenance that matched. About six feet, two inches tall, with a craggy face and large ears, when he donned his Stetson fedora, Rogers bore a passing resemblance to Lyndon B. Johnson.

Politically, Rogers was the former president's opposite. He had chaired the Orange County Republican Central Committee in the early 1960s and was active in anti-tax efforts. When plans for the San Joaquin Hills Transportation Corridor were released, it became apparent that a portion of Rogers' backyard would likely be affected to accommodate the toll road's connection to I-5. His opposition to the toll road, blended with his general anti-tax attitude, coalesced into his decision to lead the charge against the county's plans in 1984 to increase the sales tax for transportation improvements.

Standing David-like against the Goliaths of county real estate development and their major-employer allies, Rogers enjoyed a sweet victory when Measure A was rejected by county voters in a landslide in 1984. Rogers became the de facto leader of the county's anti-growth advocates.

In June 1987, Rogers and his compatriots, organized under the banner of Orange County Tomorrow, presented the Orange County Board of Supervisors with a copy of their proposed growth management initiative, which tied project approval to traffic speed during rush hour.

In a long analysis, *Los Angeles Times* reporter Jeffrey Perlman laid out what he described as an unprepared landscape of elected officials and land developers struggling to decide how to react to the measure. Rumors flew that individual supervisors might offer countermeasures to head off a signature drive (66,000 were needed to qualify). Irvine Company executives Thomas H. Nielsen and Gary Hunt had a private meeting with Rogers and others in his camp, including "Irvine Mayor Larry Agran and Irvine Councilman Ray Catalano, both of whom were involved in the effort to put the proposed initiative together, also participated in the discussion."[21] The meeting produced no results.

Eileen Padberg and Harvey Englander, both veteran political consultants in the county, told the *Times*,

> They believe that the initiative would pass if it were put to a vote right now. But as popular as the concept of limiting growth may seem, they say, politicians fear being linked to anything that may cause an economic downturn.[22]

A year later, after the politicians, developers, and others did find their way and opened their wallets, it was Rogers and his allies who were asking themselves how their proposal had been rejected by voters 56 to 44 percent.[23]

Perlman once again provided an analysis, which was succinct: "In the end, it all came down to money. The nearly $2 million spent by opponents of the slow-growth initiative was the major factor in the measure's . . . defeat Tuesday, both sides agree."[24]

It was the last vote on development affecting the entire county. A few years later, the county would become consumed in a series of ballot box battles over the fate of the El Toro Marine Corps Air Station (see Chapter 18). And in Irvine in 1991, the Irvine Company faced a referendum on the zoning of the second phase of its Westpark development but beat it back using the same tactics that bested Rogers: spend a lot of money in opposition.

There would also be pockets of anti-growth sentiment that cropped up in places like San Juan Capistrano, Costa Mesa, and Seal Beach. But by the time El Toro's fate was revised into what would become the Great Park and surrounding development, only an extension of one of the county's three toll roads would spark a major environmental conflict. And in Irvine, it would be 2025 before growth and development once more sent communities to the barricades (see Chapter 26).

Notes

1 Brad Hellis, *Oral History* (DeGraff Center for Oral and Public History at California State University Fullerton, 1985).
2 Daniel Yankelovich, *New Rules: Searching for Self-Fulfillment in a World Turned Upside Down* (New York: Random House, 1981), 15–17.
3 Irvine City Council Minutes, November 1975.
4 Herman Wong, "Growth Is the Issue in Irvine Election," *Los Angeles Times*, April 2, 1978.
5 Larry Agran, The Cancer Connection (New York: Anchor Books, 1977).
6 "Councilman Attacks Irvine Company," *Los Angeles Times*, April 24, 1980.
7 Mark Baldassare, *Trouble in Paradise: The Suburban Transformation in America* (New York: Columbia University Press, 1986), 45–48.
8 Mark Baldassare, "Orange County Annual Survey," University of California, Irvine, 1982.
9 Baldassare, *Trouble in Paradise*, 45–48.
10 "Newport Center Development Limits Cut," *Los Angeles Times*, March 19, 1981.
11 "Fashion Island Renaissance," *Los Angeles Times*, January 27, 1985.
12 "Newport Center Remodel Cost $60 Million," *Los Angeles Times*, November 23, 1986.
13 "Irvine Company Files $300 Million Expansion Plan," *Los Angeles Times*, August 17, 1986.
14 "Newport Center Referendum Defeated," *Los Angeles Times*, October 1986.
15 Greenlight Initiative, Newport Beach City Charter Amendment, 1990.
16 "Irvine Company Builds 20-Story Tower," Orange County Register, 1995.
17 "Banning Ranch Saved by $50 Million Gift," *Los Angeles Times*, March 2022.
18 "Jean Watt, Newport Environmental Icon, Dies at 95," *Daily Pilot*, December 2023.
19 "Allan Beek, Ferry Heir and Growth Foe, Dies at 96," *Newport Beach Independent*, February 2024.
20 *Orange Juice Blog*, "The Yellow VW Deathtrap," August 2025.
21 Jeffrey Perlman, "Slow-Growth Initiative Stirs County Debate," *Los Angeles Times*, June 21, 1987.
22 Perlman, "Slow-Growth Initiative."
23 Orange County Registrar of Voters, "Measure A Election Results," November 1984.
24 Jeffrey Perlman, "Slow-Growth Measure Defeated," *Los Angeles Times*, June 9, 1988.

3 Committee of 4000

In mid-1981, the owners of two homes on leased land in Newport Beach received notices that the value of their land had gone up.

Way up.

There were thousands of homes in the bayside city, as well as in nearby Irvine, where residents had purchased the house that became their residence but not the land.

The reason was the deal the Irvine Company had offered at the time: buy the land with the house, at a higher price and resulting in higher mortgage payments, or lease the land for 50 or 75 years and keep payments lower.

It was in the fine print of the lease that homeowners found that was later described as a "ticking time bomb"; the provision that 25 years into the term of the lease, the value of the land would be revised, and with it, the annual lease payment.

Then the news began to circulate about the astounding increase in annual rent the revaluations would cause for the homeowners.

Harry Baker became the poster boy for what would become an uprising of (largely) upper-middle-class leaseholders, primarily in Newport Beach, who challenged the Irvine Company and its new owners on a dollars-and-cents issue many feared could cause them to lose their homes.

Baker owned a large house on Bayside Drive, located on a quiet saltwater channel across from Balboa Island. Baker was a successful business owner—he had purchased the home in 1972 for $200,000. He took the land lease option, thinking at the time the $1,600 annual fee was a good deal.

In 1980, his land had been reappraised, rising to $480,000. He thought the new number was fair and paid his new annual rent payment.

Then came 1981.

Baker was astonished when, in July, he received word that the value of his land had more than doubled to $1.2 million, raising his annual lease payment to $67,666. Accusing the Irvine Company of using "real estate boom" values to justify the increase, he refused to pay, forcing him into default on the lease.

As word of Baker's dilemma began to spread through Newport Beach, worried homeowners dug out their lease documents and began to calculate what a significant increase in land value might mean. In 1982, just 60 homes were scheduled for reappraisal, with more in line in the following years.

DOI: 10.1201/9781003683087-3

Baker was clearly at the top of the valuation pyramid, with a 3,600-square-foot home fronting on the waterfront. His neighbors, had they taken the lease option, were likely in the upper 5 percent of Orange County income earners.

But many residents living on Irvine Company leases were not wealthy. They were teachers or aerospace workers who had purchased a home in the newly planned communities developed decades earlier. And they were scared.

On September 27, 1981, a handful of leaseholders gathered in a local park to discuss the impending crisis and determine what, if anything, to do about it. These were largely well-educated professional people—attorneys, bankers, real estate agents, and the president of the Los Angeles branch of one of the nation's largest advertising firms, Foote, Cone & Belding.

A few days later, the group took out a full-page ad in local papers declaring their concerns about the leasehold situation, signing it with their new name: The Committee of 4000 (there were an estimated 3,600 leases on Irvine Company land).

***

The concept of leasing land on the Irvine Ranch dated to the early part of the 20th century and the business philosophy of James Irvine II.

The owner of the huge ranch was determined to retain as much of the property under his ownership as possible. The way to do it, he reasoned, was to avoid selling. By leasing land to those who would till and graze it for him, he would receive an annual income and retain ownership of the land.

The philosophy continued after his death, and in 1960, when plans were announced for the development of a new city adjacent to the forthcoming University of California at Irvine, leasing the land instead of selling it was to be a principal element in development.

A lengthy story in the November 13, 1960, edition of the *Los Angeles Times* laid out the plan: "William Spurgeon III, vice president of the Irvine Company in charge of developing the new city, said all land is to be leased and improved by the owners."[1]

"The concept of leasing rather than selling land for homesites and other buildings is probably one of the most unusual departures of all in the development of this planned city," the *Times* story continued.

Spurgeon noted that land leasing for homes had been underway in Newport Beach since the 1930s, "and the idea has worked for all concerned." The story went on: "present plans call for 55-year leases with reappraisals every 25 years (which permits both the land user and the company to determine whether the lease is equitably priced in accordance with economic conditions at the given intervals").[2]

Almost a year after Spurgeon's interview, Ray Watson—well into his new job as a junior planner—wrote a memo on leasing after a meeting with a Mr. DeVault, identified as an employee or consultant to Kaiser Industries in Hawaii.[3]

DeVault counseled that both commercial and residential properties were fair game for leasing. However, he said, it is difficult to determine a precise lease payment at the outset of the lease. "The lessor cannot, with fairness, expect to reap all the benefits of the increased value of the property at the point of re-evaluation,

Figure 3.1 Full-page advertisement by Committee of 4000

since the lessee has contributed materially to the increase by his own investments in the improvements therein," Watson wrote in his memo.[4]

He went on:

> If, at the point of re-evaluation, the Irvine Company determined the new rent by taking the entire increase in the land value for themselves, we would, in

> fact, be penalizing the lessee for his past performance in the upkeep of the property. In Mr. DeVault's opinion, a fair formula would be a policy set forth at the beginning of the lease and clearly understood by the lessee . . . that an independent appraiser would determine the new appraised value at the point of re-evaluation, and that only fifty percent of the difference between the original and new appraised value would be used in the determination of the new ground rent.[5]

Watson then noted a key point from the De Vault meeting—the consultant's recommendation that

> the Irvine Company develop a clear and equitable policy on (the company's) method of determining the ground rent . . . as well as unalterably clear (Watson's emphasis) on what basis the re-evaluation will be made at the end of the fixed lease rent period.[6]

A final section of Watson's memo highlighted how wrong he and the other well-educated planners and executives at the Irvine Company would be about future rising land values. His memo laid out the example of a homeowner leasing land worth $15,000 instead of buying it. If the owner instead invested it in stocks or mutual funds, the value would grow with inflation. In some cases, Watson believed, "the interest may well be greater than the lease payments and the growth of the stock will be greater than the inflated value of the land."[7]

Watson had been gone from the company for four years when the lease controversy erupted. It would be up to the new owners and managers to handle it; whether he remembered his memo from nearly 20 years earlier isn't known.

***

The Irvine Company's response to the dramatic language of the Committee of 4000 ad was nonplussed. What was the big deal, company public relations officials asked? "There is nothing to fight about, because the terms of the lease were spelled out to homeowners," one PR representative noted. But in a lengthy timeline analysis of the controversy, the Committee noted that "just to be on the safe side, the company, in an advertising campaign, announced that it is not greedy and the matter is out of its hands."[8]

Company public relations were the domain of Martin Brower, who had been hired eight years earlier, and Jerry Collins, a former *Daily Pilot* reporter with more than fifteen years of service at the company. Collins was an extremely talented writer and had honed his skills in the past as the author of regular full-page ads sponsored by the company in the *Pilot* and *Orange County Register* and, later, in the Orange County edition of the *Los Angeles Times*.

Collins and Brower may not have been surprised when the Committee of 4000 turned the full-page strategy around on them. After all, one of the committee leaders was Louis E. Scott, head of the Los Angeles office of Foote, Cone & Belding, a national advertising firm.

(Brower, in his memoir, said he was tipped off to the impending controversy by one of the company's appraisers weeks before the notices of increased payments went out. He said he developed an outreach plan to contact all company lessees, but company president Peter Kremer nixed it. Kremer, Brower explained, figured that with just a handful of leases affected, the blowback would be minimal.)[9]

With a large headline reading "An Open Letter to The Irvine Company from the Committee of 4000," the ad laid out a litany of objections and claims that covered virtually every aspect of the brewing controversy, from increases in land values and subsequent lease payments to how appraisals were conducted. "We have no desire to go to litigation, but we are prepared to do so," the ad stated, concluding with a call for "a reasonable basis to resolve this matter and work together."[10]

Readers were asked to fill out a card and return it to Barbara Young, identified as "president of the Committee of 4000."[11]

Young owned an escrow service in Newport Beach. At first, she didn't feel any immediate concern because her home in Bayshores had another ten years on its lease, she told the *Los Angeles Times*. However, she did have concerns about her friends, so she attempted to find someone at the Irvine Company to talk to about the issue. The company, she said, ignored her.

Big mistake.

Young described herself as "combative" and quickly channeled that element of her personality into leadership of the Committee of 4000. She relished the fight, although she realized taking on the Irvine Company was like "fighting God. In this community, there is God and there is the Irvine Company, or maybe it is reversed."

Young swept into motion at the Committee as the first of hundreds, then thousands of dollars in support began to roll in. A meeting was held at Corona del Mar High School, attracting 800 people, including both lessees and property owners. The story was running regularly in the local press, stoked by Young's attacks and the Irvine Company's responses. Young proposed a budget of $150,000 to support the Committee's efforts, not an insubstantial sum for a grassroots effort.[12] ($150,000 in 2025 dollars would equate to $588,000.)

While Brower and Collins attempted to fend off the Committee's relentless assaults in the press, the internal debate at the company was heating up.

Kremer, whose real estate experience was founded mainly on dealmaking, initially adopted what he considered a reasonable position: the lease language is clear, the homebuyer agreed to it, and the new payment amount was therefore owed in full.

Irvine Company board chairman Alfred Taubman backed up Kremer. The veteran of dozens if not hundreds of negotiations of store leases, construction agreements, and financings, Taubman told a small meeting of owners and senior executives, his view was simple: a contract is a contract. (Taubman was quoted in the *Los Angeles Times,* expressing disbelief that "multimillionaires" in Newport didn't understand the terms of their leases.)[13]

Kremer and Taubman aside, the man who owned the largest stake in the company, Donald Bren, was worried that the lease situation had the potential to poison the well of community attitudes the company needed to cultivate for future entitlement efforts. Bren had also been concerned about Kremer's leadership style for months—his management of the leasehold issue reinforced Bren's attitude that Kremer wasn't up to the task of running the company.

As Christmas approached, the company announced a rent deferral plan for leaseholders coming up for adjustment in 1982. It proposed that 50 percent of the new rent be paid immediately, the remainder due when the property was sold. And the future increase in annual rent would be limited to the CPI index or 8 percent, whichever was higher.

Not enough, the Committee said. The real issue is the value of the land.

Both sides paused for the holidays. It was the Committee that launched the first volley a few days after the celebration of the birth of the Prince of Peace.

On Sunday, January 10, more than 800 people jammed the ballroom of the Marriott Hotel in Newport Center, which overlooked Irvine Company headquarters at 550 Newport Center Drive. The Committee unveiled their "Leaseholder Bill of Rights," its headlines composed in Old English type to reinforce its intended resemblance to what had been written at the Constitutional Convention nearly 200 years earlier. It laid out a series of demands favorable to leaseholders.[14] (The event was black-tie, prompting some local cynics to start circulating the phrase "rent control for the rich.")

A week later, the company was back with another variation of its previous offer but now with a significant concession—a purchase plan with a 7-year loan on a down payment of 20 percent of the land value. The interest rate would be variable, from 12 to 16 percent.[15]

And while the details of the financing plan would continue to be fine-tuned, the company drew a line in the sand: it would deal with individual leaseholders on a case-by-case basis but would not negotiate with the Committee on any sort of comprehensive settlement.[16]

The following month, the promised legal action took place as four leaseholders filed a class-action lawsuit against the company in Orange County Superior Court.[17] It was a bold move—the company had virtually unlimited resources and retained some of the county's most powerful and brilliant legal minds. The leaseholders and the Committee had public opinion on their side but would have to raise money to pay their attorneys. However, by moving the battle into court, discovery could be made and depositions taken—things the company would not welcome.

The lawsuit accused the company of a variety of offenses, from moral ("unconscionable to raise rents beyond the cost of living"); to contractual ("the company breached provisions of the lease . . . by not excluding from land appraisals certain contributions of value allegedly made by lease holders"); to basic legal ("a contract of adhesion . . . lease presented on a take it or leave it (basis)").[18]

For the remainder of 1982, the conflict settled into something of a siege, with the Committee of 4000 lawsuit waiting in line at Orange County Superior Court with other business disputes for its day in court and the Irvine Company occasionally tweaking offers for lease buyouts or land purchase.

In late spring and early summer, the Committee made two legal jabs at the Irvine Company to try to apply pressure on the company's continuing refusal to negotiate with the Committee itself on behalf of homeowners. Both motions—seeking to prevent the company from even communicating with leaseholders—failed, with Committee lawyers conceding that the maneuvers were, at best, a long shot.

The Committee did score a victory in late December, when Superior Court judge Claude Owens ordered the company and committee to "use their best good faith efforts to attempt a negotiated settlement of some or all of the issues raised" in the legal action that had been filed earlier in the year.[19]

Internal currents at the Irvine Company were continuing to swirl during the same timeframe, however.

Debate within management was sharpening. Gary Hunt, who had risen to a senior position in the public affairs department, recalled in 2022 going to the board of directors with both polling and anecdotal evidence that convinced him "if the leasehold situation is not resolved, the Irvine Company will never get another favorable zoning decision in Newport Beach."[20]

Bren remained distressed about the leasehold stalemate, both for how it was damaging the company's reputation and because it might impact his secret plan to buy out his partners in early 1983 when the management agreement the owners had written at the time of the original purchase came up for renewal.

And while Bren's plan was closely held, Peter Kremer sensed that his days at the company were drawing to a close. Kremer knew Bren had lost faith in his management of the company, and if Bren purchased a controlling interest, there would be no place for him. Kremer took the initiative and, in October, announced that he would resign effective January 1, 1983.[21] His ownership of a handful of Irvine Company shares would fully vest in December, making him—at least on paper—a multimillionaire.

Kremer's departure set the stage for the Bren buyout, which took place in April 1983, and just days later, Bren and his newly installed company president, Thomas H. Nielsen, were promoting a new plan that would allow leaseholders to purchase their land.

Like many real estate transactions structured by or with the Irvine Company, the various elements of the proposal were complex, so the letter than Nielsen sent to all homeowners boiled down the details to the bottom line: for those who wanted to purchase their land and forever say goodbye to lease payments, "the net effect of the agreement would be to reduce the price to purchase the land . . . as a result, more than two-thirds of our leaseholders will be able to buy their land for less than $50,000."[22]

For leaseholders who wanted to continue to lease, the company proposed a deal that "under the amended lease, more than 60 percent of all leaseholders would have a rent under $100 a month and more than 80 percent would have a rent under $200 per month."[23]

And along with the carrot came a stick: "The settlement agreement requires that at least 75 percent of the leaseholders in any one development must accept the settlement package before any of the leaseholders in that development may take advantage of the terms of the settlement."[24]

Two months later, at a park in the Irvine Terrace neighborhood along Pacific Coast Highway across the street from Newport Center and Fashion Island—the

Irvine Company's flagship commercial and retail center—a celebration took place where Nielsen and Barbara Young buried a hatchet in the soil to mark the conclusion of the conflict.[25]

Young declared victory, while Nielsen acknowledged that the company "got to the table a little late" but was pleased that it was all over.

The Los Angeles Times story that described the afternoon celebration contained several comments that reflected some of the attitudes that had circulated around Newport Beach after the 1977 sale of the company to Taubman, investors from Detroit and New York, as well as locals, Bren and Joan Irvine Smith.

The *Times* story said:

> Several of the 250 picnickers said their feelings about the Irvine Company warmed as soon as developer Bren purchased the firm in April. Residents hailed what they called the new regime led by Bren, who grew up in Los Angeles and Newport Beach, and spoke bitterly about the 'out of towners' who took over the company in 1977.[26]

Another resident was quoted as saying, "I would have gone to the carpet before I would have paid Taubman and gang what they wanted."[27]

It is doubtful that Taubman or any other members of his "gang" cared much about the opinions of anyone in Newport Beach. Few of them spent any time there except for board meetings, and in a few short months, they would realize an extraordinary return on the investment Taubman had lured them into.

For Bren, it was a short-lived moment of glory. Within a few months, his plans for the Irvine Ranch would proceed in the face of almost continuous opposition, in Newport Beach and in other communities.

Notes

1 "Irvine Company Plans Land-Lease Development." *Los Angeles Times*, November 13, 1960.
2 "Irvine Co. Land-Lease Plans."
3 Raymond L. Watson Papers, MS-R001, Special Collections and Archives, University of California, Irvine Libraries.
4 Raymond L. Watson Papers, MS-R00.
5 Raymond L. Watson Papers, MS-R00.
6 Raymond L. Watson Papers, MS-R00.
7 Raymond L. Watson Papers, MS-R00.
8 Committee of 4000 Records, 1979–1983. MS-R160. Special Collections and Archives, The UC Irvine Libraries, Irvine, CA.
9 Martin A. Brower, *The Irvine Ranch: A Time for People: A 50-Year Overview of the Development of the Irvine Ranch in Orange County, California* (Bloomington, IN: AuthorHouse, 2013).
10 Committee of 4000. "An Open Letter to The Irvine Company." Promotional advertisement, circa 1982. Box 1, Folder 3, *Committee of 4000 Records, 1979–1983*. MS.R.160. Special Collections and Archives, The UC Irvine Libraries, Irvine, CA.
11 Committee of 4000, "Open Letter," Box 1, Folder 3, MS.R.160.

12 Committee of 4000. "Corona del Mar High School Meeting Materials," Box 1, Folder 4, *Committee of 4000 Records, 1979–1983*. MS.R.160. Special Collections and Archives, The UC Irvine Libraries, Irvine, CA.
13 "Taubman Defends Lease Terms Amid Newport Uproar." *Los Angeles Times*, July 12, 1983.
14 Committee of 4000. "Leaseholder Bill of Rights Event Materials," Box 1, Folder 5, Committee of 4000 Records, 1979–1983. MS.R.160. Special Collections and Archives, The UC Irvine Libraries, Irvine, CA.
15 Committee of 4000, "Financing Proposal," Box 1, Folder 6, MS.R.160.
16 Committee of 4000, "Company Response," Box 1, Folder 6, MS.R.160.
17 "Leaseholders File Suit Against Irvine Co." *Los Angeles Times*, February 1983.
18 "Leaseholders File Suit Against Irvine Co."
19 "Legal Proceedings and Court Orders," Box 1, Folder 7, Committee of 4000 Records, 1979–1983. MS.R.160. Special Collections and Archives, The UC Irvine Libraries, Irvine, CA.
20 Warren Fix, interview by H.P. Oliver and C.M. Stockstill, June 7, 2023.
21 United Press International. "Peter Kremer, President of the Irvine Co., Announces Resignation." UPI Archives, October 12, 1982.
22 Committee of 4000. "Settlement Proposal and Nielsen Letter," Box 1, Folder 8, *Committee of 4000 Records, 1979–1983*. MS.R.160. Special Collections and Archives, The UC Irvine Libraries, Irvine, CA.
23 Committee of 4000, "Nielsen Letter," Box 1, Folder 8, MS.R.160.
24 Committee of 4000, "Nielsen Letter," Box 1, Folder 8, MS.R.160.
25 "Leaseholders, Irvine Co. Bury Hatchet in Newport Celebration." *Los Angeles Times*, April 11, 1983, Metro section, B1.
26 "Leaseholders Bury Hatchet," *Los Angeles Times*.
27 "Leaseholders Bury Hatchet," *Los Angeles Times*.

4 The passion of Joan Irvine Smith

For decades, Joan Irvine Smith had burned with the desire to control the Irvine Ranch.

Now, in late 1977, it appeared that while absolute control was still out of reach, the potential to wield genuine influence was possible; she had been invited as a partner with the group of men who now owned the ranch.

For a woman whose birth name was not Joan, who lost her father while a toddler, and by 1977 had married and divorced three men and remained (barely) wed to a fourth, it had been a contentious journey.

Her roles as advocate, opponent, and general mischief-maker in the corporate drama that would define her adult life were created in 1947, when her grandfather's death made her a very wealthy teenager. James Irvine II willed control of the Irvine Ranch to trusted friends and family members who sat on the board of the foundation bearing his name, but the largest gift of stock to a family member–20 percent–went to his granddaughter.[1]

As Athalie Anita Irvine (her real name) grew into a young woman, her public persona was defined by her beauty and her money, as well as her seemingly untiring willingness to litigate, challenge, and defy her fellow owners of the Irvine Company.

Joan joined the company board of directors in 1957 when her mother, who also owned shares in the company she inherited when her husband died, gave up her board seat in favor of her daughter. Decades younger than board members who had been contemporaries of her grandfather, Joan found that her youth and dearth of business experience allowed them to treat her as little more than family window dressing at meetings.

But Joan had plenty of ideas and, as it turned out, a mind for boardroom politics and strategy that soon became evident. Labeling the aging board members (including Loyall McLaren, who had advised her mother on business affairs and taxation for years) as "old bozos" and worse, Joan's accusations at her first board meeting sent ranch general manager Brad Hellis into an early retirement.

She hadn't done it alone. By the time she joined the board, Joan and her mother had retained the venerable Los Angeles law firm of Loeb and Loeb to advise them. A young partner at the firm, Howard Friedman, would build a bridge of trust with mother and daughter that would last for decades.

DOI: 10.1201/9781003683087-4

Throughout the 1960s, Joan's image was shaped by professional public relations assistance as well as the advice and counsel of her legal retainers. Smartly dressed in Dior and Adolpho, she would turn up in San Francisco, Orange County, and Washington, D.C., on her quest to find a path to taking control of the Irvine Company.

Perhaps her finest hour came with the decision of the company and the James Irvine Foundation to donate 1,000 acres of land to the University of California to build a new campus in Orange County. Joan had been a fierce advocate for the land gift inside the boardroom as well as in public. She joined local newspaper publisher Walter Burroughs to form a support coalition to pressure the company, rallying hundreds of other residents who believed in her vision of a university campus as the intellectual and physical center of what would become the New Town of Irvine.

Then it was on to Washington, D.C., to join forces with Representative Wright Patman to reform the nation's tax code as it applied to charitable foundations. Joan's advocacy was hardly grounded in good government; forcing foundations to divest all or part of their holdings of private companies could make it possible for her to leverage her 20 percent holding of Irvine Company stock into majority ownership.

Passage of the Tax Reform Act of 1969 would eventually convince the James Irvine Foundation to sell its 54 percent holdings in the Irvine Company at auction, with Joan now a part of the new ownership.[2]

***

As the only direct descendant of the men who had owned the Irvine Company, Joan was big news in the heady weeks and months that followed the September 1977 sale of Irvine Company stock to the new owners for $337.44 million. Not yet 50, she wore her blond hair shorter and had maintained her youthful figure. And after years of public and court appearances, she offered opinions about the future of the ranch in thoughtful sound bites that local and national news outlets reported in detail.

But behind the smiling facade of Joan Irvine Smith, darker forces were at play.

It is difficult to pinpoint an exact date or time frame, but in the late 1970s, Joan began to change. Her family members said she became steadily more withdrawn and suspicious, not so much of them individually, but of enemies seen and unseen.

And it wasn't just family members who were impacted.

Joan was invited onto the board of directors of the Irvine Company. However, with just over 11 percent ownership, she had almost no ability to shape strategy or operations without other members in her corner. Within weeks, Joan was forging her own path at the board, taking positions and actions that soon began to alienate her fellow owners.

Moreover, Joan took it on herself to begin attending city council and water district meetings. She would stake out a position at the back of the room with her attorneys and furiously take notes. She was often the last person in the audience. Her presence made waves at Irvine Company headquarters when confused elected officials began to ask just who was making the big decisions and speaking for the company—its management or Joan?

Figure 4.1 Joan Irvine Smith and her mother, Athalie Irvine Clarke in the mid-1960s
Source: Orange County Archives

Board chairman Al Taubman patiently explained that Peter Kremer, the company president, spoke for the company, and Joan spoke for herself. As she continued her lone wolf actions, the board told her not to bother coming back for any meetings. Forced out of the action, she began flooding Kremer with hundreds of memos, position papers, and copies of newspaper clippings. Joan also showed up unannounced one day at the offices of Irvine Pacific, the company's internal homebuilding arm, demanding to see the financial records. After reviewing them, she deemed herself satisfied that they were in order.

Finally, as had been her go-to strategy for decades, she sued her own company over a decision to help build a regional water line. As usual, she lost the case.

For reasons unknown, after her flurry of public attendance at local government meetings, she did a U-turn and submerged.

Paranoia was the term often used to describe Joan's outlook on life at the time, reinforced by her decision to retain armed bodyguards to stand watch over her home and to accompany her to both public and private events on the increasingly rare occasions she ventured out of self-imposed seclusion. (Joan's desire for secrecy didn't extend to her automobiles. For months, her bodyguards made almost daily deliveries of her messages to Irvine Company headquarters in Newport Beach, driving one of two baby blue Mercedes-Benz sports coupes with personalized license plates JIS 1 and JIS 2.

More than one Irvine Company executive who found themselves in close quarters with Joan and her bodyguards described the experience as unpleasant. "These guys were big and they wore their suit jackets open so anyone could see the pistols hanging under their arms," one recalled. "We were in laid-back Newport Beach, where nothing like this was part of everyday life. It was troubling."

Peter Kremer recalled in his memoir that while Joan was still attending board meetings, her suspicions never receded. "When we held board meetings in New York City, she traveled with bodyguards and moved furniture in front of her hotel room to block entry. Also, she never once ate with the board members or directors for fear that someone might poison her."

One company executive who did dine with Joan was Richard Reese, the vice president of planning. He recalled joining a dinner with Taubman, Donald Bren, and Joan. He thought he had been invited, he said 40 years later, because neither man wanted to sit next to Joan. During dinner, she told him her uncle, Myford Irvine, had been murdered by the Mafia, and she might be next.[3]

Irvine city councilman Basil "Bill" Vardoulis recalled that he once received a phone call from Joan asking him to meet late at night at a remote spot for a discussion. Intrigued, he showed up on a dark roadway where Joan and her bodyguards were waiting. Vardoulis had a brief conversation that seemed to be going nowhere; increasingly nervous about the location and the company, he cut it short and departed.[4]

Late-night phone calls were something a number of people who dealt with Joan at the time said was standard procedure for her. One woman who shared an interest in equestrian sports with Joan found herself on the phone for hours late at night, listening to Joan's thoughts, fears, and opinions. Peter Buffa, a city councilman from Costa Mesa, had a similar experience.

Two excerpts from a book and a magazine article written in 1980 and 1984 offer insights into Joan's state of mind at the time.

From *California Rich* by Stephen Birmingham:

> Now she has more or less settled in her California camp—an armed camp, literally—to fight the fight she says she will never give up while there is any fight left in her.
>
> The battle . . . has caused a deep rift between Joan and her ever loyal mother and the rest of the Irvine family . . . In the process, she had become more antisocial, and people who once considered themselves her friends complain that she never answers their letters (or) telephone calls.
>
> Says her cousin, Linda Irvine Gaede, "she's become convinced that everyone is out to get her, that people are trying to do her in, that the whole family is trying to do her in. Why else the bodyguards? She thinks people are trying to kill her."[5]

From an *Orange Coast* profile of Joan by Larry Petersen:

> With occasional interruptions, the Irvine Ranch heiress had locked herself into an 18-hour-a-day grind. Sometimes she talks herself hoarse . . . mulling with

> her attorneys over the points of her latest lawsuit, conferring into the early hours of the morning. Often she has stayed up all night, poring over dog-eared news clips . . . or typing the latest of thousands of letters and memos she has fired off to (Irvine) company officials . . . she has almost no personal life. Recently she told a local reporter that she hasn't "kept company" with a man since 1978. And except for occasional court appearances, she is a virtual recluse. She has always been suspicious of almost everyone and is quick to see conspiracies where others see coincidences.
>
> Frightened by a rash of burglaries and convinced her enemies want to "do me in" she sometimes has shuttled back and forth from several homes she owns in Laguna Beach and Newport Beach in the hopes of fooling them by being a moving target. And in case anyone wants to initiate her demise, the would-be assailant will have to stare down the barrel of one of her two 12-gauge shotguns.[6]

Joan's personal passion play would continue into the mid-1980s, when a decision by Donald Bren would set the stage for her final act as an owner of the Irvine Company.

Notes

1 U.S. Congress, House, Committee on Ways and Means, *Subject of Tax Reform* 473–504 Tax Reform Act of 1969, Pub. L. No. 91–172, 83 Stat. 487.
2 Richard Reese, interview by author, Newport Beach, CA, 2023.
3 Basil Vardoulis, interview by author, Irvine, CA, 2022.
4 Stephen Birmingham, *California Rich* (New York: Simon & Schuster, 1980), 215–217.
5 Larry Petersen, "Joan Irvine Smith: Heiress in an Armed Camp," *Orange Coast Magazine*, May 1984.
6 "Joan Irvine Smith: Heiress in an Armed Camp."

5 Donald

Donald Bren was just 31 years old when he set out to build a new master-planned community in Orange County.

With less than a decade of experience as a homebuilder, Bren had teamed up with the former city attorney and planning director of the City of Santa Ana to purchase and develop 11,000 acres of hilly grazing land owned by the O'Neill family.

Like the Irvines, the O'Neills were descendants of Gold Rush–era Californians who purchased thousands of acres of land from Spanish rancheros who had fallen into economic distress due to drought, mismanagement, and bad luck.

The Spanish name for the property was Mission Vieja; Bren and the new owners changed it to Mission Viejo and named their company the Mission Viejo Company. The land was located about ten miles south of the Irvine Ranch; its potential for development had opened up when Interstate 5 punched through to San Clemente at the San Diego County line.

James Toepfer had been Santa Ana planning director for just two years when city attorney Phil Reilly approached him in late 1963 with an offer. "Don Bren has this idea of creating a city. He wants me to be the attorney, and I'd like you to be the planner and engineer."[1]

A few weeks later, Toepfer, Bren, and Reilly met in Bren's office—the O'Neill family had agreed to a deal:

> Bren looked at Reilly and threw him a checkbook, as I recall. Reilly threw it at me and said, "Plan it." I threw it back and said "Hey, I think I need some help here. How many people have created a city?"[2]

While Reilly and Toepfer searched for money and engineers to lay out a basic infrastructure plan, Bren spent his time, in Toepfer's words, dreaming—visualizing what the new community would look like, how potential homebuyers would view it. Bren, Toepfer recalled, "thinks in a very refined way. He's got a great mind when it comes to the delicate things."[3]

Bren created a theme for streetlights and other signage designed around the Mission-style bell emblematic of Old California, spending what was at the time a great deal of money to have them custom-made. Toepfer also recounted an incident that future Irvine Company employees would find all too familiar: Bren unloading

DOI: 10.1201/9781003683087-5

dozens of pallets of adobe bricks in his company's Newport Beach parking lot, then carefully examining each color variation until he found the precise shade that would be used for the walls throughout the new community.

According to Toepfer's memory, Bren's vision extended far beyond the physical look. He conceived Mission Viejo as a complete community, built by a single company that would blend residential and commercial elements in a complementary fashion, an antidote to the typical subdivisions that cropped up in the same general vicinity without any real thought given to how they related to one another.

John Martin, who Bren had hired to do initial market surveys for Mission Viejo, said that Bren's youth and entrepreneurial outlook were another key part of his approach to the new community. "He had business acumen, but also the artistry . . . design and authenticity were important to him."[4]

The first years of the operation were tough, Toepfer recalled. Mission Viejo was a considerable distance from urban Orange County, and builders were hesitant to purchase lots there. The hillside environment ate up funds necessary for grading, water, and sewer. Finally, Bren persuaded the Deane Brothers to build homes at the entrance to the community. With a good design and a favorable price, the homes sold well—Mission Viejo was on its way.[5]

After four years at the helm of the company, Bren decided to pull out. Martin believes "he must have known that the relationship with the O'Neill's . . . wasn't going to work out over time." Indeed, when Bren confided in Martin about his plans to depart, he told Martin, "I can't make any money selling lots to the Deane Brothers."[6]

Bren turned his attention back to his homebuilding company—The Donald Bren Company—which had continued to operate while he worked at Mission Viejo. Within a few years, he had sold more than 20,000 homes all over southern California, and in 1970, he sold his homebuilding company to International Paper (IP) for $34 million. When the real estate market went into a slump, IP had a corporate change of heart: Bren bought it back for $22 million.[7]

***

Bren was born in 1932. He enjoyed a privileged postwar childhood in Beverly Hills and Newport Beach.

His father, Milton Bren, had started in the movie business as an agent, later moving up to the role of producer. The most famous of his productions was the *Topper* movies made in the 1930s. Milton Bren also had a sideline business in real estate development.

Donald's mother was Marion Newbert, who, in addition to him, had given birth to his brother Peter. Marion came from a wealthy family, making life in the most affluent city in greater Los Angeles more than comfortable. The family also had a home in Newport Beach, with neighbors like Dick Powell and June Allyson, Humphrey Bogart, and Lauren Bacall. Bren told a reporter in one of his rare interviews that a favorite boyhood pastime was rowing around Lido Island and the Newport Harbor.[8]

The Bren brothers, despite being raised in luxury, attended public schools and held summer jobs on their father's rental properties or construction sites. Time off was spent rebuilding cars, with Peter recalling that Don was meticulous about every detail of the reconstruction, a trait that would continue in later life.[9]

Most biographies give little detail about his childhood and especially the life stories of his parents. But research into the newspapers of the late 1920s and 1930s reveals a rich, colorful, and detailed portrait of his father's career in the motion picture business as well as his mother's wartime volunteer efforts.

Milton Bren's beginnings were humble. According to an extended profile written about him at the height of his Hollywood fame in 1939, he grew up in near-poverty but showed pluck by rounding up empty beer bottles for the deposit. A few years later, Milton talked his way into an attorney's office and from there into a movie studio where his enthusiasm was noticed by the legendary producer Irving Thalberg.[10]

Thalberg began feeding Bren small assignments, where his creativity blossomed; soon, he was writing movie scenarios and even entire scripts. One of the first he sold was a movie about a flea circus. Before long, Bren had graduated into the role of agent, and his name began appearing with increasing frequency in the Hollywood gossip columns of Los Angeles newspapers (dozens were publishing at the time).[11]

As his success grew, Milton Bren was recognized as an attendee at galas and parties alongside some of the biggest movie stars, writers, producers, and Hollywood insiders, including Hal Roach Jr., who ran a productive movie studio that churned out dozens of films. Milton Bren had, in a few short years, developed a deep understanding of how Hollywood worked and signed on with Roach as an in-house producer. It was there that Bren produced the *Topper* films that are best remembered nearly 90 years later. But in addition to *Topper*, Bren was involved in dozens of others, rounding up actors and actresses, commissioning scripts, and arranging for sets, lighting, and the myriad details involved in making a film.

An essential element of Hollywood business was socializing, and based on the dozens of news items of the time, Bren and his wife were active participants. They attended dinners, dances, receptions, and film openings at legendary Tinseltown spots like the Coconut Grove, Ciro's, and others. When Don and Peter were just toddlers, they accompanied their parents to a birthday party attended by, among others, the Jack Bennys and the Mervyn LeRoys.

They could afford the social whirl. Given the studio hierarchy and the success of *Topper*, it's likely Milton Bren was making $25,000 to $50,000 annual salary in the mid-1930s, about $550,000 to $1.1 million in 2025 dollars.[12] They purchased a house in Beverly Hills for $25,000. They likely employed a maid for around $75 per month, possibly a cook for $100 a month. Newspaper reports show that Bren owned a number of racing yachts and motorboats, often competing in races to Catalina and Ensenada. The Brens spent weekends in Palm Springs with other movie business friends and went skiing in Sun Valley, where Marion took a nasty spill on the slopes.

In 1942, along with the rest of the nation, Hollywood went to war. Milton Bren, at that point nearly 40 years old, received a commission in the Coast Guard Reserve as a lieutenant commander. An August 1943 news report said he was on duty "in the East."[13]

With her husband away, Marion Bren took on an active volunteer role, along with the wives of other stars and film executives, in the Naval Aid Auxiliary (NAA), a group formed to support the wives and children of Navy servicemen on duty overseas. Marion led the NAA three times and also played a key role in organizing a canteen to provide refreshments and entertainment for Navy personnel and their dependents.

As World War II was about to end, Milton Bren was at Columbia Studios to act as an advisor for the film *Tars and Spars*, which he had written and sold to movie boss Harry Cohn. It was apparently based on Bren's own experiences in the Coast Guard, where the tars were the men, and the SPARs were members of the Coast Guard auxiliary staffed by women.

The film was a musical, and while it did well at the box office, controversy ensued afterward, resulting in Bren testifying in a court battle involving Cohn, whom nearly everyone in Hollywood feared. Bren's brush with Cohn didn't impact his reputation—after the war ended, he was back in business making films, this time as an independent producer.

The Bren brothers, while raised in luxury, were no strangers to work. Their father outlined his expectations for his sons in an interview with the *Los Angeles Times*:

> When they are nine or ten, they are going to do regular work in their off-school hours. Probably selling papers. A boy who works can't find enough trouble—I don't know how the idle ones ever keep out of it a second.[14]

Bren was asked, "What do you think is most important for them, or for any boy?" "Principle," he replied. "Nobody ever amounts to anything until he acquires it, and a job is more important than the job's money—a chance to express yourself, that's living."[15]

Donald was 14, a student at Beverly Hills High School, when Milton and Marion Bren divorced. Milton subsequently married movie star Claire Trevor. Marion Bren remarried to steel magnet Earle Jorgenson. (In a strange coincidence that was not apparent at the time, Bren and Trevor were married by then-Superior Court judge Thurmond Clarke, who had wed Athalie Irvine, the widow of James Irvine Jr. and mother of Joan Irvine Smith.)

When Claire Trevor married Bren, she moved into the family home with her 6-year-old son, Charles. Trevor offered her views on child rearing in an interview with the *Los Angeles Citizen-News*:

> the boys aren't allowed "soft living." They must keep their rooms clean and pick up after themselves. In the summer, the older boys work. If parents spoil their children no one suffers but the children. They have to learn to live in a world that is not theirs.[16]

The article continued, "Perhaps the most rigid rule in the Bren home is, never be late for meals. It isn't fair to the cook to straggle into the table, says Claire. In fact, late arrivers are penalized by having their allowance docked."[17]

Figure 5.1 Donald L. Bren circa 1990

Source: Courtesy of David Dmohowski

Bren remained close to his mother and her new husband and credits them, along with his father, with instilling in him a strong work ethic and discipline. His step-mother exposed and educated him to the world of artwork and design. "That was such an exciting period," Bren told the *Los Angeles Times* in 2011. "What I learned from Claire was to see a different dimension, the depth, the symmetry, relationship, color. That carried over to architecture for me."[18]

Bren attended the University of Washington on a skiing scholarship and was good enough to be considered for a spot on the U.S. Winter Olympics team, but an injury prevented it.

Looking back on the experience in his *OC Metro* interview in 2003, Bren said, "It was a wonderful opportunity to be able to participate at that level in any sport. It was character-building."[19]

Bren recovered well enough to gain a reputation in future years as a skilled master of the downhill, bordering on daredevil. In the early 1980s, he nearly broke his neck skiing at Sun Valley in Idaho and, for weeks thereafter, conducted business in a steel frame halo brace around his head.

(It wasn't the first time Bren cheated death or major injury. In August 1969, a helicopter he was piloting made a crash landing in the ocean off Laguna Beach. Bren and his passenger escaped with back injuries and cut scalps. The aircraft, owned by Bren's homebuilding company, sank.)[20]

After college, Bren joined the U.S. Marine Corps and attained the rank of lieutenant. He taught survival skills to other Marines. Bren told *OC Metro* in 2003 that he was inspired to join the military by his father, a Coast Guard lieutenant commander who served in World War II. "I followed that tradition," he said. "Wearing the Marine uniform, I just felt I could wear that with special pride."[21]

Mustering out of the Marines, Bren considered law school but, instead, found an empty lot on Lido Isle in Newport Beach and obtained a $10,000 bank loan to build two homes on speculation. They sold, and Bren's career as a homebuilder began.

Within a few years, Bren was competing against builders like Macco, which had been an early buyer of Irvine Company raw real estate on the west side of the Upper Newport Bay. Others in the game were John Lusk, a former banker who saw that the loans he was making builders were making them a lot more money in profits than the interest he was collecting. The Holstein Company was building large homes, and the William Lyon Company, headed by a former Army Air Force general, was also churning out products.

Bren focused on design and execution and was in regular contact with Ray Watson and other Irvine Company executives in the hopes of landing a project. Bren must have made a good impression, as he accompanied Watson, Irvine Company planner Al Trevino, and several other urban planning notables (including famous *New York Times* critic Ada Louise Huxtable) on a 15-day tour of New Towns in Europe in the summer of 1965.[22]

Bren's break with the Irvine Company came in 1968 when he was selected to build a large tract in Harbor View Hills, on the slopes of a gentle hillside just above MacArthur Boulevard. The homes were set along extensive greenbelts similar to what had been created in Eastbluff, and the interior designs—reflecting Bren's artistic touches—were well-received by homebuyers. The project turned out to be a winner for Bren and the Irvine Company.

During these years, Bren developed a taste for politics, and while it is not chronicled anywhere in print, it is likely he met Ronald Reagan through his mother and her second husband. Earle Jorgenson, who, along with auto dealer Holmes Tuttle and retail mogul Justin Dart, was part of Reagan's initial "kitchen cabinet" of supporters who guided the former movie star's successful path to politics and the governorship of California in 1966.

Bren became involved in Republican politics, supporting candidates and the GOP. It was something of an odd affinity for Bren, who avoided crowds and disliked the glad-handing of retail politics. A 1990 *Fortune* magazine profile of Bren described him as "a loner with a small circle of close friends."[23]

Bren continued to build houses, but the taste of large-scale development was never far from memory. So, when reports began to circulate about the potential sale of the Irvine Company, he was more than just interested. Bren began contacting banks and savings and loan institutions in Southern California to inquire about loans—he wanted to purchase the Irvine Ranch.

Notes

1 James Toepfer, Oral History, Mission Viejo Library and Mission Viejo Heritage Committee, January 22, 2012.
2 Toepfer, Oral History.
3 Toepfer, Oral History.
4 John Martin, interview by author, Irvine, CA, 2023.
5 Martin, interview.
6 Martin, interview.
7 "Donald Bren: How America's Richest Real Estate Mogul Built His Empire," *CRE Analyst*, accessed September 8, 2025, https://www.creanalyst.com/insights/donald-bren-how-americas-richest-real-estate-mogul-built-his-empire.
8 Scott Kraft, "The Bren Legacy," *Los Angeles Times*, February 13, 2011, Business section, accessed September 8, 2025, https://www.donaldbren.com/2011/donald-brens-legacy.
9 Gary Hector, "The Invisible Empire Builder," *Fortune*, April 27, 1990.
10 *Los Angeles Times*, March 5, 1939.
11 Bonnie Brennen, "Malice in Wonderland: Hedda Hopper and Louella Parsons in Hollywood," *International Journal of the Image*, Temple University, accessed September 8, 2025, https://www.ijpc.org/uploads/files/Hedda%20Hopper%20and%20Louella%20Parsons%20-%20Gossip%20Columnists.pdf.
12 Thomas Schatz, *The Genius of the System: Hollywood Filmmaking in the Studio Era* (New York: Henry Holt, 1988), 215–220; Dollar Times Inflation Calculator, accessed September 8, 2025, https://www.dollartimes.com/inflation/.
13 *Los Angeles Times*, August 10, 1943.
14 "Milton Bren Profile," *Los Angeles Times*, March 5, 1939.
15 "Milton Bren Profile," *Los Angeles Times*, March 5, 1939.
16 Los Angeles Citizen-News, October 29, 1949.
17 Los Angeles Citizen-News, October 29, 1949.
18 Kraft, "The Bren Legacy."
19 *OC Metro Magazine*, March 6, 2003.
20 *Daily Pilot*, September 1, 1969.
21 *OC Metro Magazine*, March 6, 2003.
22 Raymond L. Watson Papers, MS-R001, Special Collections and Archives, University of California, Irvine Libraries.
23 Gary Hector, "Donald Bren: The Invisible Empire Builder," *Fortune,* April 27, 1990, pp. 62–70.

6 The new world of politics and media

With new owners and management, the Irvine Company's engagement in politics intensified. New lobbyists were retained in Sacramento and in Washington, D.C. Political contributions increased. Political consultants were retained to help shape strategy. Public opinion polling was initiated that focused on the attitudes of local residents on topics like growth, development, and transportation.

Personal relationships began to play a larger role in the company's political operation, with four individuals evolving into key supporters of the company's business and political desires.

****

Marian Bergeson had four children; she was a former teacher and was concerned for their education in the Newport-Mesa Unified School District. She was elected to a seat on the school board and quickly displayed an ability to grasp issues and influence her colleagues. Petite, blond, and blessed with wrinkle-free skin long before Newport Beach women had ever heard of Botox, local Republican leaders soon saw Bergeson as a potential future candidate for higher office.[1]

In 1976, the Assembly district covering Newport Beach opened up. Bergeson was the establishment choice, but the owner of the local Mercedes-Benz dealership jumped into the race and poured hundreds of thousands of dollars into his campaign. Jim Slemmons had a sleazy reputation but won the Republican primary, which in any other year would have guaranteed election in November. But Bergeson and an army of outraged local women mounted a write-in campaign that split the party vote and elected Democrat attorney Ron Cordova. Two years later, normalcy was restored, and Bergeson won the seat.

Bergeson served in the Assembly and Senate for 17 years. Although Republicans never controlled either legislative body during that time, Bergeson managed to pass significant legislation, particularly in the field of education. And she could always be counted on to advise Irvine Company lobbyists and executives about goings-on in the state capitol.

(Bergeson's demure manner masked a tough constitution. While a state senator, Bergeson retaliated when Assembly speaker Willie Brown killed a bill she was sponsoring. Using a procedural maneuver, she stopped a bill in the Senate that Brown was eager to see pass. A few minutes later, Bergeson was talking to a

DOI: 10.1201/9781003683087-6

Figure 6.1 Marian Bergeson

Source: Orange County Archives

colleague on the floor of the Assembly when a visibly livid Brown stalked into the chamber and confronted her, his face just inches from hers. His voice rising, Brown shook his finger at her, a scowl creasing his face. Bergeson—perhaps channeling her experience as a teacher—calmly stared down the enraged Brown as she might a child throwing a tantrum.)

Bergeson returned to Orange County to win a seat on the Orange County Board of Supervisors. The homecoming was marred a few months after her election when the county declared bankruptcy following a scandal that took down the elected county treasurer. Bergeson limped through two years of her term and then resigned to become secretary of education in the cabinet of then-governor Pete Wilson.

***

Pete Wilson had a personal connection to the Irvine Company—Donald Bren was a friend and fellow Marine. Equally important, three key company executives who came to operate political and media relations in the critical decade of the 1980s to early 1990s had strong links to Wilson and his team of advisors.[2]

Wilson was a California transplant who was practicing law when the 1964 "one man one vote" decision upended the traditional map of state politics. He won a seat in the state assembly in 1966, then in 1970 vaulted to the mayor's post in fast-growing San Diego.

Moderate in outlook, Wilson governed carefully and wisely, building a reputation as a competent administrator. While faulted for being charisma-challenged, as one admirer put it, he nevertheless won the 1982 Republican nomination for U.S. Senate and beat former governor Jerry Brown in the general election.[3] Wilson had just been re-elected to the Senate in 1988 when he decided to return to California and seek the governorship. He won and served two terms.[4]

Wilson's 14 years in office ensured the Irvine Company's ability to influence federal and state legislation at the highest levels. Public affairs executives Gary Hunt and John "Jack" Flanigan had known and worked for or with Wilson for years; they rarely found it difficult to get his ear.[5] Larry Thomas, who by then had been recruited to oversee the company's public relations, had been Wilson's press secretary when he was mayor.[6] Topping it off, long-time Wilson chief of staff Bob White had deep affection for Hunt, Flanigan, and Thomas, in some ways looking at them as the sons he never had.[7]

***

Thomas Riley's political career was launched on a gruesome tragedy.

In June 1974, Ronald Caspers had just finished his first term as a county supervisor representing all of the Irvine Ranch and the southern part of the county. To celebrate, Caspers and his sons set off from Dana Point on a boat owned by Fred Harber, Caspers' closest political advisor. They enjoyed a sunny journey to Baja California but, on the return trip, were overtaken by a hurricane that sank the boat.

A frantic search for the ill-fated *Shooting Star* was undertaken, but when the boat's Boston Whaler lifeboat washed ashore, hope faded to resignation. All aboard were declared dead.

Weeks after the accident, *Fullerton Tribune* reporter Gary Granville dug deep into the story of the sinking and its cause, determining that at its last landfall, the boat had taken on fuel in 50-gallon drums that were stashed—perhaps not securely—into the hold. The rough seas, Granville theorized, broke loose the drums, which punched holes in the side of the boat. The vessel was never recovered.[8]

California governor Ronald Reagan filled the empty seat on the board with a surprising choice: former Marine Corps brigadier general Thomas J. Riley, a complete newcomer to Orange County politics.

Riley had been the inspector general of the Corps, a one-star general with a distinguished career as a Marine, including combat duty in the South Pacific during World War II. Riley had been working quietly in Orange County for a decade, unhappy—it was later reported—with his role as an administrator for an aerospace firm.[9]

Riley dove into his new job with gusto. Soon known not as "supervisor" but "the general," it was not unusual for Riley to answer his own phone, especially early in the morning when he habitually started work. He had a raspy voice and was uncomfortable in speaking extemporaneously in his early months in office, so he usually spoke from a script. In private, he was warm and welcoming, especially with men who had served in the military. He embraced growth in the Fifth District, which he proudly proclaimed as the "Fabulous Fifth." Developers and builders, including the Irvine Company strongly supported Riley. He would serve 20 years as supervisor, during which time more than one million new residents moved to the county to take up new homes and jobs.

Riley rarely wavered from his support for development in his district but also saw the wisdom of creating open space and parkland. He was also a leader in expanding and modernizing the Orange County Airport and naming it in honor of John Wayne.[10] His backing of a set of toll roads in the county was another key element in the success of the Irvine Company and other major landowners.

***

Bruce Nestande had also served in the Marine Corps, including a stint in Vietnam. Like thousands of others, Nestande fled the Midwest for a brighter future in California, where he found both the weather and political climate more to his liking.

Nestande was an avid Republican whose personal and professional life revolved around politics. Eager to attain elected office, Nestande ran and lost three times against Kenneth Cory in a west Orange County assembly district that had been so tortuously redrawn to favor Cory that critics labeled it the "Coryidor." In between elections, Nestande worked for the Republican Party, Governor Ronald Reagan, and on various political campaigns. He even spent a few months in the employ of arch-conservative Walter Knott, who, with Nestande's help, constructed a replica of Independence Hall at his Knott's Berry Farm amusement park.[11]

Finally, Nestande won a seat in the state assembly, serving six years in the Republican minority. It was a decision by Jerry Brown that gave Nestande his final and most influential office as an Orange County supervisor.

Brown appointed a former Marine Corps (Vietnam) prisoner of war, attorney Edison Miller, to a vacant seat on the board of supervisors. Miller, a Democrat, had been accused of cooperating with his captors—the county Republican establishment exploded in anger at Brown's action. Nestande was drafted to run against Miller, which he did, winning the seat in a brutal, mud-slinging campaign.[12,13,14]

Nestande was well-known to Hunt, Flanigan, and Thomas. Along with dozens of other business executives in the county, they had a love/hate relationship with Nestande, who could be edgy, crude, and occasionally offensive in both public and private settings. He also had a sensitive ego and was ever alert to slights and plots against him, some real, most imagined.

Figure 6.2 Thomas Riley

Source: Orange County Archives

But Nestande understood power and quickly elbowed his way into a leadership role on the board of supervisors. He even wrangled an appointment to the California Transportation Commission while serving on the board, a reflection of how critical transportation funding issues were to local developers and his desire to influence them.

Despite his rough exterior, Nestande was bright and shrewd and had a strong staff to support his efforts. Like Pete Wilson, Nestande was a moderate on most matters. The Irvine Company and others also counseled Nestande on issues but as often as not would end up debating them to the point of exhaustion and frustration. In the end, Nestande almost always sided with business interests.

Nestande left the board of supervisors mid-term in 1987 under a cloud of suspicion related to campaign contributions. He pivoted to a political position with developer George Argyros and played both a public and behind-the-scenes role in the county's airport wars for the next decade.

***

The first major statewide issue the new owners of the Irvine Company faced came in 1978 with the passage of Proposition 13.[15]

A growing political tidal wave of fear and resentment over skyrocketing property taxes swept the issue onto the June primary ballot. Crafted by crusty gadfly Howard Jarvis, the measure promised to cut back property tax levels to 1 percent of assessed value and to drastically limit future annual increases.

Like virtually every other establishment business in the state, the company opposed the ballot measure. However, it was not lost on the new managers how beneficial it would be for the company if Proposition 13 passed; the Irvine Company paid more property taxes than any other business in the county.

Counterbalancing the situation were the consequences of what would happen if local governments—especially school districts—had their funding slashed. It was not the big aerospace firms or established manufacturers that would be hardest hit but land developers and builders who depended on roads, bridges, and schools to be built with public funds.

Basic infrastructure was the foundation of the new communities built on the Irvine Ranch. Regional and neighborhood systems of water and transportation had to be installed up front. As the specter of Proposition 13 loomed larger, the company-controlled Irvine Ranch Water District took the unprecedented action of authorizing (but not issuing) nearly a billion dollars in bonds a few weeks before the election.[16]

Passage of Proposition 13 and how it upended public financing gave the Irvine Company a bitter taste of how statewide issues could impact its future; four years later, it was water supply that came before the voters, but by that time, the company had evolved to a point that it was prepared to take a leadership role in meeting the challenge.

The legislature and governor Jerry Brown, shell-shocked by the success of Proposition 13, moved quickly to staunch the flow of lost revenue to local governments and school districts. The state was sitting on a substantial surplus at the time, which helped ease the immediate loss of local funds.

Equally important to developers, Proposition 13 created a new standard for the passage of any "special taxes," a two-thirds majority of those voting. It was a nearly impossible number to achieve, even in the best of economic times, and in the aftermath of a voter revolt, a complete non-starter.

It was not until 1982 that a lasting and sophisticated funding alternative was created that would allow developers and homebuilders to obtain the necessary revenue for projects. It was especially helpful to owners of large, unpopulated areas like the Irvine Ranch.

Henry Mello was a senator from the central coast of California, and Michael Roos was an assemblyman from the mid-Wilshire area of Los Angeles. Both were Democrats, moderate, and eager to solve problems. Out of their collaboration was born the Mello-Roos Act, which created a mechanism for funding infrastructure needed to serve new developments.[17]

The Irvine Company and its lobbyists were deeply involved in the creation of the Mello-Roos legislation. William Geyer and David Booher were intelligent and policy-oriented, each more likely to spend time engrossed in books or white papers than prowling the Capitol watering holes after hours.[18]

The Mello-Roos legislation created Community Facilities Districts (CFDs), a defined area that could be drawn across city, special district, or county jurisdiction. It provided that the landowner and residents of the area could choose to levy taxes within the district for 20, 25, or 30 years, using the proceeds to build schools, roads, fire stations, water and sewer lines, or any other public facilities.[19]

What made Mello-Roos attractive to developers was the fact that if there were no residents living within the district boundaries, the landowner would have all the votes necessary to form it—an election involving people would take place only if 12 humans were residing within the district.

In 1982, vast areas of the Irvine Ranch remained undeveloped and uninhabited, perfect candidates for the creation of Mello-Roos districts. As homebuilding and commercial development progressed in the future, dozens of CFDs would be created on the Irvine Ranch and elsewhere in Orange County, raising hundreds of millions of dollars. In the years after 1982, the Mello-Roos Act was regularly amended, sometimes to the direct or indirect benefit of the company. For example, the ability to issue variable-rate debt in addition to debt with a fixed interest rate came about in one of those amendments, likely saving the Irvine Company and other developers millions of dollars.

Funding for infrastructure became the foundation of Irvine Company's political decision-making at every level of government.

***

During the initial years of Taubman-Allen-Bren-Irvine ownership, Robert Shelton oversaw the company's political operations.

Shelton had transitioned from consulting for the company to becoming an employee. Widely respected for his integrity and measured approach to dealing with politics and politicians, Shelton had positive relations on virtually all sides of the political spectrum. While registered as a Republican, Shelton was more moderate—even liberal—than most people thought (Shelton confided to an associate that his first vote in a presidential election went to Socialist Norman Thomas).

Shelton's bipartisan reputation paid an unusual dividend on a trip to Sacramento with Kremer. The company had sold the Irvine Coast property to the state for $32 million. With interest rates high, Kremer was eager to get the windfall invested. He and Shelton stopped off at the office of then-state controller Ken Cory, a former Orange County assemblyman. (The controller pays the state's bills). Without an appointment, Shelton asked to see Cory, who greeted him like an old friend. Informed of the situation, Cory advised Shelton and Kremer to return to his office later before flying home, which they did—with a check for the full amount tucked into Kremer's suit pocket.[20]

In 1982, Shelton decided to employ a lobbyist in Washington, D.C. to monitor and lobby on federal issues.

There was plenty of legislation that could potentially impact the company, with transportation funding at the top of the list, followed by the impact of the two airfields at either side of the company property—Lighter Than Air Station Tustin, a World War II blimp base converted to host a Navy and Marine Corps helicopter fleet, and Marine Corps Air Station El Toro, where fighter jets screamed across company lands day and night (see Chapter 18).

Shelton selected William Ferguson for the job. Ferguson had just started building a firm at the time, focusing on cities, counties, and special districts—the Irvine Company would become his first private sector client. Ferguson had substantial experience in the nation's capital, having worked as a staffer to a senator and for the National League of Cities. Compact and intense, Ferguson raced bicycles in his spare time and was a near-fanatic when it came to working out.[21]

Shelton liked the fact that by representing primarily cities and other government agencies, Ferguson didn't fit the manner or mores of the high-powered, politically connected K-Street firms that dominated the lobbying scene in Washington, D.C. Ferguson, he thought, was more process-oriented, someone who worked behind the scenes, paying more attention to the legislative sausage-making than preening at a cocktail reception and handing out checks.

Ferguson also had a strong personal connection to an obscure analyst within the land use section of the Department of Defense, who would prove extremely valuable to the company's relations with the Marine Corps and the El Toro airbase.

The Irvine Company made political contributions directly and indirectly.

In 1978, the first full year of new ownership, the company donated $178,000 to state races—constitutional offices, assembly, and senate. Those numbers rose steadily until 1984, when the company put $271,000 into the attempt to levy a one-cent sales tax in Orange County for transportation improvements. By 1985, total giving for all political races was just over $300,000 (about $900,000 in 2025 dollars), with a significant portion of it going to political action committees (PACs).[22]

There were hundreds of PACs in California at the time (today, there are thousands). A PAC was a collection of contributors, usually with a common connection related to their business interests but also for other reasons—pro- or anti-abortion supporters, victims of crime, labor members, environmental advocates—the list was long.

The Irvine Company had many business interests that justified contributing to PACs. Chief among them was the California Building Industry (BIA), the California Housing Council (apartments), Western Growers (agriculture), National Association of Office and Industrial Parks, California Roundtable, California Chamber of Commerce, Orange County Business Council, and others.

Membership didn't guarantee that a single donor could make a contribution to the PAC war chest and direct the same amount to a specific candidate—that was known as money laundering and was illegal. But if an individual donor had a position on the PAC board where the money allocations were made, it was possible to sometimes influence specific contributions.

To support elected officials at the federal level, the company formed a PAC composed solely—as the law dictated—of owners and employees and whimsically called the Irvine Company Employees Political Action Committee or ICEPAC. During each two-year election cycle, it raised and distributed around $30,000.[23] On occasion, the company gave $1,000 honoraria to federal elected officials for visiting the company, taking a tour of the property, and meeting with executives to discuss issues.

PACs or not, the other factor that was always considered at the Irvine Company when it came to political contributions was public perception. Large checks, especially in the company's backyard of Orange County, were sure to draw press attention or criticism from activists.

(As this is written, political giving in California and nationally has exploded—hundreds of millions of dollars from donors large and small are flooding into elections. And those are the ones reported—equally large amounts of money issue forth from "dark money" funds.)[24]

In 2023–24, the Irvine Company made $1.3 million in political contributions.[25]

***

The new owners of the Irvine Company inherited three media companies: a weekly newspaper, a semi-monthly magazine, and a cable television operation.

The newspaper, the *Irvine World News*, had been established years earlier by the company as a monthly publication listing community events. It transformed into a weekly paper that sometimes printed news with varying degrees of frankness about its owner along with a significant amount of city, school district, and local sports news. By 1980, it was a thriving community paper.

The magazine was *New Worlds,* published every other month. Initially little more than a slick collection of high-quality photography and light editorial topics, it had shifted slightly in the mid-1970s to carrying more substantial stories, tackling subjects with more depth and complexity, while still emphasizing beautiful photos of new homes, buildings, and natural areas on the Irvine Ranch.

Donald Bren, Al Taubman, and the other new company owners had no interest in becoming media titans, but they recognized that the publications and the cable system had potential value that could be exploited to pay off the massive debt they had incurred to purchase the company. Finance department personnel began exploring the sales value of all three, but the rapid and unexpected payoff of the loans after just over a year of new ownership took the pressure off for a fire sale.

Nevertheless, in 1980, *New Worlds* was sold to a publishing company in Laguna Beach—the price was not disclosed. An effort was made a few years later to revive the publication, but it went nowhere. And, by then, two regional magazines were publishing in the county, leaving no room for a third; *New Worlds* was dead.

Community Cablevision continued to expand as new development took place in Irvine and Newport Beach. In late 1992, it was acquired by Times-Mirror Cable. The price was not revealed, but Richard Sim recalled in a 2024 interview that it was $100 million (equivalent to approximately $230 million in 2025 dollars.)[26,27]

The *Irvine World News* held on until 2000, when it was sold to the *Orange County Register*. The countywide paper was experimenting with highly localized editions at the time, trying to stem the national trend of daily newspaper decline in the face of the onslaught of the internet.[28]

The *Irvine World News* became a weekly feature of the *Register* and, for a brief time, was even turned into a daily before returning to its weekly status. For a community institution that had once been delivered free to every home in Irvine, to a four-page insert in a paper with a countywide circulation of less than 100,000, it was a sad decline.

In the years before the *Irvine World News* was sold, within the Irvine Company, executives were considering going on the offensive with a new media strategy aimed directly at Irvine residents.

The concept came from the results of the regular public opinion polling the company undertook to take the pulse of Irvine and Newport Beach. Both communities were generally tolerant of new development—Newport less so than Irvine—but there were strong undercurrents of dissatisfaction with the pace of growth, traffic, and availability of schools.

Reading the polls, it became apparent that a single element of Irvine Ranch history resonated strongly with a large majority of residents: the belief that planning, made manifest in the city's General Plan, was good for the community.

The decision was made to create a communication tool and place it directly before residents. The result was a newsletter called *Planning Ahead*, written by the company's public relations team with substantial direction from technical experts in planning, engineering, and transportation.

In direct, persuasive prose, *Planning Ahead* was sent to Irvine households four to six times per year, examining issues in either a generic or specific manner but always with the underlying theme that almost everything the Irvine Company proposed, planned, or built was in accordance with the General Plan.[29]

Planning Ahead became a fixture in the company's array of public relations efforts, accomplishing two goals: it provided a credible way to communicate technical policy information and policies in a format that was easy to understand and generally viewed as credible. Second, it became a factor that company opponents had to be wary of if they considered launching an anti-company campaign. Before *Planning Ahead*, the company had been reluctant to engage publicly when criticized or attacked. But a communication that arrived in the mailbox of every resident—potentially in multiples—gave pause to opponents.

For a variety of factors, as the new century dawned in Irvine and Newport Beach, the fire of concern about growth and development burned itself out, ushering in a new era of relative peace on overriding issues. And while there was neighborhood opposition to individual projects, on a macro level, the need to address the issues *Planning Ahead* specialized in faded.

In 2018, the Irvine Company went full circle in the history of creating its own publications, with the introduction of the *Irvine Standard.*[30]

Formatted as a tabloid, the *Standard* initially ran just eight pages, but it was published in full color, making it hard to miss when it arrived in mailboxes throughout Irvine. After a couple of years of publication, it became clear that the content of the *Standard* could be predicted. The overriding theme was that everyone who had moved to Irvine had made the right choice because it was a great place to live.

The *Standard* explained why: Irvine was a community with a master plan that had been carefully created and implemented since the early 1960s. Homage was paid to architect/planner William Pereira, whose firm created the plan, and to men like Ray Watson, who implemented it.

The sub-theme of the master plan was open space and recreation. Irvine had created many of its own parks and recreational facilities, and the expansive donations of open space by Donald Bren both in and around Irvine were repeatedly celebrated in *Standard* editions, complete with color photos and maps.

Equal printed space was also regularly given to the Irvine Spectrum, where goings on at the Irvine Entertainment Center were chronicled in detail; new restaurants opened every year in the giant shopping area, and each one received favorable coverage. In the commercial/industrial section of the Spectrum, entrepreneurs and established executives were featured as companies were opened or expanded.

Education got its fair share of *Standard* attention as well, with students excelling on the athletic field and in the classroom recognized for their achievements.

In the early 2020s, internet-based blogs focusing on neighborhoods in Irvine and Newport Beach began to gain greater traction. The city and especially the police department began to use the blogs to promote events and post warnings about crime or explain when a crime took place and how it was dealt with.

Notes

1 "Marian Bergeson, Former State Senator and Education Leader, Dies at 90," *Los Angeles Times*, August 17, 2016.
2 Pete Wilson, Oral History interview by Lisa Rubens, *California State Archives State Government Oral History Program*, 2001, accessed July 2025, https://archives.cdn.ca.gov/oralhistory/wilson_pete.pdf.
3 "Pete Wilson Wins Senate Seat," *New York Times*, November 3, 1982.
4 "Pete Wilson," Wikipedia, accessed September 6, 2025, https://en.wikipedia.org/wiki/Pete_Wilson.
5 Author's recollection.
6 Author's recollection.
7 Author's recollection.
8 Gary Granville, "Fuel Drums May Have Caused Fatal Boat Sinking," *Fullerton Tribune*, July 1974.

9 "Thomas J. Riley, Longtime OC Supervisor, Dies at 85," Orange County Register, September 12, 2004.
10 "John Wayne Airport History," County of Orange, accessed July 2025, https://www.ocair.com/about/history.
11 Bruce Nestande, Oral History interview by Susan Douglass, Center for Oral and Public History, California State University, Fullerton, 2008.
12 "Nestande Wins Bitter Supervisor Race," *Los Angeles Times*, November 6, 1980.
13 The aftermath of the campaign was equally bitter. Miller sued Nestande for libel; he lost at the Superior Court level and on appeal. (Edison Miller v. Bruce Nestande, Orange County Superior Court Case No. 81-CV-1123, affirmed on appeal, 1983).
14 Miller continued to assert his innocence against charges of conspiring with his captors and years later won a Pentagon decision affirming his contentions. Miller died in 2022 and was buried at Arlington National Cemetery with full military honors. ("Edison Miller Dies at 84," *Stars and Stripes*, March 14, 2022).
15 "Proposition 13: How It Changed California," *Sacramento Bee*, June 6, 2018.
16 "Irvine Ranch Water District Authorizes $1 Billion in Bonds," *Orange County Business Journal*, May 1978.
17 Henry Mello and Michael Roos, *Mello-Roos Act Legislative History* (California State Assembly Archives, 1982).
18 Author's recollection.
19 "Community Facilities Districts: A Primer," *California Debt and Investment Advisory Commission*, accessed July 2025, https://www.treasurer.ca.gov/cdiac/publications/cfds.pdf.
20 Author's recollection.
21 William Ferguson, interview by C. Michael Stockstill and H. Pike Oliver. February 12, 2024.
22 "Irvine Company Political Contributions, 1978–1985," *California Secretary of State Campaign Finance Archive*, accessed July 2025, https://cal-access.sos.ca.gov.
23 "ICEPAC Formation Documents," Federal Election Commission Archive, 1980–1986.
24 "Dark Money and Political Giving in California," *CalMatters*, October 2024.
25 "Irvine Company Political Contributions, 2023–2024," *California Secretary of State Campaign Finance Archive*, accessed September 12, 2025, https://cal-access.sos.ca.gov.
26 Richard Sim, interview by C. Michael Stockstill, 2024.
27 "Times-Mirror Cable Acquires Community Cablevision," *Broadcasting & Cable*, December 1992.
28 "*Irvine World News* Sold to *OC Register*," *Editor & Publisher*, March 2000.
29 Author's recollection.
30 "*Irvine Standard* Launches with Eight-Page Tabloid," *Orange Coast Magazine*, February 2018.

7 Parks and recreation

The creation of parkland and open space for public use on the Irvine Ranch began in 1897 and continued for more than a century.

The first park was a long swath of land adjacent to Santiago Creek that hugged the lower reaches of the Lomas de Santiago. The modest foothills rose gently above the flat, fertile fields that were being turned into a vast farm. But this particular parcel was full of trees, mostly California Oaks, spreading their long limbs covered with serrated leaves and studded with acorns.

For many years, the land had been known as the Picnic Grounds, where residents of nearby Santa Ana, Orange, and Tustin gathered on weekends to spread a tablecloth and enjoy the quiet splendor of the shady groves and verdant, grassy ground.

James Irvine II made the generous gift of 160 acres of the Picnic Grounds to the County of Orange, with a series of conditions that Joan Irvine Smith enumerated in her autobiography:

> A road that bisected the park had to be relocated . . . the grounds were to be fenced and an 'inspector' appointed to keep out sheepherders and woodchoppers, no intoxicating liquors were to be sold on the premises, admission would be free, and above all . . . the trees were to receive good care and the grounds be kept as natural as possible.[1]

In 1920, county supervisors commissioned a master plan for the park, retaining Florence Koch, a member of a pioneer family from Laguna Beach.[2] Koch was a talented designer who later prospered in the movie business, creating sets for such screen classics as *Gone with the Wind* and *The Grapes of Wrath.*

James Irvine II would make two other substantial land gifts in his lifetime: a narrow strip of coastline in Newport Beach between Newport Bay and the northern city limits—about half a mile—restricted for recreational and non-industrial uses (much of the property was later leased to the Balboa Bay Club by the city of Newport Beach) and a small parcel of land in Laguna Canyon that became the grounds of the Laguna Festival of the Arts. Beyond the Irvine Ranch, Irvine gave the Save the Redwoods League close to $100,000 in the early 1940s, equivalent to nearly $2 million in 2025 dollars. One of the most popular groves in Prairie Creek Redwoods State Park is named for Irvine.

DOI: 10.1201/9781003683087-7

From the turn of the 20th century until the mid-1960s, James Irvine II and those who followed him as owners and managers of the Irvine Company gave little to no thought about parks, playing fields, swimming pools, or any of the modern recreational amenities homeowners enjoy today.

As parcels of land on the Irvine Ranch were sold off to developers, primarily in Newport Beach, the decision to place a park within a tract of homes was left to them. If a pleasant area filled with trees and playground equipment would promote home sales, the builder might consider it. But it was not until 1965 that state law mandated that 2 to 5 acres of land within a development be set aside for every 1,000 homes built.

As the Irvine Company was gearing up in the early 1960s to move out of agriculture and into large-scale urban development, the fate of two large (mostly) natural features on the ranch was being examined by company managers and planners.

Both government and private planners envisioned Upper Newport Bay as a potential marina and watersports center. However, those plans were thwarted by growing concerns about environmental preservation, resulting in the creation of one of the largest and most heavily used recreational areas on the ranch: the Upper Bay Ecological Reserve, a state wildlife sanctuary acquired by the State of California from the Irvine Company in 1975 for $3.4 million. The western and eastern bluffs above the Upper Newport Bay were gifted to the County of Orange by the Irvine Company in 1989. The county dubbed their new property the Upper Newport Bay Nature Preserve.[3]

A short distance from where San Diego Creek enters the Upper Newport Bay, a large inland estuary had existed from the time James Irvine I and his partners purchased the 110,000 acres that would become the Irvine Ranch.

The San Joaquin Marsh was created by the meandering San Diego Creek, which originated in the Lomas de Santiago foothills 16 miles from the connection point to the Upper Bay. Along the way another natural runoff channel, Peters Canyon Wash, joins San Diego Creek. Between the two, thousands of gallons of natural and (in later years) agricultural runoff flowed into the San Joaquin Marsh and then into the Upper Bay. The marsh was ill-defined, ebbing and flowing into surrounding acreage depending on rainfall and irrigation.

Around the early 1900s, two duck clubs were built in the marsh by hunting enthusiasts. They created ponds to attract migrating waterfowl, erected shooting platforms and blinds, and set about to bag ducks and geese that found their way along the Pacific Flyway to the Irvine Ranch.

In the 1950s, portions of the marsh were drained for agricultural use; in the Upper Bay, a salt production facility was established. The continued human use of the entire area resulted in its gradual degradation. In 1968, the Orange County Flood Control District dug a channel along San Diego Creek adjacent to the marsh, improving flood protection but also depriving the marsh of a regular source of water.

By this time, ownership of the marsh property began to splinter. The Irvine Ranch Water District, created and at that time controlled by the Irvine Company, found marsh property a good location to build a water treatment plant. The portion of the marsh nearest the Upper Bay was wasting away, but the University of

California saw potential to use it as a teaching tool, and in 1969, paid the Irvine Company $200,000 to establish the San Joaquin Marsh Reserve and began to restore its 200 acres.[4]

The IRWD also began an effort to improve water quality and access to the portion of the marsh it owned, and today there are multiple opportunities for public use of the marsh, including 12 miles of walking trails and a small complex of relocated farmhouses operated in partnership with the Sea and Sage Audubon Chapter of Orange County.[5] The houses have been rehabilitated and are available at a modest cost for community events, weddings, and conferences.

***

Eastbluff was planned and underway before the 1965 law took effect, so there is only one formal park in the community in the upper section where homes were built in a variation of the familiar grid pattern. But in the lower part of Eastbluff, homes were interspersed with greenbelts, both broad and narrow, with occasional nodes along the way that functioned as small parks. The greenbelts were well received by the new residents, so the Irvine Company duplicated the concept in the first village in Irvine, University Park, as well as in the hillside area of Turtle Rock. As both communities expanded, formal parks were constructed to meet the new state law requirements.

Irvine was incorporated in late 1971, but it was not until 1977 that the city adopted a General Plan that was substantially similar to the one written by the

Figure 7.1 San Joaquin Marsh

Source: Courtesy of Irvine Ranch Water District

Irvine Company. However, in 1974, the young city proposed a $16 million park development bond, accompanied by a $2 million bond to fund bikeways. Both passed by a 74 to 26 percent margin ($16 million in 2025 dollars is nearly $106 million; $2 million equates to about $13 million).[6] As a result, large-scale community parks around the city were built as well as an aquatics complex adjacent to Irvine High School. At the same time, city planners tasked with creating other parks partnered with the Irvine Unified School District to locate local parks immediately adjacent to new schools, blending the school playgrounds and parks into a single facility.[7]

Since incorporation, Irvine continued to build parks as development occurred and now claims a total of 302 parks, including Aldrich Park in the center of UC Irvine, as well as two parks in the city owned by Orange County: Mason Regional Park and Irvine Historic Park.

Perhaps the most unique parkland in the city is the Jeffrey Open Space Trail (aka the Jeffrey Spine), conceived as a key element in the Irvine Open Space Initiative in part as a response to citizen concerns that the southern and central sections of Irvine were benefiting unduly from park development, leaving the older, eastern side of the city wanting. (See Chapter 14).

The Jeffrey Open Space Trail is 3.5 miles long, running from Portola Parkway at the northern edge of town in a village of the same name and currently terminating at the Interstate 5 Freeway. The trail is consistently 80 to 120 yards wide, landscaped with both active and passive areas, including a faux stream of cobblestones that drains the area during the rare Southern California rains.

Designed by the SWA Group, which has a long history of creating landscape projects in Irvine, the 76-acre park features three bridges and two stone tunnels along trails shared by bikers and walkers.[8] Local history was incorporated into the park, with a series of plaques in both tile and metal that trace the property's timeline from the days of Indigenous residents, Spanish explorers, and into the era of the rancheros, the Irvine family, and later the Irvine Company and city. There is also a Stonehenge-style monument that marks the site of the last produce and citrus packing house on the Irvine Ranch, the Frances Packing House. Fourteen weathered wooden beams stand silently encased in concrete blocks, forming a ghostly reminder of the city's agricultural heritage.

In 2024, the city extended the trail with a bridge over Interstate 5, continuing the park to a new terminus in Woodbridge at Barranca Parkway.[9]

Similar to the Jeffrey Open Space Trail but on a much smaller scale is the Venta Spur Park, a linear park about a mile long built along an abandoned railroad spur line that served the Frances Packing House. When city leaders in the 1980s proposed turning the decrepit right-of-way into a landscaped trail, it met with surprising resistance from some of the Northwood homeowners whose properties backed up to it. After numerous contentious hearings, the city council approved the improvements.[10] Today, the Venta Spur is a serene and safe splinter of greenery. And along Hicks Canyon Wash, also in Northwood, a trail system, both paved and natural, was created as part of that village's development.

Irvine's park system proved to be fertile ground for the growth of a youth sports culture. Within months of the opening of neighborhood and city parks, hundreds,

Figure 7.2 Jeffrey Open Space Trail

Source: Author's image

then thousands, of children had taken to the field as part of the American Youth Soccer Organization (AYSO) program that introduced what at the time was a largely foreign game to boys, girls, and their parents.[11]

Irvine eventually had two AYSO regions and hundreds of volunteer coaches and referees. Dads, Moms, and other family members were learning that offside in soccer was much different than in American football and that footwork was the primary manner of moving the ball and scoring.

Baseball was not far behind, as boys and girls as young as 5 or 6 started their journey in the game invented in the United States more than a century earlier, with T-ball. This variation of the game proceeded with a ball placed on a pole at home plate, allowing a stationary target for young batters. T-ball was followed by slow-pitch (adults pitching) and finally into gradations of Little League. In 1987, a team from the village of Northwood won the Little League World Series, besting a team from Taiwan.[12]

The city's neighborhood pools, along with a strong Southern California tradition of competitive water sports (and fair weather), combined to stimulate a robust youth swimming culture. The leading organization that reflected and encouraged Irvine youth to take to the lanes was Novaquatics.[13] It was helped along immensely by the construction of a twin competitive pool center adjacent to Irvine High School and Heritage Park. Soon, the state-of-the-art facility was hosting dozens of swimming and diving competitions annually. Proudly displayed on the walls of the center are the names of several Irvine swimmers who medaled in the Olympics.

Currently, reflecting both changing sports attitudes and the city's ethnic diversity, sports such as lacrosse and cricket are played regularly on city fields. It has also been reliably reported that the occasional Quidditch match[14] has been seen at UC Irvine.

(Irvine's park system expanded dramatically with the development of the Great Park. Details in Chapter 20).

Five parks in Irvine honor men whose lives touched the city, and a sixth is related to Irvine Company owner Donald Bren.

At UC Irvine, Aldrich Park graces the center of the William Pereira–designed campus, honoring the university's first chancellor, Daniel Aldrich.[15] He was trained in agriculture and treated the park with the same care a new homeowner would lavish on a backyard. Only half-jokingly, Aldrich once told a compatriot he would happily return to the campus as a gardener after retirement.

The large community park adjacent to the Irvine Civic Center honors the memory of William Barber, a city resident who won the Congressional Medal of Honor for his heroic actions during the Korean War.[16]

William Mason Regional Park, which meanders between University Drive and an intermittent streambed, was originally called University Park after it was donated to the County of Orange by the Irvine Company in 1969. After the untimely death of company president William Mason in 1973, it was renamed in his honor.[17]

David Sills Lower Peters Canyon Community Park is in Northwood, across the street from the Racquet Club neighborhood, where Sills first lived in the city.[18] An attorney, Sills led a community protest against the first Northwood zoning, which paved the way for his election to the city council. Appointed to the Superior Court bench in 1985, Sills was elevated five years later to the appellate court in Orange County. He died in 2011.

In the center of Woodbridge, the Mike Ward Community Park is situated adjacent to a senior center and overlooks South Lake in the Village.[19] Ward was the first—and, to date, only—member of the giant Woodbridge Community Association Board of Directors to be elected to a city council seat. Ward's life was tragically cut short in 2013 after a short battle with ALS.

The aquatics center in Heritage Park, which the city shares ownership with the Irvine Unified School District, is named for Irvine's first city manager, William Woollett Jr. He and his first wife, Diane, were active supporters of youth aquatics and were regularly seen on the pool sidelines during swim meets.[20]

In Turtle Rock, a quiet, four-acre park ascends a gentle slope above Turtle Rock Drive, separating two tracts of homes. The Charles R. Bren Park is named in memory of Donald Bren's stepbrother, who died in a plane crash in San Diego in 1978.[21]

***

While parks and open spaces in urbanizing communities of Irvine and Newport Beach were being planned and developed, the huge swath of coastal property between Corona del Mar and Laguna Beach remained much the way it had been for centuries. The exceptions were Crystal Cove, an assemblage of cottages where Los Trancos Canyon met the Pacific Ocean, and a smattering of mobile homes on the sand and above the beach at Moro Cove. Both had been developed in the 1930s.

In 1969, a group of Irvine Company executives traveled to Europe to gain inspiration from communities along the coasts of Spain, France, and Italy. Ray Watson was particularly taken by the charm and beauty of Portofino, believing it might be possible to create the tiny coastal village's La Dolce Vita vibe at Crystal Cove.

But as he pondered the future of the coast, Watson concluded that whatever was built there, the coastal bluffs that ran almost uninterrupted from one end to the other would have to be sold or dedicated to the public. His viewpoint became reality in the early 1970s when the Irvine Company released its first draft of a coastal plan and approached officials at the California Department of Parks and Recreation about purchasing the coastal strip. A price of $7.5 million was proposed (about half the appraised value of $14 million the company had arrived at), but Joan Irvine Smith quickly filed a lawsuit to quash the offer, labeling it far too low.

Thus began a decade of discussion, negotiation, and bargaining among individuals and organizations with a stake—official or not—in the future of the coastal property. Looming over every conversation was the Irvine Company's continuing desire to fashion a plan for coastal development that would be acceptable to the South Coast branch of the California Coastal Commission.

Equally concerning was an effort by two local congressmen to make the Irvine Coast (and possibly other lands to the south) a national park. The plan had gained traction and even had a placeholder in the federal budget for land purchase (the plan never came to fruition).

In the end, after what Irvine Company president Peter Kremer called a decade of "planning, compromise, and hard negotiations" that concluded with the state paying $32.6 million for 1,896 acres of the coast strip, plus a giant patch of land around Moro Canyon, to establish Crystal Cove State Park. At the time, it was the highest price the state had ever paid for a park.[22,23]

And while it was never memorialized in any legal document, it was clear to both the company and environmental interests that the coastal open space sale was the unspoken quid pro quo that would allow coastal development to proceed (see Chapter 16).

Notes

1 Joan Irvine Smith, *A California Woman's Story* (Irvine, CA: The Irvine Museum, 2006).
2 Florence Koch, referenced in Orange County Parks Commission Archives, 1920.
3 California Department of Parks and Recreation, "Upper Newport Bay Nature Preserve Dedication," 1989. See also discussion of the Irvine Company's role in the original sale in earlier volumes of Irvine Ranch history.
4 University of California, "San Joaquin Marsh Reserve Acquisition Summary," 1969.
5 Irvine Ranch Water District, "San Joaquin Marsh Restoration Overview," internal report, 1990.
6 City of Irvine, "Park Bond Election Results," November 1974.
7 Irvine Unified School District and City of Irvine, "Joint Use Parks Agreement," 1975.
8 SWA Group, "Jeffrey Open Space Trail Design Narrative," 2002.
9 City of Irvine, "Jeffrey Open Space Trail Extension Plan," 2024.
10 City of Irvine, "Venta Spur Park Development Hearings," 1980–1982.
11 American Youth Soccer Organization (AYSO), Region 213 Archives, Irvine, CA.
12 Little League International, "Northwood Team Wins World Series," 1987.
13 Novaquatics, "Heritage Park Aquatics Center History," internal publication, 2000.
14 Boomers who've been out of touch will have to look it up as they get trained for the new world of AI.
15 University of California, Irvine, "Daniel Aldrich Biography," Chancellor's Office Archives.
16 U.S. Department of Defense, "Congressional Medal of Honor Citation: William Barber," Korean War Records.
17 Orange County Parks Department, "William Mason Regional Park Dedication," 1973.
18 City of Irvine, "David Sills Park Naming Resolution," 2012.
19 City of Irvine, "Competitive Aquatics Center Dedication: William Woollett Jr.," 1995.
20 Woodbridge Community Association, "Mike Ward Memorial Statement," 2013.
21 City of Irvine, "Charles R. Bren Park Dedication," 1978.
22 California Department of Parks and Recreation, "Crystal Cove State Park Acquisition Summary," 1979.
23 Peter Kremer, quoted in "State Buys Irvine Coast," *Los Angeles Times*, December 12, 1979.

8 UCI grows up

To build a great city, build a great university and wait 200 years.

Attributed to Daniel Patrick Moynihan

The sale of the Irvine Company in late 1977 had little impact on UC Irvine. Just 12 years old, the campus was spreading out and growing its student body under the leadership of its founding chancellor, Daniel Aldrich.

Aldrich had been exhilarated at the prospect of building a new campus literally from the ground up. The administrators and other academics he had recruited shared the vision and were busy fulfilling it with him by hiring both seasoned and junior faculty members. The initial resistance to the new campus, looked upon with suspicion by some of Orange County's extreme right-wingers, had faded away. Aldrich had a firm hand on his beloved campus and was confident in its future success.

But there was one decision made in 1967 that Aldrich had no hand in, a decision that would provoke a dramatic and contentious dispute between town and gown a few years later.

The UC Regents voted to move a medical school in Los Angeles to the new university campus in Orange County. The California College of Medicine had operated since 1896, beginning as an osteopathic teaching institution. Relocating to the new UCI campus the following year as the UC Irvine College of Medicine, the new medical school quickly achieved an affiliation with the Orange County Medical Center, providing it with a teaching hospital.

(In a 1989 oral history, Aldrich described the political machinations behind what was essentially a shotgun marriage of the medical school and university. UCI officials at the time joked that of the 88 medical schools in the nation, "California College of Medicine was probably ranked 89").[1]

Another seminal event in UCI's history would occur after Donald Bren took total control of the Irvine Company in 1983. Bren's appreciation of the potential power of a partnership with the campus, plus the arrival of a new chancellor, would combine to lay the foundation for millions of dollars in gifts and the creation of an even more valuable connection between the giant landowner and the growing university that would pay future dividends to both.

***

DOI: 10.1201/9781003683087-8

Fourteen years after admitting its first class, enrollment at UC Irvine had passed the 10,000 student mark. The Brutalism-style buildings favored by William Pereira were the dominant feature but by then had softened as the landscaping at the campus took hold and filled in. Pereira occasionally visited the campus to review his handiwork, but by 1980, he had slowed down considerably; he died in 1984.

The same year Pereira died, Aldrich retired. He had served 19 years in the leadership post, and his stellar record as an academic, administrator, and, in some ways, the spiritual leader of the campus would be cited again and again as the history of the campus was written.

Tall, formal, but with a dry sense of humor and a smile waiting just beneath his patrician demeanor, Aldrich had steered the new campus through the turmoil of the late 1960s, sparked by the Vietnam War and marked by the burning of a branch of the Bank of America just across the street from the campus.

And, as he revealed in a 1989 oral history interview, he had essentially stared down the president of the university, Charles Hitch, when he presented Aldrich with a new study forecasting the number of students to expect at UCI in the future, lowering the initial estimate of 27,500 to just 7,500. Aldrich argued strenuously that the forecast was wildly inaccurate. As it turned out, Aldrich was right—growth on the campus continued to blossom, and he and his academic colleagues simply kept pressing for new buildings and faculty to meet the demand.

Aldrich envisioned the new campus as a modern land-grant university. Those historic colleges had focused on agriculture, mining, and engineering, reflecting the needs of a 19th-century nation still expanding and modernizing. Aldrich saw UCI as a place "that would be developed in the tradition of the land-grant philosophy, but essentially geared to meet the needs of a society as we face it now and in the future."[2]

One of his proudest achievements at UCI, he said, was the creation of the Program in Social Ecology led by Professor Arnold Binder. Social Ecology was grounded in an experiential philosophy, "the application of Social Sciences, Biological Sciences, Physical Sciences, and Engineering and Humanities to the solution of problems confronting individuals, as well as the organization of individuals."[3]

Social Ecology got its students out of the classroom and into real-world situations. Its success on and off the campus helped to burnish UCI's reputation in its first decades as a place where innovative approaches were welcomed.

(Social Ecology was helped along the way by Assemblyman John Vasconcellos, chair of the influential Ways and Means Committee in the state assembly. Vasconcellos was an unusual combination of a streetwise political operative and a humanistic seeker, intrigued by New Age philosophies, meditation, and the occasional hallucinogen. The vision that Social Ecology presented appealed to his counterculture instincts, leading Vasconcellos to play a key role in assuring funding for a new Social Ecology building on the campus, Aldrich said.)[4]

Social Ecology was just one example of how Aldrich and the other administrators he had hand-picked to lead the new campus were willing to take new—sometimes radically new—approaches to organizing academic programs. Citing the drama and dance department, and the decision that it would be performance-oriented,

Aldrich (1989) said, "We concluded that if we wanted to attract top people to Irvine, how (academic departments) were organized and how they worked together varied very much with the field."

Breaking with tradition was,

> to me, an important opportunity for a campus that was starting from scratch, had an opportunity to strike out in new directions, exploiting the best of work we had in the way of information, in a field as well as an organization, and as I used to say, "we'll make mistakes, but let them be our own."

In the 1989 interview, Aldrich was asked about the high points of his nearly two-decade tenure as well as the disappointments. He identified the worst experience without hesitation: the high-level contest of wills, corporate and political factions, and personal animosities that combined into a bitter controversy over the creation of a new hospital in the City of Irvine.[5]

***

The 1961 Master Plan for the Irvine Ranch, prepared by William Pereira and Associates, contains, among many other elements, a location for a zoo. It omits any reference to a hospital or medical center.

Most of the details in that plan focused on Newport Beach, with emphasis on Newport Center. At that time, Newport Beach had a new hospital, Hoag Memorial Presbyterian, which opened in 1952 on a bluff overlooking the intersection of Newport Boulevard and Pacific Coast Highway, near the western boundary of the city.

Irvine, including the site for the new university campus, was nothing more than vast fields and citrus orchards. Pereira's plan assumed it would be decades before large numbers of homes and businesses would be built. Moreover, it was not until 1972 that California law required cities and counties to create general plans; within the elements required to be addressed in the plan, hospitals or medical centers were not specified. Among the elements—housing, commercial, industrial, schools—was the catch-all designation "institutional." Presumably, a hospital would fit in that category.

But Irvine grew much faster than Pereira anticipated, as did the rest of the county. The new city was incorporated in late 1971 with a population in excess of 20,000 and unmistakable signs that more people were coming (by 1980, Irvine had 60,000 residents).

For the next decade, individuals and organizations began to coalesce around the concept of building a hospital to serve the growing community. Two distinct alternatives emerged: a hospital located on or near the UC Irvine campus affiliated with its medical school or an independent hospital built in the center of Irvine.

UCI Medical School had a hospital, having made a deal with the County of Orange in 1968 to take over the Orange County Medical Center in the city of Orange. The hospital was old and run down, tracing its roots to 1905 as the county Poor

Farm. The aging facility gave the new medical school a ready supply of patients to train its future physicians—most were poor, and the emergency room was always busy. But medical school administrators knew that bringing the hospital up to modern standards would require a great deal of money. They dreamed of a new facility, fully equipped and designed as a teaching hospital.

After several false starts, a campaign for the community hospital gained traction in 1975 with the creation of the Western World Foundation, backed by Hoag Hospital, some of its doctors, and the esteemed Dr. Arnold Beckman, founder of Beckman Instruments. The Irvine Company—still controlled by the James Irvine Foundation but preparing to sell its shares to comply with new federal law—offered 18 acres of land near the university as a hospital site.

Leading the university forces with an uncompromising goal of a hospital on the campus was Dr. Stanley van den Noort, the dean of the medical school. Described by the *Los Angeles Times* as "a compact, combative man of strong will,"[6] van den Noort was also well-connected in the state capital and had the support of Joan Irvine Smith, whose wealth and cache as the best-known member of the Irvine family made her a formidable ally.

Dean van den Noort was suspicious of both Hoag and its allies. He continued to oppose any hospital that was not on university property. Meanwhile, the UC Regents, viewing the battle from their system-wide perspective, undercut van den Noort in 1976 by purchasing the Orange County Medical Center and renaming it the UC Irvine Medical Center.

Between van den Noort's political connections and the inability of a competing medical entity to take the high ground in the battle, the issue stalemated. It was the formation of a new, community-based entity in Irvine that also attracted the support of the Irvine Company and other corporate interests, which proved to be a turning point in the contest.

A handful of Irvine activists formed People for an Irvine Community Hospital (PICH) to promote a local hospital. Casting about for someone to lead the group, they found one in David Baker, an attorney and former UCI basketball star, ambitious for elective office. Baker had just lost a close race for Irvine City Council when he was approached by PICH in late 1980 and asked to lead the effort to make a community hospital in Irvine a reality once and for all.

Baker's imposing presence (well over three hundred pounds on a six-foot, nine-inch frame), along with an engaging personality, gave PICH the boost needed to galvanize the fight. The Irvine Company also viewed Baker as someone who could win a seat on the city council in the next election and maintain its majority pro-development stance. But it wasn't just the Irvine Company who promoted Baker—Arnold Beckman, commercial developer Don Koll, and Robert Fluor all liked what they saw in Baker as well as the hospital and joined to support PICH.

Within a few months of Baker's ascendancy to PICH leadership, money began to flow into the group, used for public relations and community outreach. Hoag Hospital gave at least $3 million, as well as a $5 million matching challenge from Arnold Beckman's foundation.

Perhaps the most comprehensive chronicle of how PICH prevailed was written by Mark Pinsky in the *Los Angeles Times* on October 13, 1985. What follows is excerpted from that article:

> [After van den Noort blocked a proposed hospital site selected by PICH] in stepped Irvine Company President Thomas H. Nielsen, who in February 1983 offered . . . a gift of 15 acres adjoining the Golden Triangle. The idea was the community hospital . . . could form the nucleus of the company's proposed bioscience complex.
>
> Van den Noort and his supporters, already caught off guard by the growth of PICH, were blindsided by the land offer. In the meantime, the dean . . . had lost some of his clout in Sacramento with the arrival of the Dukemejian administration . . . Chancellor Aldrich began to sense that the time was near to end the estrangement with the Irvine Company, its corporate allies and the Irvine community, all of which he perceived to be solidly behind the proposed Irvine Medical Center. [While Aldrich publicly still supported an on-campus hospital], he was also beginning to listen more to his new vice chancellor for advancement, John Miltner, an advocate of reconciliation.
>
> In the Spring of 1983, Aldrich had lunch with Irvine Company chairman Bren and president Nielsen. [He] outlined what he thought were the mutual interests of the company and the university. There was, Aldrich said, much interest on the part of the Irvine Company to locate the hospital out there to be the catalyst for its high-tech area. The chancellor brought up his own aspirations for the development of the inclusionary land . . . The thrust of the discussions, Aldrich said, was that all the ingredients were present for successful cooperation . . .
>
> Still, however, there was the troubling dispute over the hospital. The PICH and UCI plans for a medical center were competing for county and state approval." [At this point, the Irvine Company made an offer to move the UCI Medical School to the Irvine Spectrum area and spend $50 million to pay for it. Van den Noort thought it would take $300 million to make it work and rejected the company's offer.]
>
> If the rapprochement between the university and the Irvine Company was going to go anywhere, something had to give. In August 1983, something did. According to [van den Noort], Aldrich called [him] to his office and announced he was withdrawing the university's request to the UC Regents for campus land to build the hospital.

(According to the *Times* story, van den Noort quoted Aldrich as saying his decision was linked to the potential for getting the Irvine Company to lift restrictions on the 550-acre Inclusion Area. Aldrich denied any connection between the hospital decision and the potential for easing the limits. Another person present at the meeting quoted Aldrich as saying, "The way things were going, the community was not going to end up with any hospital.")

The article continued:

> The weeks that followed Aldrich's announcement were the start of a new period of cooperation. Miltner and Gary Hunt, Irvine Company vice-president, joined Beckman and Baker on the board of (the new entity created to bring the hospital into reality: the Irvine Medical Center, or IMC), along with five other UCI representatives.
>
> Pinsky, "Irvine Hospital Dispute Ends in Compromise."

In late 1983, the Irvine Company and UCI announced a joint planning process that would allow development in the Inclusion Area in a way that produces future income for the university. In 1984, Irvine Medical Center, Inc., broke ground for a 177-bed hospital to serve the Irvine community on Sand Canyon Road, just north of the San Diego Freeway, approximately 3.5 miles southeast of the university campus.

Some academics and community members viewed the outcome of the hospital battle with cynicism and bitterness, heaping criticism on Aldrich in particular, accusing him of making a Faustian bargain with the Irvine Company, compromising the university's integrity in the process.

Chancellor Aldrich retired in 1984. It would fall to succeeding chancellors to turn the blueprints for a partnership with the Irvine Company into reality.

***

Figure 8.1 Aerial view of the center of the UC Irvine campus

Source: Courtesy of UC Irvine

Figure 8.2 UC Irvine Chancellor, Jack Peltason

The UCI campus had a new nickname when Jack Peltason arrived in 1984 to assume the position as chancellor: "UCI, Under Construction Indefinitely" was the not-entirely-joking moniker students and faculty were using to describe the day-to-day difficulty of moving about a campus that had dozens of new buildings under construction to accommodate the growing student body. "One school year," Peltason said in an oral history, "we actually had more buildings under construction than existed on the campus."[7]

There was also development underway across the street at University Town Center, the village William Pereira had envisioned as the first element in the new community of Irvine, one emphasizing mixed uses linked physically and intellectually to the new university.

Despite the focus in the Pereira plan of the early 1960s, the Irvine Company initially focused on developing University Park and Turtle Rock, both of which were well underway by the early 1970s. A few years later, following the grand opening of the ambitious Village of Woodbridge in 1976, people lined up for a chance to buy a home. This company continued to ignore University Town Center and focus its staff and finances on maintaining the pace of sales in the nearly three-square-mile Village of Woodbridge.

By the early 1980s, University Town Center was the locus of just a couple of commercial buildings. When the first phase of significant development finally got underway in the early 1980s, the focus was residential. The planned commercial center was once slated to feature a hotel (never built) but struggled to find market support for a strong town center identity. The City of Irvine abandoned University

Town Center when it decided to relocate its city hall several miles away to a site in the Village of Westpark. And students at UCI voted to place their student union complex on campus rather than in University Town Center.

Jack Peltason was more concerned about his burgeoning campus and its fragile relationship with the business community and the Irvine Company. For Peltason, becoming chancellor was a homecoming—he had been one of the nucleus of administrators Aldrich had recruited to plan and open UCI in the early 1960s. After serving as a vice chancellor, Peltason moved to Illinois to head up the state university system. He left Illinois for Washington, D.C. as president of the American Council on Education, then he returned to UCI.

Peltason's casual manner and avuncular appearance masked a shrewd mind. Trained in political science, Peltason was the co-author (with James McGregor Burns) of a text that was widely used in high school civics classes, *Government by the People*. He also co-authored *Understanding the Constitution* with Edward Corwin. Like Aldrich, Peltason enjoyed mingling with students both informally and with cause. His hours-long debates with student protestors in Illinois during anti-war protests may have helped to keep the student unrest there largely nonviolent.

The medical school/hospital contest was over, but bad memories lingered on campus and in Irvine. It fell to Peltason to drop the ax on Dr. Stanley van den Noort as medical school dean. And though the move was anticlimactic, it was a reminder of how deep the fissure between the campus and community had become. In an oral history Peltason gave in 2001, he said the aftermath of the hospital wars, plus the continuing operational and budget problems at the UCI Medical Center, "took up half my time." (In the eight years Peltason was chancellor, the medical center went through seven directors.)[8]

Peltason came to his new position with the attitude that UCI had to develop stronger ties to the local community and, more importantly, to the business community, with the unabashed goal of raising private funds for the public university. At the top of his list of potential partners was the Irvine Company.

Donald Bren made the first move, meeting with Peltason when he was still in Washington, D.C., to discuss what was happening on university property as well as on the 500 acres of land the university had purchased from the company in the late 1960s. That property, the Inclusion Area, came with deed restrictions limiting land uses to those related to the university's academic mission. Peltason learned that Daniel Aldrich had been testing the limits by allowing two businesses to locate on the campus, on the condition that the buildings be turned over to the university 50 years later. This activity had not gone unnoticed by the Irvine Company.

Over lunch, Bren described a vision for an Irvine Company–UC Irvine partnership that was very similar to what Peltason wanted to achieve:

> He was anxious to be helpful. And from that grew a friendship. When I became chancellor, I spent a great deal of time negotiating with the Irvine Company about freeing up the constraints on the inclusionary land. Don Bren and I both have the same interests, his (was) to be sure we used this land to promote the university, so we negotiated a major amendment to the

> agreement . . . which removed the constraint from the university's right to develop the land, and the constraint against leasing it or using it for some for-profit enterprises.[9]

Peltason's discussions came to fruition in late 1985, when the university and Irvine Company made a preliminary agreement for

> UCI to grant the Irvine Company exclusive rights to develop 292 acres of campus land, together with 123 acres of adjoining company property, for a bioscience research and development park that both parties hope will be a magnet for high-technology firms.[10]

The announcement that day did not include details on the sharing of development revenue. But Peltason had privately made a handshake deal with Bren that the money would endow faculty positions at UCI that would become known as Bren Fellows. It was the best way, Peltason said, to meet Bren's desire that the money remain at the campus and not find its way into the UC system for other uses.[11]

As significant funds from the new development would not arrive for several years, Bren primed the pump with initial funding of $2.5 million to endow the first chairs, allowing the university to select biologist Francis Ayala as the first Bren Fellow in 1989, followed by Sherwood Rowland, a physicist.

As was the case when Aldrich capitulated on the proposed hospital on campus, there was grousing from faculty and others about the partnership with the giant landowner. Peltason dismissed the complaints: "Don Bren is a visionary. He doesn't need the university land to fulfill his vision, but if it grows, and gets distinguished professors and becomes an outstanding university, it will attract people to the neighborhood."

Peltason also saw the Irvine Company's single ownership as an advantage for the university, describing the complexities he would have faced had there been dozens or hundreds of landowners to negotiate with on a deal like the one he had just completed. He also compared the staying power of the university to that of the Irvine Company. Both institutions, he said, would be around for another century, providing potential donors to UCI the assurance that their donations would benefit the university over the course of several lifetimes.[12]

Peltason had looked beyond Bren and the Irvine Company after he became chancellor, recognizing that with the hospital battle behind the campus, it was an opportune time to seek connections with the larger county business community. He created the UC Irvine CEO Roundtable, inviting the leaders of major companies to offer advice, friendship, and funding to the university.

Peltason was not shy about asking for money. He viewed it as one of the chancellor's primary roles. He had inherited John Miltner from Aldrich and soon found that the charismatic vice chancellor had a remarkable talent for attracting funds and cultivating those who could raise funds from peers in the business world.

Peltason also created the UCI Medal, an honor to bestow on individuals within the university as well as outside it who made significant contributions to the well-being of UCI (the medal substituted for an honorary degree, which the UC system stopped awarding in 1970).

Another initiative that helped the university make a stronger connection to the larger Orange County community was the creation of the UCI Annual Survey, created and managed by urban planning professor Mark Baldassare.

Soft-spoken and modest, Baldassare was passionate about governance and election issues and saw the survey as a tool that could help shape the county's future. Just a few years after its inception, the survey had become a key barometer of county concerns, drawing the attention of both elected officials and community leaders.

While Peltason was busy with problems at the medical center and strategizing on how to raise money for programs and facilities, on the campus itself, the signature stark architectural style of William Pereira was giving way to something more diverse.

***

According to the 2015 publication celebrating the first 50 years of UCI's birth, growth, and expansion,

> by the early 1980s, amid enormous physical expansion, planners and architects sought to free themselves from design constraints. Campus leaders felt that the Pereira buildings were too much alike. The campus hired architect David Neuman, who was passionate about putting Irvine on the map. Also, every building designed in the 1980s won architectural prizes.

Diverse aesthetic styles and structures defined the postmodern phase of architecture: McGaugh Hall, built by Canadian architect Arthur Erickson and dubbed 'the Emerald City' for its green hue; Charles Moore's Italian Renaissance-style University Extension and Alumni House; Robert Stern's Studio Four of the Claire Trevor School of the Arts; the Paul Merage School of Business 'decorated shed' building designed by Robert Venturi; and celebrated British architect James Stirling's Science Library with its central courtyard.

Neuman was followed in 1986 by Rebekah Gladson. According to the anniversary publication, "Gladson espoused a design aesthetic called contextualist architecture. Future buildings, she said, would have a distinct bottom, middle, and top would be sensitive to their neighbors, and would convey a sense of permanence and quality."[13]

During Gladson's tenure, UCI decided in 2007 to demolish a building designed by Frank Gehry in 1986. Gehry's signature California architectural creation, the Walt Disney Concert Hall in downtown Los Angeles, had opened just four years earlier, so the razing of what was known as the "310 Building" and an adjacent structure garnered attention and comment from throughout the nation.

Gladson and other campus planners explained that the 310 Building was never meant to be used for very long and had some serious maintenance issues, including dry rot. Moreover, they argued, additional space for a new engineering classroom building that would replace 310 was badly needed.

A variety of architects and critics panned the decision. Perhaps the most pungent comment came from famed architect Robert A.M. Stern, who at the time was the campus architect at Yale University. Noting that the Yale campus had buildings from the 19th and 20th centuries in active use, Stern said that (in a curious Yoda-like fashion) "Irvine had no buildings older than 40 years, and it's the plain vanilla capital of the world, in my opinion, Orange County is. This is a death wish. In my humble opinion."[14]

While there might be controversy about one building or another, there was virtually universal agreement about the beauty and ambiance of the 16-acre park at the center of the first section of the university. Designed by Laguna Beach landscape architect Fred Lang, the park had a craggy rock formation at the center, dotted with drought-resistant plants long before such vegetation became de rigueur in southern California—large grassy areas dotted with eucalyptus and California Oaks. A series of sidewalks connected the buildings ringing the park to one another. Daniel Aldrich, whose academic background was in agriculture, loved the park, and after he died in 1990, it was named Aldrich Park in his honor.

***

The 1985 agreement between the Irvine Company and UCI to ease land use restrictions in the Inclusion Area proved to be a significant milestone in their relationship, with an even more fulfilling marriage of interests coming 13 years later.

The campus had moved slowly to take advantage of the newly available Inclusion Area land. After considering several approaches to development there, Peltason said in his 2001 oral history that leasing the land to a developer was best. The model that became standard was that the university would collect a lease payment on a building or laboratory the developer built and financed and receive use of a portion of the building space. Then, after a term of 50 or 60 years, ownership of the building would revert to the university.

This model would be repeated on an 85-acre parcel of university land adjacent to 99 acres of Irvine Company property already being developed for what would become known as the University Research Park.

Peltason had retired, and his successor, Laurel Wilkening, had concluded her term as chancellor by July 1998, when Chancellor Ralph Cicerone announced an agreement between UCI and the Irvine Company that would lead to the expansion of the University Research Park.

The Irvine Company, he explained, would receive a 65 to 75-year lease on 85 of acres of university land, upon which

> it will develop a total of about one million square feet of investment property . . . the company will continue to focus its marketing on companies involved in biomedical research, biotechnology, information technology, and health care—all areas in which UCI researcher excel.[15]

In addition to taking all the risk for obtaining tenants in the more than 40 buildings that it expected to construct in the area, the Irvine Company agreed to fund infrastructure and other improvements for the project at an estimated cost of $20 million, as well as $3 million in fees for the nearby San Joaquin Corridor toll road, and another $3 million to help finance two buildings in the campus' Irvine Biomedical Research Center.

Finally, the most valuable aspect of the agreement for UCI is that, at the end of the lease term, all properties built on the 85 acres will become the property of the university.

Twenty-seven years into the lease agreement, the University Research Park is thriving, with 36 buildings occupying 2.2 million square feet of space. As of late 2024, just over 249,000 square feet of various office footprints were available for lease, giving the complex close to a 90 percent occupancy level.

The park website invites prospective tenants to "unlock the power of UCI's innovative ecosystem, . . . that's connecting businesses with direct STEM talent, partnership opportunities and more."[16]

And perhaps most significantly, the research park is producing revenues from lease payments that accrue directly to the UC Irvine campus.

In 2019–20, according to university records, UCI received $5,565,545 in lease revenue. By the 2022–23 fiscal year, lease revenue had increased to $7,089,729, thanks in part to a Consumer Price Index catch-up clause in the lease agreement. Considering that there are at most 49 years left on the lease, using a modest inflation factor, it appears feasible that, in that timeframe, UC Irvine could receive in excess of half a billion dollars in lease revenues from University Research Park.

***

The positive news of the University Research Park agreement was a welcome conclusion to the decade of the 1990s. University officials hoped it would overshadow a series of scandalous events in 1995 that dealt a blow to the reputation of the system and UCI's beleaguered medical school.

The *Orange County Register* broke a blockbuster story in May 1995, accusing two doctors affiliated with the university medical system of enriching themselves by selling eggs from women who had come to them seeking fertilization treatment, without the knowledge or the permission of the women.

Doctor Ricardo Asch had developed a breakthrough procedure for in vitro fertilization and had built a thriving practice at a clinic affiliated with the UCI Medical School. Couples trying to conceive a child had flocked to his practice with hope and a willingness to pay for treatment, sometimes in cash.

In a series of articles that would win the paper a Pulitzer Prize, the *Register* detailed how Asch and Dr. Jose P. Balmaceda sold eggs from unsuspecting donors to other couples seeking to become pregnant. Four months after the story broke, both doctors fled to Mexico and Chile; a series of criminal charges followed as well as dozens of civil lawsuits by couples who were outraged and numbed by the prospect that children had been born to other families from their reproductive tissue.

Asch and Balmaceda were declared fugitives. Asch was briefly detained in Mexico a few years after the scandal but never extradited to face the charges.

Balmaceda pleaded guilty to tax evasion in May 2022 and coughed up about $37,000 in back taxes in a settlement with the federal government. The University of California paid an estimated $27 million to settle the lawsuits filed by the unknowing egg donors.

***

While controversies were an element in the history of UCI's first four decades, during that time the campus steadily accumulated academic honors and what would over time total hundreds of millions of dollars in donations.

One of the most unusual gifts came early, in 1972.

Marianne McDonald, who had studied classics at UCI, donated $1 million to create a digital encyclopedia of Greek writings from the time of Homer (8 BC) to well into the 20th century: *The Thesaurus Linguae Graecae* (*Treasury of the Greek Language*) "The concept was extraordinary, since no one until then had considered the marriage of Classical scholarship with the rapidly emerging new technologies."[17] Today, the database is used by students and academics from around the world.

A similar gift transcending centuries came in 1986 from alum Patrick Hanratty, who donated one of the few existing *First Folios* of William Shakespeare to the UC Langson Library Special Collections center. In remarkable condition, some pages show signs of use by people who likely read the Bard's comedies and tragedies while eating. The library brings the *First Folio* out once a decade or so for public viewing.

In 1991, the university partnered with the City of Irvine to construct a small performing arts center on a parcel of UCI land adjacent to University Town Center. With a gift of $1 million from Richard and Marjorie Barclay to seed the effort, the Irvine Barclay Theater was born. The facility's shared nature requires its operator to reserve part of the season for student use, while the non-profit entity that runs it sponsors the remainder for public performances.

Dr. Arnold Beckman had invented the pH meter, which became a foundational tool for chemists around the world. While he spent most of his business career in Pasadena near the California Institute of Technology, where he studied, he and his wife had a home in Corona del Mar. One of his companies, Beckman Instruments, was located in Fullerton. By 1988, Beckman was nearly 90 and focused on giving away much of his fortune. With a seven-acre gift of land adjacent to UCI from the Irvine Company, Beckman's $20 million contribution to the National Academy of Science and Engineering lured the prestigious institution to establish a western presence, building a sleek, low-slung center that hosts academic symposia and other meetings.

As the new century dawned on UCI's 35th year, a few of Orange County's multimillionaires were considering their legacies.

Paul Merage perfected a process to bake or microwave a hearty calzone-style sandwich combining frozen pastry and a meat or vegetable filling into an edible delight, giving birth to the Hot Pocket. As a result, millions of dollars flowed into the pockets of the immigrant from Iran. In 2003, Merage gave $30 million to the university's graduate school of business, and the university renamed the school in his honor.

In 2008, a little-known Orange County businessman named M. A. Douglass died at age 97. Working in the shadow of county real estate titans such as Henry Segerstrom, Don Koll, George Argyros, and others, Douglass had quietly amassed 10,000 apartments. Two years after his death, his estate gave $21 million to the UCI medical system, which named its City of Orange hospital after him.

A decade later, the *really* big dollars began to roll into UCI.

First was bond guru Bill Gross. An eccentric genius who once supported himself by counting cards at Las Vegas casinos, Gross transformed the bond business while at Pacific Life (then Pacific Mutual) insurance company. Pacific Life created a subsidiary business for Gross called PIMCO, which made billions of dollars. In January 2016, the family foundation Gross and his wife, Sue, created gave $40 million to establish the Bill and Sue Gross School of Nursing.

The Gross gift was eclipsed in 2017 by another power couple, Henry and Susan Samueli. Henry was a co-founder of Broadcom, which had taken the Orange County high-tech landscape by storm after the company relocated there in 1995. The company's wild success drove its stock price into the stratosphere, making the Samuelis billionaires.

In 2017, they made two staggering gifts: $30 million to kickstart matching funds from the UC system to build a 100,000-square-foot science building and an unprecedented $200 million to create the Susan and Henry Samueli College of Health Sciences (in 2001, the couple gave $5.7 million to establish the Center for Integrative Medicine, named in honor of Susan).

The $200 million gift was the largest ever received by UC Irvine. In 2023, the pair gave $50 million to the school of engineering, which had been named in Henry's honor in 1999 in recognition of his gift of $20 million.

And in 2024, the waterfall of donations continued, led by a $50 million donation from biotech entrepreneur Charlie Dunlap. The gift led to the decision to name the UCI School of Biological Sciences in his honor.

And in 2024 as well, the UCI Foundation reported assets of $900 million.

***

After paying the symbolic $1 dollar in 1960 to purchase 1,000 acres from the Irvine Company for the new campus that would become UC Irvine, university officials purchased an additional 550 acres in 1963 that would become known as the Inclusion Area.

That land was for "university-related" purposes, principal among them, housing for staff and faculty. UC administrators were specific in their desire for housing—they had seen what had happened at UCLA in Los Angeles, where land values adjacent to that campus had soared, pricing out students, faculty, and staff within a few years of the creation of the campus.

The initial class of students arrived at UCI in 1965, but it would be 19 years before the university built faculty housing at the far southeastern corner of the Inclusion Area. Dubbed University Hills, the university established a public benefit non-profit organization, the Irvine Campus Housing Authority (ICHA), to develop the residential community and manage a community association.

The first of what would eventually total 1,600 units on 300 acres of land was completed and sold in 1985. In addition to single-family homes, ICHA added apartment units over time.

University Hills was the perfect neighborhood for those who taught at and administered the growing campus. It was a short walk to classrooms or offices; there were recreation facilities and several small parks. The only shortcoming was the absence of a neighborhood school—children could either enroll at the Farm School operated on campus or go to nearby Turtle Rock Elementary School.

But University Hills also came with several restrictions.

First, all of the lots remained property of the Regents of the University of California. Homes built on each lot would be the property of the owner but not the land. Owners would pay an annual lease payment to the UC Regents for the privilege of living in University Hills. Second, when a resident decided to sell the home, it had to be offered first to, a faculty member, or a staff member, second to an individual associated with the university, and finally to a member of the general public. Throughout the process, the Regents have the right to exercise their option to purchase the property.

Finally, a formula spelled out in the initial lease agreement between the buyer and the UC Regents limits sales prices. The ICHA made it clear to its residents that the purpose of the community was to provide housing that faculty and staff could afford. At the initial sale and when resales occurred, buyers who wanted to make a significant capital gain on the sale of a home were advised to look outside of ICHA.[18]

***

The Long Range Development Plan (LRDP) governs development of the UCI campus.

The first LRDP, written in 1963, was a collaboration with William Pereira and Associates, the firm selected by the UC Regents to create the plan, and senior UC and UCI administrators. Among the foundational decisions made at the time was UC president Clark Kerr's recollection of a book by a German author that inspired the circular design of what became known in the LRDP as the Academic Core, the ring of buildings around a central park where the university located the fundamental undergraduate schools (social sciences, physical sciences, biological sciences, humanities, fine arts, and engineering).

Beyond the Academic Core, the plan envisioned a radial mall of buildings primarily for graduate education and specialized institutions . . . where learning could be applied to "real-world problems."[19] From that section of the campus, over the decades, the campus expanded, amoeba-like, filling virtually every available space on land, flat and sloping.

The most recent LRDP was written in 2007 and is due for an update in 2027. When planners sit down at their maps and computers to begin that process, they will see a campus that is jammed with 170 buildings that total 4.8 million square feet of space for academics, administration, and support. There are 10,822 units of student housing. The current enrollment at UCI is 36,000.

The most prominent new development sponsored by the university system is at the far boundary of the Inclusion Area at the intersection of Campus Drive and Jamboree Boulevard in Irvine. In late 2024, two new medical treatment facilities opened there, followed in mid-2025 by a 144-bed hospital.

The flagship facility is the Chao Family Comprehensive Center and Ambulatory Care building, named in recognition of the $50 million the family has contributed to UCI health programs and facilities since 1999. The center combines cancer research and treatment.

Adjacent to the Joe C. Wen and Family Center for Advanced Care is a 168,000-square-foot building that, in addition to providing healthcare, has dedicated the second floor to the county's well-respected Children's Hospital of Orange County (CHOC). The Wen family donated $20 million for the facility.

Notes

1 Daniel G. Aldrich, interview by Samuel C. McCulloch, May 24, 1989, UCI Oral History Project.
2 Aldrich, interview.
3 Aldrich, interview.
4 Aldrich, interview.
5 Aldrich, interview.
6 Mark I. Pinsky, "Irvine Hospital Dispute Ends in Compromise," *Los Angeles Times*, October 13, 1985.
7 Jack W. Peltason, interview by Ann Lage, Oral History Office, University of California, Berkeley, 2001.
8 Peltason, interview.
9 Peltason, interview.
10 "UCI, Irvine Co. Plan Research Park," *Los Angeles Times*, October 22, 1985.
11 Peltason, interview.
12 Peltason, interview.
13 Sharon Roan and John Westcott, *UCI: Bright Past, Brilliant Future* (Regents of the University of California, 2015).
14 Orange County Register, February 11, 2007.
15 University of California, Irvine, "University Research Park Lease Agreement Announcement," July 1998, in Office of the Chancellor Records: Ralph J. Cicerone (1998–2005), Special Collections and Archives, UC Irvine Libraries, https://oac.cdlib.org/findaid/ark:/13030/c828069d/.
16 Irvine Company Office, UCI Research Park, accessed January 14, 2026, https://www.irvinecompanyoffice.com/locations/orange-county/uci-research-park/uci-research-park.html
17 "Thesaurus Linguae Graecae," University of California, Irvine, accessed September 9, 2025, www.stephanus.tlg.uci.edu.
18 "University Hills," Irvine Campus Housing Authority, accessed September 9, 2025, www.icha.uci.edu.
19 University of California, Irvine, *UCI Long Range Development Plan*, 2007.

9 Ray Watson returns

It had been nine years since Ray Watson walked out of the Irvine Company as its president, and the day he returned as one of three vice chairmen reporting to the man who owned or controlled almost every share of the company—Donald Bren.

Watson had arrived at the giant Irvine Ranch in late 1960, hired as a junior planner. He joined a handful of other young men who had been given the task of transforming 93,000 acres of land into a new city that would radiate into the flat, fertile fields from the site of a new campus for the University of California.

Watson rose quickly, especially adept at convincing community and government officials that the Irvine Company's plans were in everyone's best interests. Watson was also friendly, approachable, and engaging—people liked him. It was a winning combination. Watson formed a strong partnership with William Mason, the company's chief engineer, who was promoted to president in 1966 and took Watson along with him into the upper levels of executive management. Together, the pair implemented an acceleration of both planning and construction of new homes and businesses, and the infrastructure to support them.

In 1974, Mason died unexpectedly of a heart attack. Watson was devastated. Mason was much more than an associate, Watson told others who mourned the loss—"he was like a brother." Still, Watson was prepared to move up, and he was named president of the company shortly after Mason's death.

Watson was raised in a boarding house. His mother had died when he was just 5 years old—his father and grandmother had moved to Oakland, California, from the family home in Seattle.

Watson had to grow up quickly. He had a paper route and did chores around the boarding house with his sister, absorbing the atmosphere of struggle and sadness reflected by the residents, describing them in an oral history decades later as "hard of luck."[1]

The boarding house years made a lasting impression on Watson, shaping his outlook on life. He developed a deep and abiding sympathy for ordinary people. While attending the University of California at Berkeley, Watson also discovered the writings of Carl Jung.[2] More than once, when confronted with business issues, Watson balanced public and private interests, trying to achieve something fair as well as profitable. A key example was his attitude during the planning of the vast, undeveloped coastal property the company owned.

DOI: 10.1201/9781003683087-9

Without hesitation, Watson had advised the senior executives and board members who controlled the company that almost all the land between Pacific Coast Highway and the shoreline should remain in public ownership. Watson knew intrinsically that no matter the tremendous value of property looking directly over the Pacific Ocean, the political reality argued for simply acceding early on and concentrating on using the goodwill of the "donation" to leverage greater development opportunities elsewhere on the coastal shelf above the highway.

After the passage of the Tax Reform Act of 1969, the James Irvine Foundation, which controlled the Irvine Company by virtue of its holding of a slim majority of company stock, continued to support accelerated housing and commercial development. The giant village of Woodbridge broke ground; Newport Center continued to expand, as did the industrial complexes next to the Orange County Airport and the "Golden Triangle" where Interstates 5 and 405 met. Throughout this period, Watson, first as a senior vice president and then as president, guided the giant land developer.

When the company was sold to new owners in 1977, Watson made the tough decision to leave. In his oral history, he did some self-analysis, acknowledging that had he stayed, he would have made a great deal of money both from an increased salary as well as ownership of company stock. But in the end, he said that by leaving, "I stayed true to my conscience. I did what I believed was right for me and them (new company owners)."

Watson and three other former Irvine Company executives, along with two former business associates, formed a development company and achieved some immediate success by building a condominium project and a handful of other small developments. However, Lanny Eberling, one of the owners, recalled in a 2021 interview that Watson became much more cautious in taking on new ventures. "Ray was not an entrepreneur, which he admitted," Eberling said. "He was very slow to make a decision."[3]

Watson had been with the new firm for six years when an opportunity presented itself that would set his career on a second trajectory of intense business conflict and notoriety.

In 1965, Watson had been invited by Walt Disney himself to look over the plans for EPCOT, an experimental "city of the future" Disney was planning in Florida. Watson recalled reviewing a roomful of maps, with Disney making the presentation and fielding Watson's questions as a cadre of silent Disney executives looked on. During a subsequent meeting, Watson walked through a part of the vast Burbank studio and got a glimpse of early prototypes of the Pirates of the Caribbean ride.

For the next few years, Watson was asked to review Disney's real estate plans. The company's executives were impressed by Watson's knowledge as well as his family-oriented reputation, which they saw as a good fit for Disney. In 1974, Watson was offered and accepted a seat on the Disney board of directors.

He and Shirley Temple Black were the only two outsiders on the board, which Watson said was essentially a rubber stamp for company owners and executives.[4] That changed in 1983, however, when a radical upheaval occurred in the upper tier of Disney management.

Ron Miller, Walt Disney's son-in-law, was being promoted to CEO after working in the company as president and chief operating officer. But Miller didn't have a deep understanding of business (he had dropped out of USC and spent a year with the Los Angeles Rams as a receiver) was not self-assured, and was uncomfortable making public presentations. He admitted to Watson that he needed someone more experienced to help him.

Watson had been elected vice chairman of the company in February 1983. According to John Taylor's history of that time at Disney—*Storming the Magic Kingdom: Wall Street, The Raiders and the Battle for Disney*—as Miller turned to Watson more and more for advice, Watson "was soon immersed in a staggering array of decisions that needed to be made daily at a company as large and diversified as Disney."[5]

The new role was taking so much time Watson decided to depart from the Newport Development Company and come aboard full-time at Disney:

> He immediately had his hands full . . . and one of the reasons why, Watson discovered, was that Walt Disney Productions, a company with revenues of more than $1 billion a year, had no business plan. It had never had a business plan.[6]

But Disney's internal management challenges were eclipsed by a prolonged, bitter, and public battle to take control of the company through a hostile takeover. Watson was plunged into the midst of a conflict with corporate raiders like Ivan Boesky, Saul Steinberg, and Michael Milken. Giant New York investment houses joined the cast, along with private investors like the Bass brothers. Adding intrigue and an element of personal animus was Roy Disney, Walt's nephew. Roy had worked at Disney for decades and owned more Disney stock than any single individual or business entity.[7]

He had become disillusioned about the direction the company took after Walt's death and had earned the animosity of senior Disney executives, who either ridiculed him or ignored him. When Roy resigned from the Disney board of directors, he began to conspire with Stanley Gold, his investment advisor, to gain control of the company.[8]

As Disney stock swooned in the backwash of the takeover tides, Watson said in his oral history that his priority quickly became trying to save the company instead of managing it. For nearly a year, Watson found himself like the captain of a storm-tossed ship who had been lashed to the helm in ever-rising seas. And while he had served on corporate boards in the past, none had experienced the turmoil of a hostile takeover.[9]

Storming the Magic Kingdom chronicles several tense meetings and confrontations between Watson and the cast of corporate vultures circling the beleaguered Disney organization. For Watson, it was both exhilarating and frustrating. However, in his oral history, he said he never felt intimidated or outmatched in the contest as it played out on a national stage.[10]

In the end, Disney offered corporate raider Saul Steinberg a $25 million "greenmail" payment to go away, paving the way for Michael Eisner and Frank Wells to assume control of the company under revised ownership. Watson gave up the board chairmanship in the transition but continued to fill a board seat as well as a position on the firm's executive committee. He left both positions in 2004.

***

Watson's return to the Irvine Company in 1986 was the antithesis of his experience at Disney.

While there were few familiar faces in the employee ranks nine years after his departure as president, he was pleased to renew his relationship with Donald Bren. The two men met decades earlier when Bren was angling to build homes on Irvine Company property. Highly focused and detail-oriented in his choice of design for his homes, Bren made a favorable impression on Watson. Bren accompanied Watson and a handful of company executives on a trip to Europe to study New Town developments.

Figure 9.1 Ray Watson in 1978

Source: Author's image

The public explanation for Watson's appointment as one of the three vice chairmen reporting to Bren was that he would take an active role in ongoing planning for the company's coastal properties. And while he did become part of the team of company employees, lobbyists, and lawyers shaping the coastal strategy, he also immersed himself in the company's array of operations and projects.

Watson kept his copies of the company's executive committee meetings and donated them to the Langson Library at the University of California, Irvine, where they are available for public review.[11]

Examining the meeting minutes and other documents Watson retained, it becomes clear that he was deeply attentive to the strategic and financial implications of the myriad development efforts that had been ramped up after Bren took complete control of the company in early 1983.

In addition to attending those meetings and providing private advice to Bren and other senior executives, Watson resumed one of his favorite activities—management by walking around.

As had been his habit during his presidency of the company more than a decade earlier, Watson began to roam the halls of the company offices at 550 Newport Center Drive, stopping to say hello or more, often, inquiring about what the employee was working on at the moment.

At the time, it was increasingly unusual to see the company's most senior executives anywhere outside the ninth floor of the building. Bren never left his office there, engaging a chef to prepare lunch for him and other executives in his private dining room. The only time most employees caught a glimpse of Bren was if they happened to walk into the ground-floor elevator at the same time he did. The result was usually an awkward silence on the way up for both riders.

So, Watson's accessibility and friendly demeanor were a refreshing experience for employees, most of whom happily engaged with Watson on matters large and small. Watson also enjoyed being mildly provocative if he thought the employee had the intellect and strength of personality to argue a point with one of the most senior and influential executives at the Irvine Company.

A division vice president stuck his head out of his office one afternoon after hearing raised voices and recognized that they belonged to a subordinate as well as Watson. Alarmed, he interceded with a joke and suggested Watson join him behind closed doors. Watson expressed no qualms about the incident—he viewed such intellectual jousting as just another way to get to the heart of issues and create dialogue.

During these years, it was not unusual for Watson's name to appear in the press as either an official or unofficial spokesman. Watson was always comfortable speaking to reporters, and they liked his habit of eschewing talking points or corporate nonstatements for straightforward commentary.

And Watson was back on the speaking circuit, often appearing at building industry or civic groups to discuss environmental or housing issues.

One news columnist described Watson as something of a "corporate grandfather," playing the elder statesman role. Another called him "the spiritual keeper of the flame for the company's planners."[12]

But Watson saw it somewhat differently. As was often the case, he engaged in a bit of self-deprecation when asked to explain what his role was as an Irvine Company vice chairman. With a grin, he said, "I go in, I pontificate, I leave."

Notes

1 Ray Watson, oral history interview by Ann Lage, *Planning and Developing the New Town of Irvine, California, 1960–2003* (Berkeley: Bancroft Library, University of California, 2005), accessed July 2025, https://digicoll.lib.berkeley.edu/nanna/record/218181/files/watson_ray.pdf.

2 Carl Jung, *Modern Man in Search of a Soul*, trans. W.S. Dell and Cary F. Baynes (New York: Harcourt, 1933), 112.

3 Lanny Eberling, interview by C. Michael Stockstill and H. Pike Oliver, March 2021.

4 Shirley Temple Black, oral history interview, Stanford University Libraries, Department of Special Collections, accessed July 2025, https://oralhistory.stanford.edu.

5 John Taylor, *Storming the Magic Kingdom: Wall Street, the Raiders, and the Battle for Disney* (New York: Knopf, 1987), 38.

6 Taylor, *Storming the Magic Kingdom*, 38.

7 Taylor, *Storming the Magic Kingdom*, 40.

8 Taylor, *Storming the Magic Kingdom*, 41.

9 Taylor, *Storming the Magic Kingdom*, 42.

10 Taylor, *Storming the Magic Kingdom*, 43.

11 Irvine Company Executive Committee Meeting Minutes, 1986–1990, Ray Watson Papers, Langson Library, University of California, Irvine.

12 "The Spiritual Keeper of the Flame for the Company's Planners," *Los Angeles Times*, December 30, 1999.

10 The buyout

From late 1977 to early 1983, Donald Bren bided his time.

As the owner of just over a third of the newly formed Irvine Company, Bren was the logical choice to lead the new company's first board of directors. But he had been outvoted by supporters of Alfred Taubman, who had—at the instigation of Allen and Company—rounded up the other investors to complete the purchase of the company from the James Irvine Foundation. Bren agreed to Taubman's advice to incorporate the new Irvine Company in Michigan. And he acquiesced to Taubman's desire to chair the board.

It was no surprise to those who knew and worked with the Detroit-based retail magnet that Taubman would insist on the top leadership post of the new company. In a profile of Taubman in the August 7, 1983, edition of the *Detroit Free Press,* former Taubman Company employees, tenants in his malls, and industry observers recalled working with a man they described as "a taskmaster," "demanding," "passionate," and "an exciting person to be around." A man identified as "a leader in the shopping center industry" summed it up by saying, "he's very forceful. No one has a higher opinion of Al Taubman than Al Taubman himself. He wants to be the top dog; he wants to be Number One. He has an immense ego."[1]

The other Irvine Company owners Taubman had recruited were his close friends and business associates: Max Fisher, Milton Petrie, and Henry Ford II. Each one had done business with Taubman for years; Fisher was especially close, sharing Taubman's commitment to revitalizing Detroit. Petrie and Ford were more passive investors, as was Allen and Company, the family-owned, low-profile Wall Street investment bank. Rounding out the investors were Irvine heiress Joan Irvine Smith and her mother as well as agribusiness executive Howard Marguleas. (Three Irvine family members also had very small ownership positions.)

Taubman had wasted no time finding opportunities to leverage his Irvine Company ownership for his own gain. Taubman installed his management company to operate Fashion Island, the company's flagship retail center across the street from company headquarters in Newport Center. Bren had the opportunity to move his homebuilding company to first in line to build new homes in developments like

DOI: 10.1201/9781003683087-10

Woodbridge. And in his book, Martin Brower—who was director of public affairs at the time—wrote that

> Irvine Center (at the intersection of the 5 and 405 Freeways, dubbed the Golden Triangle) announced with letters of intent from May Company and Bullocks back in 1976, before the sale of the company, had been set aside immediately after the sale. Taubman . . . did not like the deals struck with the department stores and wanted to restudy the entire center.[2]

Files from Taubman's papers at the University of Michigan include a letter dated September 12, 1980, that he sent to his fellow Irvine Company owners. "You may recall," he reminded them, "when the Taubman-Allen-Irvine group initially entered into the bidding for the acquisition of The Irvine Company, one of our objectives was to seek direct ownership of real estate assets."[3]

Taubman went on to explain that the banks that financed the purchase objected to the owners double-dipping in the company's assets and vetoed the idea. But now, he went on, "with the bank loan paid off, however, it seems appropriate to reconsider the concept."

Taubman's letter then identified the target for the owners to purchase—an 18.7-acre site at the western edge of Newport Center that had been zoned for 5 two-story office buildings yielding 213,000 net leasable square feet. It was called Civic Plaza. He said as long as the deal would return fair market value for the land to the company, the lawyers had blessed it:

> Based on the enclosed pro forma for the project (valued at $34 million), the partnership would fund a $1 million equity investment with the remaining capital provided through the land loan (value $10 million) from the Company and a construction loan and eventual permanent mortgage loan to be arranged by the partnership. A deferral of a portion of the interest payments under the Company land note is assumed during the initial five years of operations. This will provide a positive cash flow from the project and a reduced equity requirement.

This is the only documentation about the Civic Plaza deal that is public. The project was built and is occupied today.

Also found in Taubman's files at the University of Michigan is an analysis of a bizarre proposal to divide the Irvine Company into two separate entities—a concept that has never been reported in print or otherwise. The idea was outlined in a memo from the company's principal legal advisors, Latham and Watkins.

Dated March 7, 1980, it noted that

> under the reorganization proposal, at least two new subsidiaries of Irvine would be created. One would confine itself to agricultural purposes and is hereinafter referred to as "Agrisub." . . . The other subsidiary (herein after

referred to as 'Dealersub') would perform all land development activities with respect to the agricultural land.[4]

The object of the restructuring was all about taxes, specifically capital gains. The lawyers acknowledged that no matter how convoluted an ownership structure might be patched together, "the capital gains tax will have to be paid only when the developed land is sold or exchanged outside the affiliated group. However, the character of the gain will, it is hoped, be changed for federal tax purposes." The bottom line: the attorneys estimated that at an 18 percent capital gains tax rate, the two-company structure could save $4.7 million annually on the assumption that 350 acres of land were developed per year.

Agrisub and Dealersub were apparently discussed—if at all—in the abstract at the management levels of the Irvine Company, which remained a single entity.

According to a source who was close to Taubman during the bidding war, the Taubman-Allen-Irvine partnership agreement had a provision that after five years in existence, there would be the potential for what has been described as a "capital event" It may also have been a trigger date to allow for ownership shares to be sold or exchanged among the owners.

In any case, in Al Taubman's view, things were going quite well at the Irvine Company, and he was enjoying his role as board chairman, setting policy and making the big decisions as he cruised into Newport Beach on his private jet from Detroit or Palm Beach. The five-year mark made no difference to him—he saw the Irvine Company as a long-term investment that was just starting to provide him with substantial returns.

Donald Bren, however, was increasingly unhappy with how the Irvine Company was being run. According to a profile of Bren in *Fortune* magazine, Bren had increasingly clashed with Taubman on business strategy. Taubman, Bren said, wanted to sell off not just land for residential development but for retail, commercial, and industrial. Bren wanted to keep it.[5] And even though Bren had recruited Kremer to run the company, after seeing how Kremer's style was being received in Irvine and Newport Beach, Bren began to ask himself if his instincts about Kremer had been correct. Also, Kremer usually sided with Taubman at board meetings.

The company had tightened up its relationships with tenants and was making tougher deals with builders and other business partners. And while Kremer and senior managers always found time to meet with local political leaders, Kremer often made it clear to subordinates that he could not spend the amount of time his predecessor, Ray Watson, had to hold hands with council members, planning commissioners, and other influential community types.

One meeting with Kremer, Irvine city councilman Bill Vardoulis, Irvine city manager William Woollett Jr., and Irvine Company vice president Tom Wilck (as well as a couple of junior Irvine Company staff on hand to take notes) exemplified how Kremer's style sometimes shaped opinions about him.

Woollett and Vardoulis had arrived with an agenda of concerns about a variety of issues they believed required the Irvine Company's attention.

Kremer welcomed the pair into his office and motioned them to chairs in front of his desk. Kremer sat behind the desk. Ten or fifteen minutes into the meeting,

Vardoulis was becoming visibly agitated as he pressed Kremer, who showed little reaction. A moment later, Vardoulis started to come up out of his chair; Woollett silently reached behind him, grabbed his belt, and pulled him back into the chair.[6]

Kremer smiled passively as Woollett and Vardoulis departed. After the door closed, he turned to Wilck and said, "You know, Vardoulis made some good points." On his way back to his office, Wilck turned to his young associate and said, "Too bad Peter didn't tell Vardoulis instead of us he agreed with some of the things he said." Wilck then observed that Kremer, in many ways, viewed conversations like the one that just took place as negotiations and was almost always reluctant to show what he was thinking lest it be used against him in future dealings.

The incident was emblematic of reports that Bren was hearing from a variety of sources inside and outside the company. Those observations began to change somewhat after Kremer hired Thomas Nielsen—who he had worked with at Newhall Land and Farming—to head up the community development division of the company, which had the most interface with political leaders and community members.

However, Nielsen's appointment had come too late to change Bren's mind about Kremer and spurred his plan to obtain majority ownership of the company before things got worse.

Bren quietly began contacting banks and accountants familiar with real estate finance to determine how to value the Irvine Company and to construct a buyout plan he could afford.

Bren also spent hours with his most trusted political advisor, Gary Hunt, to anticipate what the public reaction would be to his purchase and how best to manage the inevitable questions that would follow. Another important audience to consider was the company's employees, the vast majority of whom had never even seen Bren, let alone talked with him.

As Bren pondered the buyout, he knew Taubman was the most critical investor he had to convince to make the sale. Taubman had been the leader and public face of the 1977 bidding war; more importantly, he had almost single-handedly rounded up the money for the deal from individual investors and a large consortium of banks.

The two men could have hardly been less alike. Taubman enjoyed the public arena and the attention that came from being a big player in the business world of Detroit as well as in the numerous cities where he built his retail malls—when a Taubman Company mall opened, it was big news. Taubman also had a home in Palm Beach and was well-known in the social scene there.

Bren's world was centered in Los Angeles, especially the glitzy West Side, where his homebuilding company was headquartered and where he owned a home. Bren was occasionally written up in the society section of the *Los Angeles Times*, usually when he attended galas with his mother, Marion Jorgensen. Other times Bren was reported to be squiring beautiful women at similar events. But Bren didn't seek out such publicity, and as the 1970s blended into the 1980s, he became less and less visible to the press and public.

Bren had a home in Sun Valley, Idaho, where he often spent the Christmas and New Year's holidays slashing down the most challenging slopes the ski resort had to offer. Bren was in top physical form, just over six feet tall, a slim 150 pounds, he kept in check by a diet high on vegetables and low on red meat. And Bren hated smoking. Taubman, six foot two, carried well over 240 pounds on his stout frame, and often enjoyed a fine cigar.

Bren and Taubman shared an appreciation for fine art, but Bren's collection was something he almost never showed anyone outside a small circle of intimates. Taubman, on the other hand, was never shy about describing his art and announcing a purchase.

As 1982 came to a close, Taubman was growing concerned about a new, high-profile project he had undertaken in Beverly Hills.

Beverly Center was an eight-story retail mall unlike any other that Taubman had built. Confined to a small block of land in the heart of the affluent city—whose most famous shopping district was Rodeo Drive—the new mall would have several floors of retail on top of parking spaces built into the entire building. On the

Figure 10.1 A. Alfred Taubman in 1978

Source: Detroit Free Press

top floor, a 16-screen theater complex would draw patrons who would then, it was hoped, saunter down to the retail shops after enjoying a first-run movie.

Taubman had two partners in Beverly Center. According to Robert Schout, who had been Taubman's day-to-day project manager for the purchase of the Irvine Company five years earlier, Taubman had obtained a large loan from Allen and Company to finance his share of Beverly Center. The problem, Schout recalled in a 2024 interview, was a substantial cost overrun in constructing the massive new mall.

Schout believes Allen and Company was becoming nervous about the loan (interest rates in 1982 were high; an advertisement in the *Los Angeles Times* placed by a savings and loan next to a story about Beverly Center tempted depositors into its money market account at 15 percent return). For this reason, Schout contends, Taubman was more receptive to Bren's purchase offer than he might otherwise have been.[7]

The story of how Bren consummated the purchase of nearly all the shares in the Irvine Company was relatively straightforward at the time it occurred. Subsequent events, trial testimony, and personal recollections by the principals involved add a slightly different dimension.

***

The official announcement was made on April 20, 1983, that Bren purchased the shares of Taubman, Fisher, Petrie, and Ford. At first, there was little detail to illuminate the news—Bren's comments in his press release emphasized that ownership was now vested with California owners—in addition to him, Joan Irvine Smith and her mother, Althalie Clarke, remained as owners, as did Howard Marguleas—all were original members of the Taubman-Allen-Bren-Irvine partnership.[8]

The value of the deal was placed at $400 million, but within a few days, the real number of $518 million became known.

Bren had worked diligently to keep his intentions under wraps, but a *Los Angeles Times* story a month after the sale was announced explained how his best efforts came to naught:

> Dozens of lawyers and financial advisers had been working quietly but frenetically for weeks, but with the closing date still two weeks off, the shroud of super-secrecy was starting to lift. As Bren tells the story, he was working early one evening in his Los Angeles office when the janitor strutted in and brushed my desk. As he's emptying the trash, he says to me, "Say Don, how's the Irvine deal doing? Has it closed yet?"[9]

Bren chose to structure the purchase by forming a holding company he named Newco. It would serve as a recipient of funds that would, in turn, be transferred to the other Irvine Company shareholders in return for their shares in the company.

Bren put up $900 million of his own funds into Newco, calculated at his valuation of $860 million for the shares he owned in the Irvine Company, plus the value of 816,000 shares of International Paper that he owned. Newco then borrowed

$104 million from five banks: Security Pacific National Bank of Los Angeles, Wells Fargo Bank and Crocker Bank in San Francisco, and Bankers Trust and Manufacturers Trust Company of New York.

Bren offered the other shareholders a method to remain with the company if they chose not to sell. They could trade an existing share and receive two in return. However, in doing so they would have to agree to a stipulation in the deal that Irvine Company dividends would be limited to $12 million per year. Finally, the loan plan said that Newco and the Irvine Company would merge by November 15, 1983.

As might have been anticipated, Joan Irvine Smith and her mother, Athalie Clarke, found serious flaws in Bren's approach.

Smith had not been to an Irvine Company board meeting in five years. In the interim, she had taken to expressing her concerns in writing, inundating company president Peter Kremer with thousands of documents containing her views on a variety of company and other issues. She had also sued the company over a plan to support a regional water supply pipeline but lost in court.

The element of the Newco deal that Joan seized upon as the centerpiece of her objection was the contention that Bren had assured her he would not use company assets to pay down the loan. She pointed out that after the planned Newco/Irvine Company consolidation in November, the loan would be paid off before April 16 the following year, proving her contention.

Joan went public with her opposition in April when the Newco deal was announced, threatening to sue if the Irvine Company board of directors dared to approve it. In September, she made public the details of the Newco loan plan to bolster her arguments against it. A few days after the board approved the Newco arrangement, she filed a lawsuit against the company.[10]

In reporting on the details of the lawsuit, the *Los Angeles Times* noted that "Most of the charges in the lawsuit have already appeared in the news media . . . in the course of Smith's vigorous publicity campaign against the merger proposal. But the suit is noteworthy for the harshness of the attacks on Bren."[11]

Smith argued that the debt incurred for the buyout would "diminish the company's borrowing power . . . seriously impairing its ongoing operations and profitability, all to the enormous detriment of the minority shareholders."

For those who had followed Joan's long history of litigation, the strident tone of the accusations was familiar. And for those who were betting on the likely outcome, the smart money said she would lose. And lose she did a month later in Orange County Superior Court, where Judge Phillip Petty denied her request for a temporary restraining order to halt the transaction.[12]

The next day, Joan and her mother announced they would exercise their rights as minority shareholders and come up with their own evaluation of what their shares were worth, setting up what would become their final legal battle over their ownership in the giant ranch.

***

While Joan Irvine Smith was complaining about the buyout deal in April, the shareholders who chose to relinquish their stake in the Irvine Company were celebrating.

For Al Taubman, it was a victory lap similar to the one he proudly took after the original purchase of the Irvine Company in 1977. At that time, Taubman recounted the day-by-day strategy he undertook to outsmart Mobil Oil in the increasingly tense bidding war that was breathlessly reported by local and national media.

On April 17, after a brief story the day before on the sale, the *Detroit News* reported that Taubman "savored the sale of his and four other investors . . . with Max Fisher, Henry Ford II and their wives at Taubman's Palm Beach home."[13]

Reporter Laura Berman lobbed softball questions to the glowing Taubman, eager to verbally put a shine on his trophy sale. He modestly allowed that "the only real estate sale bigger than Friday's billion-dollar-plus sale was maybe the Louisiana Purchase—no, this is a better deal than the Louisiana Purchase."[14]

The paper went on to quote Taubman on his motivation to sell. "It takes a good idea to buy a piece of property," he said, "but very often it's a question of unique understanding to know when to sell. We felt the price offered was one we couldn't refuse."

Bren's offer, according to multiple sources that reported on the sale over the years, gave Taubman a profit of $140 million on his investment of $10 million; for Fisher and Milton Petrie, it was estimated that each made $100 million. Ford had a much smaller piece of the deal but reportedly made a profit of $19 million.

Four years later, Taubman was called to testify in the trial held to determine the valuation of Joan Irvine Smith and Athalie Clarke's shares in the Irvine Company. Facing Joan's attorney Howard Friedman under oath, Taubman's story was somewhat different.

As the *Los Angeles Times* reported in a story on October 8, 1987,

> When Bren's intentions to acquire control of the company became clear in February 1983, Taubman said, he was not happy about selling. "I felt the opportunity was still there. I wasn't interested in leaving the company."
>
> He said he was disturbed to learn that Kenneth Leventhal & Co. was preparing financial information for Bren while simultaneously serving as the company's accounting firm. Taubman said he was also "aroused" when he discovered that Bren, a fellow board member, had been quietly meeting with prominent businesspeople and politicians in Irvine during 1982 to discuss a possible buyout. The following March, after consulting with Max Fisher, Taubman said he decided he would not attempt to thwart the wishes of the other members of the ownership group who wanted to sell their shares to Bren; he agreed to participate in the sale.[15]

In the end, Taubman compared the sale to a divorce but found that "the alimony was good."[16]

Within a few months, Taubman put his money to work in a deal similar to the 1977 acquisition of the Irvine Company. Again, inviting Fisher and Ford to join his syndicate, he fronted the purchase of Sotheby's, the venerable auction house in the United Kingdom. (According to a *Washington Post* story, December 4, 2001, Taubman put $38 million into the deal, Ford and Fisher a total of $100 million.)[17]

When Sotheby's went public in 1988, Taubman and his other investors made a large profit. But Taubman was accused of colluding with rival auction house Christie's on charges of price-fixing and was convicted in federal court in 2002. He served ten months in prison.[18] Taubman died in 2015 at age 91.[19]

Notes

1 "Taubman: A Forceful Leader," *Detroit Free Press*, August 7, 1983.
2 Martin Brower, *The Irvine Ranch: A Time for People* (Bloomington, IN: AuthorHouse, 2013), 67.
3 Alfred Taubman to Irvine Company Ownership Group, letter dated September 12, 1980, Alfred Taubman Papers, Bentley Historical Library, University of Michigan.
4 Latham & Watkins, internal memorandum, March 7, 1980, Alfred Taubman Papers, Bentley Historical Library, University of Michigan.
5 Gary Hector, "Donald Bren: The Invisible Empire Builder," *Fortune*, August 27, 1990.
6 Author's recollection.
7 Robert Schout, interview by C. Michael Stockstill and H. Pike Oliver, December 12, 2022.
8 "Bren Buys Out Irvine Company Partners," *Los Angeles Times*, May 20, 1983.
9 "Bren Buys Out Irvine Company Partners," anecdote recounted by Donald Bren.
10 "Joan Irvine Smith Threatens Lawsuit Over Bren Deal," *Los Angeles Times*, April 21, 1983.
11 "Smith Sues to Block Irvine Company Merger," *Los Angeles Times*, October 19, 1983.
12 Orange County Superior Court, Smith v. Irvine Company, Case No. 83-OC-1427, ruling issued November 1983.
13 "Taubman Celebrates Sale with Fisher and Ford," Detroit News, April 17, 1983.
14 Laura Berman, "Taubman on the Irvine Sale," Detroit News, April 17, 1983.
15 "Taubman Testifies in Irvine Valuation Trial," *Los Angeles Times*, October 8, 1987.
16 "Taubman Testifies in Irvine Valuation Trial."
17 "Taubman Buys Sotheby's," *Washington Post*, December 4, 2001
18 "Taubman Convicted in Price-Fixing Case," New York Times, April 23, 2002.
19 "Generous U-M Supporter A. Alfred Taubman Dies at 91," *University of Michigan Record*, April 18, 2015, https://record.umich.edu/articles/generous-u-m-supporter-alfred-taubman-dies-age-91.

11 Show me the money

What was the value of the Irvine Company in November 1983, when Donald Bren's purchase of a majority of Irvine Company shares officially closed?[1]

Privately owned, its assets were spread across more than 77,000 acres of land in the heart of Orange County.[2]

First on the list of its holdings was land. There was a great deal of it, but not all land on the Irvine Ranch was created equal.[3]

There was land in the center of the property—hundreds of acres, almost perfectly flat, straddling two interstate freeways. It was valuable but even more valuable when it received zoning approval, allowing it to be developed for the villages that the Irvine Company had successfully created over the past two decades.

Still more valuable was the flat land that had been improved to the point where it could be sold to homebuilders—in the trade's lingo, lots that were "blue topped."

There was land at either side of the central flatlands that sloped into foothills; the San Joaquin Hills to the west, the Lomas de Santiago to the east. Development on the hillsides would require grading, at substantial cost. And to the east of the Lomas de Santiago, except for a few narrow valleys, the land was mountainous, a place where significant development would never take place.

Finally, just over the crest of the San Joaquin Hills was perhaps the most valuable single piece of property on the ranch: the Irvine Coast, 4,000 acres of hillside and coastal plateau with spectacular views of the Pacific Ocean.

Next on the asset list were commercial developments, ranging from retail centers in the residential villages to high-rise office buildings, dozens of single-story industrial buildings, and property leases.

Then apartments, thousands of them. At the high end, luxury on the coast at Promontory Point overlooking Balboa Island; at the low end, the venerable Park West Apartments on Culver Drive in University Park in Irvine, along with the Deerfield units in central Irvine.

On Newport Bay, 281 boat slips where revenue is calculated by the size of the vessels moored there. In Newport Beach and Irvine, more than 40,000 customers of Community Cablevision, a cable television service. And in the hills above the center of the ranch, thousands of citrus trees; just over the hill, a large reservoir and dam on Santiago Creek, and a pair of smaller reservoirs in the Lomas de Santiago foothills. There was also a weekly newspaper, two hotels, and a small herd of cattle.

DOI: 10.1201/9781003683087-11

Finally, the intangible value of dozens of employees with decades of institutional knowledge about every aspect of land development, finance, property management, and the intricacies of zoning law and politics in Orange County, the State of California, and Washington, D.C.

Add it all up, subtract the millions of dollars in outstanding loans that funded infrastructure like roads, sewers, and water lines, plus anticipated obligations to build schools, firehouses, and other public necessities, and *viola*, the number.

If it were only that simple.

Determining that number would lead to one of the most contentious and expensive legal battles in the history of the Irvine Company, rivaling for notoriety and media coverage the public bidding contest generated by the sale of the company in 1977.

But this dispute would tell two different stories. The first was a complex, detailed, and, at times, byzantine analysis of the intricacies of land development and property valuation, with facts, figures, charts, and arguments wielded by an array of real estate appraisers and other experts.

The second story was much more compelling; the juicy, behind-the-scenes airing of corporate and personal rivalries and motivations that allowed a glimpse into the closely guarded inner sanctum of the Irvine Company after Bren purchased it. Joining him at center stage was the leading lady of the corporate drama, Joan Irvine Smith (and her mother). For Joan, it was the last stand in her decades-long struggle to win control of the company that bore her name.

Seven months earlier, in April 1983, Bren had initiated a complex purchase of all the outstanding shares of the Irvine Company (with one exception) giving him complete control. Joan and her mother, Athalie Irvine Clarke, had agreed to sell their 11 percent interest, but made it plain, they believed the price they were offered was almost insultingly too low, by millions of dollars.

Bren had tried to accommodate the pair's concerns by offering, in addition to $114 million each, to increase the number of shares of the company, giving them 22 percent of ownership, plus two seats on the board of directors.[4] Joan and her mother declined the offer.

And as had so often been the case for mother and daughter, settling the dispute meant defending themselves in a lawsuit the Irvine Company filed against them.[5] Trusted family attorney Howard Friedman was summoned to their side, and documents were filed in Michigan, where the Irvine Company was incorporated.[6] Bren assembled a team of Orange County legal heavyweights; a war room was created on the seventh floor of company headquarters in Newport Beach, soon to be overflowing with maps and documents. Gary Hunt, Bren's most trusted employee and advisor, took charge of managing the defense.

Due to the unusual nature of the contest, a retired judge was appointed by the Michigan court system to hear the case. A temporary courtroom was built in an office building in the tony suburb of Bloomfield Hills, sparing the combatants a trek into the gritty streets of Detroit.[7]

Reporters and editors from the *Los Angeles Times, Orange County Register, Wall Street Journal,* and the *New York Times* began checking airline schedules as well as hotel accommodations in Bloomfield Hills. And Hunt's secretary Sue Kappas, put in a call to Bren's private jet company, Skybird Aviation; there would be a number of flights from Orange County to Detroit in the months and years to come—have the planes and pilots ready to go.

***

Gary Weber had come to the Irvine Company as a planner after a stint at the City of Rancho Palos Verdes. He received an unexpected welcome into the company's button-down culture soon after finding his desk:

> I walked in wearing corduroy pants, a jacket, and a plaid shirt. And one day, after I got back from a meeting, there were two cards on my desk, one for the barber in the building and another for Brooks Brothers. So I trimmed my hair and got three suits.[8]

Weber worked on a number of planning projects over his first four years at the company, then received an assignment that changed his professional life: building a database for a dispute with Orange County assessor Brad Jacobs, who had tripled the Irvine Company's tax bill after Bren bought out his partners. (The company and the county eventually settled the matter at a lower number).[9]

Although he didn't realize it at the time, the vast collection of information about the thousands of Irvine Company assets would be critical to the legal team. Weber was providing information to the attorneys who were doing a mock trial in Detroit (in preparation for the actual proceedings).

> And at one point, Bill Campbell turns around and says, "Gary, what are you doing for the next year? You are on my team." And it wasn't my decision, I don't think it was anybody's decision except his, and so it wasn't just one year, it was three years.[10]

Weber became both a technical advisor and something of the quartermaster for the team of attorneys and other Irvine Company employees working on the case:

> After the trial started, my role was to be support. We prepared graphics, and we had this huge database of land use environmental issues. A lot of the property tax stuff we were able to use that we had generated when we were doing the fight (with county assessor Jacobs).[11]

Weber added:

> And part of my duty was to maintain a travel budget and schedule. The attorneys wanted to stay in a hotel where they could get room service. Me and the rest of the team stayed at the Residence Inn so we could cook our own dinner or have breakfast without calling room service.[12]

In addition to doing his own work, Weber was backed up by Sally Smith, a company attorney; Kim Miller, a paralegal; and "five or six kids from UCI" doing research.[13]

Weber found the assignment fascinating and challenging. But the months that became years had a personal cost. "I missed a lot of my son's high school years."[14]

(Weber's post–Irvine Company career took a turn as a result of what he learned at the trial, leading him to become an expert in litigation support, including work on the defense of the *Exxon Valdez* oil spill in Alaska).

***

Four years after the lawsuit was filed, it was finally ready to be heard. The temperature was in the 90s in Bloomfield Hills in August 1987 and, like the rest of the Midwest, humid. No matter—the lawyers were dressed in their pinstripes, as were Irvine Company executives. Neither Joan nor her mother was in attendance as the judge began the proceedings.[15]

The case would likely be Howard Friedman's last one opposing the owner of the Irvine Company on behalf of Joan Irvine Smith.

Since being assigned as a junior associate at Loeb and Loeb in the early 1960s to represent her, Friedman had argued cases large and small on behalf of Joan and her mother. Many went to trial, but because Joan often sought legal redress even when it was unlikely her arguments would prevail, most cases did not go her way. However, with Friedman as her advocate, sometimes the mere threat of litigation would cause the company to retreat from the position Joan objected to.

This final case would test Friedman's legal abilities as had no other, in large part because the underlying issues were grounded less in the law than in the arcane field of real estate valuation.

Bren had successfully convinced the other owners of the Irvine Company to accept his offer to buy them out, setting a value of $1 billion for all the company's shares, which equated to $200,000 per share. Precisely how Bren arrived at the $1 billion value would be one of the primary issues Friedman faced in crafting his arguments, as Joan and her mother believed the company was worth closer to $2.5 billion. Bren's offer to purchase their shares for $114 million each was nowhere in the ballpark of their real value, the mother and daughter argued.

So in addition to attacking the value Bren had placed on the Irvine Company's multiplicity of assets, Friedman also developed a theory about motivation, asking himself (and the court) why men like Al Taubman, Max Fisher, Henry Ford II, and Milton Petrie agreed to sell their position in a company after owning it for just five years? And in the same vein, why did another owner, Howard Marguleas, reject Bren's offer so he could retain his small percentage of ownership? Did he know something the others didn't?

Friedman also wanted to illuminate the actions Bren had taken in the months and years leading up to his offer to buy out his partners. Those actions, he would argue, were less than honorable.

Facing Friedman were William Campbell and Don Morrow of the Paul Hastings, Janofsky and Walker firm. Irvine Orange County business community observers were surprised at the selection; the company had a long-standing relationship with Latham and Watkins. Gary Weber recalled that the decision to hire Paul Hastings came just two months before the trial.

The foundations of Campbell's arguments were two-fold. First, evidence from credible experts would dissect the multitude of Irvine Company holdings and place them in proper perspective, proving that the share price Bren offered was fair and valid. Second would be an analysis of the business acumen and reputation of the shareholders who agreed to Bren's deal; men like Al Taubman, Max Fisher, Milton Petrie, and Henry Ford were sophisticated, tough-minded, and shrewd business owners who didn't leave money on the table when it came to determining the value of their investments.

Bren attended the first day of the trial, explaining to reporters that it was, of course, of paramount interest to him as the owner of the Irvine Company to follow the arguments of Friedman as well as his own legal team.

But the months of preparation for the battle, plus the hours of travel time—even in the spacious cabin of his private jet—had worn on Bren. Worse yet, it was a distraction from the substantial changes at the Irvine Company he had put in motion shortly after taking control.

The management structure that Bren inherited was typical of an organization with a variety of operations. The president, Thomas H. Nielsen, reported to the board of directors, now chaired by Bren. Reporting to Nielsen were vice presidents responsible for operational sub-groups like land development, finance, property management, homebuilding, and agriculture, plus staff functions such as legal, government and media relations, and human resources.

Bren's instructions were direct and clear: make decisions faster. Increase revenues and profits. Find underutilized properties and monetize them. Within a few months, Bren had brought in new executives and deployed them in a tighter organization chart. He reinstated the planning department that Kremer had dismantled, recruiting a seasoned executive to rebuild it. Timelines for obtaining development approval were shortened. The pace of work increased, stimulated in part because bonus payments—once reserved for just a handful of top executives—were now flowing to dozens of middle managers and directors. Morale soared.[16]

The new pace and urgency, plus Bren's hands-on pressure, was paying off by the time the valuation trial was beginning. Loans to fund new projects were increasing into the tens of millions of dollars. The long-stalled coastal properties had been re-evaluated and a more feasible plan was taking shape. And in Sacramento, Irvine Company lobbyists were buoyed by Bren's willingness to substantially increase political contributions to Democrats and Republicans alike. In 1985, the Irvine Company was the largest contributor to candidates and causes in the entire State of California.[17]

But all of this success was now clouded by the potential that Bren would be forced to pay Joan Irvine Smith and her mother tens or perhaps hundreds of millions of dollars more than he first offered—plus interest, which could easily run into additional millions.

The trial began with each side exchanging its list of witnesses expected to be called. Bren's name was near the top.

***

Before Bren took the stand, two high-ranking Irvine Company officials testified about elements of geography and anticipated infrastructure costs that they believed impacted the overall value of the company's holdings.

In July 1984, at Bren's invitation, Ray Watson returned to the company he had left in late 1977. Among his first assignments was analyzing plans for the 10,000-acre Irvine Coast property. While local officials had granted permission to develop the coast, Bren believed the plans were too aggressive; he asked Watson to take a hard look at the numbers and assumptions.

Watson testified in court that he found the initial cost estimates for infrastructure to serve the coastal development much higher than anticipated, with one major road budgeted at $23 million actually more likely to cost $40 million. He also said that the extremely high initial costs needed to get development underway would increase risk, and that the 20-year timeframe for completion would yield a profit of just $40 million on $250 million expected to be spent on development costs.[19]

Watson also delivered the first attacks on Joan Irvine Smith's motivations and behavior as an owner of the Irvine Company. He reached back to 1957, when Joan

Figure 11.1 Courtroom scene. (L to R) Special master Robert Webster, Don Dorchester (RERC), Bill Campbell (PHJW), Don Morrow (PHJW), Gary Weber (Irvine Co.), Ron Hendricks (RERC), Sally Smith (Irvine Co.)

Source: Image courtesy of Gary Weber[18]

joined the company's board of directors, noting that for the next 20 years, she had voted against 80 percent of all management-backed proposals and abstained on 16 percent more. Her only positive votes, he said, related to dividends that she and other shareholders would receive.[20]

After Watson's testimony, company president Thomas H. Nielsen brought out maps and charts he said proved that about 60 percent of Irvine Company land was too rugged and steep to be developed profitably, if at all.[21] Nielsen also made the stretch argument that 27,000 acres of land could be "hampered by archeological concerns."[22] Howard Friedman scoffed at this contention, as did land developers (privately) back in Orange County. Finding and dealing with burial sites and other historic remnants of Native American culture was standard procedure; it might slow down a development, but it had never stopped one.

Nielsen also pointed out that unless public funds were used to build major regional arterials and freeways, the company could be on the hook for $300 million to do so.

Then it was Donald Bren's turn.

Over the next ten days, the famously reclusive company owner would take the stand to state his case and defend his views. For reporters Leslie Berkman of the *Los Angeles Times* and Jane Glen Haas of the *Orange County Register*, it was a once-in-a-lifetime opportunity to observe Bren for hours at a time and absorb his thinking about the company he had—depending on who was arguing the point—either strategized or schemed to purchase for his exclusive use and direction.[23]

Bren was in the prime of his life—55 years old, described in the *Times* stories as trim, tanned, and cool, as well as impeccably dressed. Berkman recalled in a 2025 interview seeing Bren taking a morning jog in Bloomfield Hills each day before his testimony. Berkman was also up early trying to determine how to get more or better information about the trial than her rival at the *Register.* The competition, Berkman said, was intense, with both reporters under pressure from editors in Orange County to deliver unique information (ironically, a few years later, Haas worked for a short time at the Irvine Company in the public relations department. She died in 2013).

For two days, responding to his attorney's questions, Bren described how he gradually lost faith in Peter Kremer's management of the Irvine Company. At first, he said, he tried to explain his frustrations and concerns in a series of confidential memos to his fellow owners. The company, he complained, was short-sighted, focused on making money over a period of months or a few years instead of investing profits that would make the company much more valuable in the decades ahead. And he accused Kremer of alienating community leaders, damaging the company's reputation.[24]

Bren also revealed in his testimony that Kremer told Alfred Taubman that he (Kremer) had come to the conclusion it was next to impossible to deal with elected officials in Newport Beach and Irvine, as well as the County of Orange. Kremer suggested, Bren testified, that the company be put up for sale and its assets liquidated. Taubman, Bren said, was interested in the idea but with a slightly different approach: sell off the vacant land while retaining the company's retail centers, apartments, and commercial properties.

Bren testified that despite the perceived weaknesses brought on by the Kremer strategies and the willingness of the other owners to take short-term profits while ignoring long-term potential, outsiders recognized the value of the company. Charles Allen of Allen and Company, who had instigated the purchase of the company from the James Irvine Foundation, was contacted by investors in the Middle East who floated a possible $1.6 billion acquisition price.[25] (Had this transpired, it would have been the height of irony: local criticism of the Taubman-led purchase of the company in 1977 was interpreted by some as thinly veiled anti-semitism directed at the Jewish owners of the company: Taubman, Max Fisher, Milton Petrie, and Howard Marguleas).

Then it was Howard Friedman's opportunity to make his case.

Bren knew Friedman's reputation well. Years earlier, when he owned a home-building company, Bren sought out opportunities to build a relationship with Irvine Company leaders. That meant keeping up with company issues and personalities, which included awareness of Joan's periodic legal clashes.

In the early weeks of his campaign to buy out his partners, Bren sent Stan Ross of the Kenneth Leventhal accounting firm to meet with Friedman. His assignment was to convince Friedman to intercede with Joan and accept Bren's sweetened offer of doubling her stake in the company, setting a guaranteed value for her stock, and adding a board seat. Bren had gambled that Joan's close relationship with Friedman could win her over but even his advice to take the deal would not budge her.

(By the time the trial started, Bren had lowered his offer to Joan and her mother from $114 million each to $88 million each, justified by arcane accounting practices relating to the timing of the initial offer and subsequent sale of the other shareholders' stock to Bren).

Friedman spent the next eight days questioning Bren's actions and motivations leading up to his successful offer to his fellow investors. As the *Los Angeles Times* reported,

> Friedman asked Bren about documents that [he] hoped would show that far from being a white knight, Bren was an opportunist who plotted to undermine the company's management and to acquire control at a bargain price.
>
> While Bren contended that he was surprised by the other shareholders' change of heart [agreeing to his purchase offer after initially declining it], Friedman accused Bren of having contributed to the turnaround by criticizing the company's management and prospects.
>
> Friedman told the judge that he intended to show that in 1982, Bren created "an orchestrated climate" in the higher ranks of the Irvine Company "designed to soften up other shareholders for the ultimate transaction which Mr. Bren was planning."
>
> In an attempt to undermine Bren's credibility, Friedman noted that over five years, Bren routinely cast his votes as a director and member of the company's executive committee in favor of the policies that he now criticizes . . . Defending his voting record, Bren said he "went along" with the majority opinion at Taubman's request to prevent discord on the board.[26]

It was nearly Thanksgiving when Bren escaped from the witness chair, and the trial went into a long recess for the holiday season. When it reconvened in early 1989, press coverage largely disappeared, with reporters and editors apparently convinced that days of testimony from appraisers, analysts, and other real estate valuation experts would put their readers to sleep.

(The papers may have missed an opportunity. The *Los Angeles Times* reported that one witness produced by Howard Friedman "testified for thirty days last year and wound up in the hospital after cross-examination by Irvine Company attorneys").[27]

It was in mid-June 1989, when the lawyers began making their closing arguments. Weary from months of tedious testimony and jet-lagged from repeated trips halfway across the country, the trial's contestants had also experienced the worst weather Motor City had to offer; during the winter months, daytime temperatures rarely rose above the 40s, and during the night, 2 degrees was not unusual. Summer brought no relief; on June 15 it was 95 degrees outside the courtroom.

So it was not surprising that when Irvine Company attorney William Campbell summed up the evidence as he saw it, he also took the opportunity to trash Joan Irvine Smith. For starters, Campbell excoriated Smith's decision not to testify at the trial, let alone to even attend it. So, in addition to labeling her as greedy, Campbell told the court that "obviously, she had nothing to say in support of her case. Just as obviously, there is no way she could explain her arbitrary, bad-faith conduct in forcing this lawsuit." In a final riposte, Campbell said the entire proceedings were conducted "to indulge the whim of a very wealthy woman."[28]

(Joan's mother, Athalie Irvine Clarke, *had* testified at the trial. She was 87 at the time.[29] Veteran Joan observers were not surprised—for years, when her fear of appearing in public was at its height, Joan often prevailed upon her mother to attend city council, water district, and other government agency meetings on her behalf. Clarke did so, accompanied by Joan's attorneys and sometimes with a family retainer. If she ever tired of the assignment, she never let on.)

After Howard Friedman made his final arguments, reprising his contention that Bren had vastly undervalued the price for the stock owned by Joan and her mother, the case went to court referee Robert B. Webster, the retired Michigan judge who had presided over the 156 days of testimony and arguments.[30]

In January 1990, the *Los Angeles Times* tried to put the case in context with a long business section article that summarized the variety of residential and commercial developments that were proving to be winners for the company (the new village of Tustin Ranch, an estimated 1,000 new apartments going up every year, continuing success of the company's travertine-clad office towers) and the losers (a retreat from a planned village in Laguna Canyon, resistance to a new village in the hills east of the city of Orange).[31]

The *Times* story also quoted Stephen Roulac of the Deloitte and Touche real estate group, who had testified for Joan. He believed that the value of Irvine Company holdings was accelerating so swiftly that the company's real price tag was upwards of $6 billion.[32]

While the business section was making its evaluation of Bren's situation, the *Times* society pages were reporting on a swank gala Joan Irvine Smith was throwing at The Oaks, her equestrian training facility and showgrounds in San Juan Capistrano. A glittering array of Orange County business and political leaders was in attendance, with Joan professing little concern about the outcome of the valuation trial.[33]

It is doubtful Webster was reading the *Times* or any other real estate publication—he had more than 27,000 pages of documents to review that the trial had generated. And he had more than one decision to make as he evaluated the value of Joan and her mother's shares. As the legal proceedings had gone on for seven years, whatever funds the pair did receive would also include interest on those sums. With the Federal Reserve rate at the time about 5.75 percent and home mortgage rates over 10 percent, it was likely the interest rate award for Joan and her mother would be in the millions.

***

The last legal dispute between the Irvine Company, Joan Irvine Smith and Athalie Irvine Clarke was settled in large part on June 25, 1990.

The retired judge who had presided over the trial awarded the two women $149 million each for their shares in the Irvine Company, $35 million more than they had been offered by Donald Bren seven years earlier but $206 million less than the value they claimed their shares were worth.[34]

Left for a later decision was the question of how much interest the pair would receive, as well as any award of legal fees that both sides sought.

Analyzing the referee's decision, the *Los Angeles Times* concluded that "in the end, the referee largely accepted the testimony of the company, whose experts appraised the company at $1.2 billion. The referee adjusted that amount upward to $1.36 billion. Using that figure, the referee determined that Smith should receive $35 million more than Bren offered her."

"But, says the Irvine Company, it has spent more than $15 million on the case and said Smith must have spent at least that much. Subtract that amount from $35 million, the company says, and she's only ahead by $20 million."[35]

Spinning the outcome as a win, the Irvine Company went on to point out the referee's comment that the aforementioned Stephen Roulac showed "a failure to display an adequate understanding of the Irvine Company or its assets."[36] And company senior vice president Gary Hunt, who had attended much of the trial, assured the *Times* that the payout, when it finally came, "won't present any hardships to the company's ongoing operations. We are carrying a range of numbers on our books, and this one falls well within that range."[37]

Three months later, the attorneys gathered once again in Bloomfield Hills to argue about interest rates. Howard Friedman demanded 12 percent interest, compounded. William Campbell countered with a figure of 9 to 10 percent, simple interest.

Referee Robert Webster was eager to put the multi-year contest behind him, ruling just two days after hearing the case that Joan and her mother would receive $102.2 million in interest on the value of their stock—they had asked for $201 million.[38] He then disposed of the other requests both sides had asked for: attorney fees, costs, and sanctions.

Webster denied Joan's request for $17 million in legal fees. Of the $658,000 in court costs she incurred and asked to be reimbursed, he awarded her $174,000. Finally, the Irvine Company asked that Loeb and Loeb be sanctioned for $750,000 as punishment for bringing forth surprise evidence at one point during the trial. Webster awarded the company $299,000.[39]

The public legal drama largely ended in September 1990, but it would be another nine months before the loose ends were tied up and a payment made.

Joan and her mother had the right to appeal the main judgment as well as the secondary rulings on legal fees, costs, and sanctions. She kept her own counsel on the subject until the early spring 1991. According to the *Los Angeles Times*, here's how the outcome was achieved:

> So this spring, when Smith began talking about a settlement, the company was interested. It eventually acceded to quickly paying her and Clarke the $256 million award—the original $150.5 million plus the $105 million in interest, rather than take the chance of losing some of it in the appeals process.[40]
>
> Negotiations between the two sides continued in secret in May and June. In return for getting the money upfront, Smith agreed to accept all the rulings in the case . . . after the quarter of a billion dollars* was wired to their accounts, Smith and Clarke each wrote checks to the company for $62,639.53, each woman's share of the sanctions minus $174,0000, the referee awarded them both for court costs.[41]

***

Larry Thomas recognized a good news story when he saw one. The immediate aftermath of the actual payment of more than $250 million in cash presented him with just such an opportunity.

In addition to responding to the local media about the myriad of issues the Irvine Company was involved in on a daily basis, Thomas—the company's vice president of corporate communications—made it his personal goal to find ways to demonstrate that the publicity-adverse Donald Bren was pretty much just a regular guy.

Bren was basically shy. He hated confrontation, was uncomfortable in crowds, and, above all, liked to be in control of his surroundings. But Bren was unfailingly courteous with both peers and his employees, had a sense of humor, wasn't pretentious, and, in the right settings, could be very personable.

Thomas had wandered into the reception area outside Bren's office and chanced upon talking to Bren's secretary, Marlys Brooks. Modest and soft-spoken, Brooks was the perfect combination of efficiency and discretion that Bren insisted on for his gatekeeper and scheduler. Brooks took Thomas aside and described an incident she found humorous. Thomas nodded, smiled, and walked back to his office. He picked up the phone and called a trusted editor at the *Los Angeles Times*. The next day an item ran on the front page in the News Briefs section of the paper:

A LITTLE LEVITY: At least somebody found some humor in the bitter legal battle between Irvine Company Chairman Donald L Bren and Joan Irvine Smith . . . when Bren returned to his office Tuesday after shelling out a quarter of a billion

* The actual amount was $127,934,845.01 for Joan, the same for her mother.

> *dollars to Smith and her mother . . . his secretary asked him how it went. Turning his pockets inside out, Bren, a billionaire, admitted: "I'm a little bit lighter."*[42]

Thomas was delighted. Bren, however, was not.

Returning to his office after Bren dressed him down for planting the item, Thomas turned to one of his staff and said wistfully, "I'm just trying to humanize him." (Thomas died in 2018.)[43]

Joan Irvine Smith also had something to say to the *Times*. A few months before she received her money, she was telling the paper her plans. In addition to making millions in charitable contributions, she would write a book about her life. Alluding to her four marriages, she also mused that a movie about her would be entertaining, "but it would have to be toned down for prime time showing."[44]

But Joan had long ago decided the script for her payday. "When the day comes, she said, she intends to ride to the bank with the check on a horse who has grown old and swaybacked, waiting for the chance. The horse's name: 'Last Laugh.'"[45]

In the end, there was no check, just an impersonal electronic transfer from one bank to another. Joan was already a multimillionaire, and when her mother died three years later at the age of 90, it was widely assumed that Joan inherited the bulk of her estate.

Joan Irvine Smith died in December 2019. Her estimated wealth exceeded $350 million.[46]

***

Three years after the valuation lawsuit was settled, Howard Marguleas found himself up against a financial wall as he was pressed toward bankruptcy. His misfortune would lead to yet another exercise in how to determine the value of the Irvine Company.

Marguleas had taken out large loans from the Bank of Scotland to invest in a new agricultural enterprise in California's Central Valley, but it had not been successful. Needing cash, he reviewed what few resources he had not mortgaged—at the top of the list was his investment in the Irvine Company. An original partner in the 1977 consortium that acquired the company from the James Irvine Foundation, Marguleas still owned 118 shares of the Irvine Company.

Marguleas began searching for a financial advisor to help him market his shares. Very few Orange County firms were interested—no one wanted to be on the wrong side of a potential fight with Donald Bren and the Irvine Company.

However, Mike Danzi was willing to step up.

At the time, Danzi was just 35 and a principal in a small investment bank that had largely focused on funding medical and tech startups in Orange County, he recalled in a 2025 interview. He remembered Marguleas as a smart and engaging man willing to sell his Irvine Company shares at a discount if that would produce timely results.[47]

Danzi did some of his own research, then approached the Irvine Company to learn more details about its financial status. "I was allowed to sit in a room and examine the books for the past three years, but I could not make copies or notes," he said. Blessed with a good memory, Danzi analyzed what was already publicly known about the company's assets, combined with what he had committed to memory, and arrived at an overall value of close to $5 billion for the Irvine Company.[48]

In a memo to potential purchasers of the Marguleas stock, Danzi said, "we can make a case for a valuation of from approximately $650,000 to $720,000 per share." Marguleas decided to value his shares at $500,000, discounted from his view that they were worth almost $618,000 per share.[49]

Within a few weeks, the first $5 million for Marguleas' shares was paid by the Claremont Colleges.[50] Soon after, other sales were made, Danizi recalls, some of them "one share at a time." Marguleas had first approached Bren about purchasing his shares, Danzi said, but the two could not reach an agreement. Eventually, by going to the public, Marguleas received more than $90 million for his holdings.[51]

Two years later, Bren bought back every share Marguleas had sold from every person or entity that had purchased them. Finally, the Irvine Company was owned by a single individual, the first time that had happened since 1947, when the death of James Irvine II resulted in the entirety of his Irvine Company ownership being distributed to the foundation he created and to members of his family.[52]

Notes

1 "Bren Buys Majority Stake in Irvine Company," *Los Angeles Times*, November 1983.
2 Thomas C. Hayes, "Developer Acquiring Control of Irvine Co.," *New York Times*, April 16, 1983, Section 1, p. 29.
3 Stephen Roulac, testimony in *Irvine Company v. Smith*, Bloomfield Hills, MI, 1987. Quoted in *Los Angeles Times*, June 27, 1990.
4 "Bren Offers Board Seats and Increased Equity," Wall Street Journal, May 1983.
5 "Joan Irvine Smith Sues Over Share Valuation," Orange County Register, April 1983.
6 Howard Friedman, legal filings in *Irvine Company v. Smith*, Michigan Court of Appeals, 1983–1990.
7 "Legal Battle Moves to Bloomfield Hills," *New York Times, August 1987.*
8 *Gary Weber, interview by authors, March 3, 2025.*
9 *"Orange County Assessor Dispute Settled," Los Angeles Times, October 1984.*
10 Weber interview.
11 Weber interview.
12 Weber interview.
13 Weber interview.
14 Weber interview, February 2025.
15 "Trial Begins in Irvine Company Valuation Case," *Los Angeles Times*, August 6, 1987.
16 "Bren's Management Overhaul," *Los Angeles Times*, September 1985.
17 "Irvine Company Political Contributions Top State List," Sacramento Bee, December 1985.
18 PHJW—Paul Hastings, Janofsky and Walker law firm, RERC—Real Estate Research Corporation.
19 Ray Watson, testimony in *Irvine Company v. Smith*, August 1987.
20 Watson, testimony in *Irvine Company v. Smith*.
21 Thomas H. Nielsen, testimony in *Irvine Company v. Smith*, August 1987.
22 "Archaeological Concerns Raised in Trial," *Los Angeles Times*, August 6, 1987.
23 Leslie Berkman, interview by C. Michael Stockstill, January 2025.
24 "Bren Criticizes Kremer's Management," *Los Angeles Times*, November 8, 1987.

25 Charles Allen, quoted in *Wall Street Journal*, October 1982.
26 "Friedman Accuses Bren of Manipulating Board," *Los Angeles Times*, November 8, 1987.
27 "Witness Hospitalized After Cross-Examination," *Los Angeles Times*, June 16, 1989.
28 "Campbell Denounces Smith's Conduct," *Los Angeles Times*, June 16, 1989.
29 "Joan Irvine Smith's Mother Testifies," Orange County Register, August 1987.
30 Howard Friedman, closing arguments, Irvine Company v. Smith, June 1989.
31 "Irvine Company Developments Reviewed," *Los Angeles Times*, January 21, 1990.
32 Stephen Roulac, quoted in *Los Angeles Times*, January 21, 1990.
33 "Joan Irvine Smith Hosts Gala at the Oaks," *Los Angeles Times*, July 5, 1990.
34 Michael Flagg. "Smith's Win Wasn't Big: Neither Was Bren's Loss," *Los Angeles Times*, June 27, 1990.
35 "Smith's Win Wasn't Big; Neither Was Bren's Loss."
36 "Smith's Win Wasn't Big; Neither Was Bren's Loss."
37 Gary Hunt, quoted in *Los Angeles Times*, June 27, 1990.
38 "Interest Awarded in Irvine Case," *Los Angeles Times*, June 19, 1991.
39 "Interest Awarded in Irvine Case."
40 "Interest Awarded in Irvine Case."
41 "Interest Awarded in Irvine Case."
42 "Bren's Secretary Offers Comic Relief," *Los Angeles Times*, June 20, 1991.
43 Author's recollection.
44 "Joan Irvine Smith Plans Memoir," *Los Angeles Times*, July 5, 1990.
45 "Joan Rides to the Bank on 'Last Laugh'," *Los Angeles Times*, July 5, 1990.
46 "Joan Irvine Smith Dies at 86," Orange County Register, December 2019.
47 Mike Danzi, interview by C. Michael Stockstill, March 2025.
48 Danzi interview.
49 Danzi interview.
50 Claremont Colleges, investment records, 1993.
51 "Marguleas Sells Irvine Shares," *Los Angeles Times*, 1993.
52 Irvine Company ownership records, 1947–1995. Special Collections, UC Irvine Libraries.

12 Planes, trains, automobiles, and schools

A critical element in the transformation of the Irvine Ranch from a giant farm into an urban environment was—and is—infrastructure. While the Irvine Company played a key role in some elements of each one explained in this chapter, in other cases, decisions—public and private—led to outcomes the company and others could only react to.

The 110,000 acres of the original Irvine Ranch straddled the middle of Orange County, making it impossible for a train, stagecoach, truck, or automobile to travel from one end of Southern California to the other without passing through its acreage.

In its earliest evolution as a giant farm, the ranch benefited from the Atchison, Topeka, and Santa Fe Railroad, whose tracks ran directly through the upper half of the property. Access to the rail system allowed convenient shipping of field crops and citrus to the rest of the state and nation. (Spur lines connected crop collection points).

Passenger service in the early days of the ranch was minimal—until the late 1970s, there were relatively few people in the area who wanted to make the journey to either Union Station in Los Angeles or the tiny depot in San Diego.

As was the case with the railroad, regional highways connecting greater Los Angeles to San Diego had nowhere to go but through the vast middle section of the ranch (Pacific Coast Highway (PCH) opened in 1926, but it was narrow and challenging to navigate).[1]

A highway originally referred to as El Camino Real (the King's Highway in Spanish) and later as US 101 found its way from Santa Ana to the lower reaches of Orange County at historic San Juan Capistrano in the mid-1920s. It was a four-lane highway, destined for expansion as the population in the county and region increased. From 1950 to 1960, the highway was constructed in increments, joining the federal interstate system as Interstate 5, the Santa Ana Freeway.

A few miles south and west on the ranch, another interstate highway, the San Diego Freeway, in the lexicon of federal highways, the I-405, which runs from Sylmar in the San Fernando Valley to a connection with the I-5 at the eastern boundary of the Irvine Ranch, was completed in 1968. The 405 provided the Irvine Ranch with two primary access points that literally connected it from Canada to near the Mexican border. Thousands, then hundreds of thousands of vehicles would fill the roads every day.

DOI: 10.1201/9781003683087-12

There was no freeway of any kind in 1923 when a young aviation enthusiast named Eddie Martin scraped out a dirt runway in Costa Mesa, leasing the property for $35 a month. He was offering airplane rides for $2 a flight up until 1939, when the need to expand a county road led to an agreement between Martin, the Irvine Company, and the County of Orange to relocate the airstrip a few miles south, creating Orange County Airport. Noting Santa Ana as the nearest town at the time, the Federal Aviation Administration (FAA) gave it the identification SNA. When World War II began, the federal government took over the airport, built a new control tower, and lengthened the runway.[2]

James Irvine II had given Glenn Martin, no relation to Eddie, permission in 1909 to bring his homemade airplane to one of the ranch's bean fields for a test flight. Martin was successful, and four years later, took Irvine up for a 20-minute tour of his land. Irvine was reportedly impressed, but according to Joan Irvine Smith, her grandfather never boarded an airplane again.

Airline service at Orange County Airport began in 1952, and as the county grew, so did air traffic: by 1967, more than 30,000 passengers a month were flying into and out of a new terminal there.[3] (There was no direct access to planes from the terminal. To board, passengers marched from a gate outside the building onto the tarmac and out to waiting aircraft.

Growing up around the eastern boundary of the airport was the Irvine Industrial Complex (IIC). Initially designed exclusively for industrial/manufacturing uses, it was inevitable that office and small commercial uses would grow there, especially after the City of Newport Beach annexed a corner of the area. The presence of an airport was, of course, a huge positive for Irvine Company marketing efforts to sell and lease land in the IIC (Irvine Industrial Complex, which later became Irvine Business Complex). But it also presented a problem.

The takeoff pattern at Orange County Airport sent planes west into the prevailing winds from the Pacific Ocean and directly over some of the most affluent neighborhoods in Newport Beach. In the 1970s and 1980s, jet aircraft were loud—so loud some residents had to flee from their patios to shelter indoors when planes passed over. It wasn't long before resistance to airport expansion had coalesced in the coastal city, beginning a long and contentious series of fights. The Irvine Company tried to avoid becoming entangled in the conflict, much as it was doing in Irvine, where the impact of jet noise from El Toro Marine Corps Air Station was raising similar objections.

County officials responded with efforts to curb noise and limit usage. An 11 p.m. curfew for landings was instituted, which sometimes caused late planes to land dozens of miles away at Long Beach or Ontario.[4]

Opposition was led by the Airport Working Group (AWG) and Stop Polluting Our Newport (SPON), both well-funded organizations able to hire sophisticated noise experts and environmental attorneys. By 1985, a tri-party agreement set limitations on airport expansion and operations, limiting the number of flights and departure times but allowing for construction of a modern 338,000-square-foot terminal for an airport renamed in 1979 John Wayne Airport, a nod to one of the county's most famous residents. (In addition to naming the airport for him, a group of fans, predominantly wealthy conservative Republicans, commissioned a 9-foot statue of Wayne in Western garb that was placed at the airport in 1982).[5]

***

As the young planners hired by the Irvine Company began to grapple with their staggering assignment to transform the giant farm into an urban landscape, the State of California was continuing its planning for freeways.

The Department of Transportation (now Caltrans) under Governor Edmund G. "Pat" Brown had been given the go-ahead to improve and expand the state's freeway system to accommodate the state's tremendous postwar growth. In Orange County, in addition to working with the federal government on the I-5 and I-405 freeways, a plan had been developed to extend a highway that would become the 55 Freeway. It ran between Costa Mesa and Newport Beach, all the way down to Pacific Coast Highway, where a massive interchange would be constructed. From there, a new freeway would be built a quarter mile or so above PCH and would extend south through Newport, across the Newport Bay, and into Corona del Mar.

All this planning activity took place in the mid-1960s, and as the maps were drawn and redrawn, lines began to appear on Irvine Company planning documents of what was to become Newport Center and Fashion Island. Looking at company records from that time, it is surprising how little mention is given to potential alignments for the Pacific Coast Freeway. It appears Irvine Company planning was in such early stages that whenever a new conceptual freeway route was identified, company plans—equally conceptual—were adjusted to accommodate the proposed alignments.

But opposition to the Pacific Coast Freeway had flickered from a spark into a full-blown conflagration as it became apparent the destructive impact the freeway's footprint would have on the blufftop location at the 55 and PCH, where Hoag Hospital had opened a decade earlier, as well as in the quiet neighborhoods of Corona del Mar, dotted with quaint cottages. The Newport Beach City Council reversed its previous support position for the freeway and amped up a local battle to stop it. By 1969, it became apparent that the state had lost its will to make the project happen; it was removed from state highway plans in 1972.

Ironically, the decision to delete the Pacific Coast Freeway planted the seed for another route from Newport Beach to San Juan Capistrano. The war over *that* route would make the Pacific Coast Freeway contest look like a mere skirmish.

In Irvine, the outline of a substantial circulation system had been clearly laid out in William Pereira's plan for both the university-adjacent village that was to be the heart of the new community as well as for the remainder of the city north of the 405 freeway. There, three large roadways were shown on the plan: Culver Drive, Jeffrey Road, and Sand Canyon Avenue. All three had been created first as dirt roads to service the agricultural fields in the vast center of the property. Each would be improved and widened as the years passed and development in the heart of the city progressed, with Culver becoming Irvine's de facto Main Street.

One of the first roadway controversies the new city council faced tested the will between concern for the environment and the hard facts of how to move autos from one side of the new community to the other.

Campus Drive extended from Culver Drive past UC Irvine and stopped at the San Diego Creek. About a mile on the other side was the Irvine Industrial Complex, at the time a significant source of property tax revenue for the city. In between was the San Joaquin Marsh, which had been sold to UCI by the Irvine Company. While the university argued for the connection, as did the Irvine Company and the IIC's business advocacy group, two members of the council objected, fearing that traffic would adversely impact birds and wildlife in the marsh.[6]

It was the first substantial infrastructure issue the city would face. Proponents pointed out that because there were only two other connections from the area near the university, pollution would increase due to the additional miles drivers would expend making a trip the long way around. The Sierra Club and another open space advocacy group complained that more study was needed to evaluate the road's impact. In the end, the fact that county and state funds would pay for the extension and an expensive bridge over San Diego Creek motivated a 3–2 vote on the council to approve it.[7]

More than a decade later, the city was embroiled in controversy over an extension of Yale Avenue from one end of the town to the other.

Yale ran smack through the middle of Irvine. But it stopped in two places: between the Village of El Camino Real and the Village of Deerfield, the impediment being the Santa Fe Railroad twin tracks. The other missing link was located two miles southwest, between Woodbridge and University Park, with the San Diego Freeway serving as the obstruction.

It took a citywide election to push through a bridge over the railroad tracks, giving much of the north/central section of the city an alternative to Culver and Jeffrey. But at University Park, resistance stiffened, and ultimately a bridge over the 405 that could accommodate bicycles was approved, leaving residents in the south half of Woodbridge with only the Culver/Jeffrey option for vehicular access.

In the ensuing years, it became increasingly evident that the railroad crossings at Culver, Jeffrey, and Sand Canyon had the potential to impact circulation dramatically. At the time, the rail line was heavily used for freight traffic and Amtrak and Metrolink passenger and commuter travel. Since being hired as director of public works for the city in 1972, Brent Muchow had been quietly salting away extra funds in his annual budget in anticipation of building underpasses on the three streets.

He got his opportunity in 1982 after city voters approved bond financing for the first underpass at Culver Drive. Several years later, a second underpass at Jeffrey Road was funded and constructed, and a third underpass at Sand Canyon as well. All of these were in place by the early to mid-2000s

Implementation of the underpasses had proceeded over the decades to relatively little fanfare. But when a group of Los Angeles-based promoters announced a plan to build a high speed train from the City of Angels to San Diego, the route map showed that it would send up dozens of trains every day whizzing through the middle of Irvine.

American High Speed Rail was the brainchild of a group of investors that included Lawrence Gilson, who had worked as a senior executive at Amtrak. Gilson became the face of a plan to build a "bullet train" modeled after the Japanese Shinkasen, famous for its feather-smooth ride at speeds in excess of 160 miles per hour.

For a few months in 1982, bullet train mania swept into Irvine, as the tentative map of the route overlaid the existing railroad tracks that cut through the heart of the city. And unlike the pioneer days when Santa Fe trains groaned through the empty fields of the Irvine Ranch, there were now hundreds of homes on each side of the right of way. Residents began to ask questions about the frequency of trains and, especially, the noise.

At the Irvine Company, meetings were convened to determine just what a bullet train might mean to the company's plans and relationship with city residents and officials. Even Bruce Young, the chair of the Assembly Transportation Committee, made a trip to a private meeting at the Fluor Corporation to discuss the potential for the project, which, at the time, had no public funding. Fluor was one of the largest construction engineering firms in the world, and the prospect of a huge public works project almost literally in its backyard was enticing.

After less than a year of meetings and hype, American High Speed Rail folded its tent and called off plans for a bullet train, citing a lack of funding. Life in Irvine resumed at a pace of less than a speeding bullet.

***

While economic conditions were favorable in the 1980s, the County of Orange and individual cities found it increasingly challenging to keep pace with the unrelenting levels of growth underway, especially in the southern parts of the county.

For the Irvine Company and other developers, a free-flowing regional transportation system was an increasingly critical factor in maintaining favorable public and governmental attitudes toward more development.

In Irvine, building new homes and businesses on both sides of two interstate freeways meant that if congestion increased on either or both, traffic would spill over onto local streets. In a perverse fashion, the company's planning for Irvine's circulation system worked against the city; the wide roads built in large part by the Irvine Company paralleled the two freeways, providing regional drivers with an attractive alternative route. The efficiency of traffic movement on roads like Alton Parkway and Irvine Center Drive was also helped along by the city, which installed computerized traffic signals on them.

The two giant employment centers anchoring the western and eastern sides of the city were designed to generate jobs for the men and women living in Irvine, which they did. But they also succeeded so well that employees from throughout Orange County, and even from surrounding counties (especially Riverside and San Bernardino), began to flood into the area, further congesting freeways and local roads.

Another traffic generator was UC Irvine, which would expand over the years from its modest beginnings in 1965 with just a handful of students and staff, to more than 10,000 students in the early 1980s. While there were dormitories on site, professors and staff all lived in surrounding communities, and many students preferred to live in the coastal areas of Newport Beach and Laguna Beach.

In Newport Beach, traffic patterns were already skewed because of geography—there were only two ways to cross the Upper Newport Bay, almost a mile long: at Pacific Coast Highway and at University Drive. When Caltrans attempted to widen the two-lane PCH bridge, SPON and other environmental groups waged a prolonged, bitter battle to halt it, infuriating locals and tourists alike.

And, of course, in a coastal community where summer weather lasted nearly six months of the year, visitors from all over southern California descended on the beaches, clogging traffic and filling parking spots. Circulation around Newport Center and Fashion Island was better than in other parts of town. Still, the retail mall and collection of office buildings attracted shoppers and employees in large numbers, spilling over into nearby communities.

Additionally, Orange County was just part of the giant Southern California human ecosystem, which had long before rejected buses and streetcars for the freedom and joy of the auto. In most families, there were at least two cars, more when the children reached 16. For Californians, unrestricted car travel was considered a God-given right; when it became increasingly difficult, Californians became very, very unhappy. It wasn't hard to look around at all those new homes being built and come to the conclusion that slowing them down, or stopping them, might make driving just a bit more enjoyable.

***

Governor Jerry Brown had crept into office with an approach to managing the state's growth that was radically different from that of his father. Edmund G. "Pat" Brown had overseen an era of building with few parallels in California history; hundreds of miles of new freeways had been constructed as well as a state water project of enormous magnitude. The university and state college system had grown by half a dozen new campuses. The economy expanded, and jobs were plentiful.

Jerry Brown's vision was reflected in the philosophy of E. F. Schumacher, author of *Small Is Beautiful*, a text extolling a world where economics were grounded more in Buddhism than capitalism, translated by Brown into state government transportation policy as less spending on concrete and more on transit.[8]

Brown's vision could not have come at a more inopportune time for Orange County. In part because its Congressional delegation in past decades had spurned federal grants for transportation improvements on philosophical grounds, the freeway system—while still relatively new—needed expansion and improvements. The county electorate, although changing, was conservative—a bond election for road improvements was unlikely. So when one county in northern California discovered a way to increase transportation funding, the Irvine Company took notice.

Santa Clara County was morphing into Silicon Valley. Its road system was lagging behind residential and commercial development. County and local government officials had convinced the state legislature to allow the county to enact a sales tax within its jurisdiction for the sole purpose of funding transportation improvements.

Tom Wilck, Irvine Company vice president for public and community relations, watched the Santa Clara experiment closely. This, he thought, was a realistic pathway to raise hundreds of millions of dollars that would benefit not only the county but could take the company off the hook for payment of road improvements needed to serve the communities it was building.

Wilck knew the company's high profile in political issues could impede taking a leadership position on the issue. The media and community activists would be quick to paint the company's motivation as selfish and profit-driven. The same would be true if the planned community builders in the south county—Mission Viejo Company and Rancho Mission Viejo—were involved.

What was needed, Wilck thought, was an organized business community effort to frame the issue as a necessity for the economic health of the entire county, with large employers taking the lead—at least publicly. Wilck pitched the plan to company president Peter Kremer, who signed off on the creation of what became the Orange County Transportation Coalition.

Finding a major employer to anchor the group was easy—the giant Fluor Corporation had hundreds of local employees. And Fluor loved to play politics, especially Republican politics. For potential Republican candidates to local, statewide, or national office, a visit to the inner sanctum at Fluor and a vetting by Tom Ellick, the chief public affairs officer, was mandatory. As a result, Fluor had a lot of friends in the state legislature and Congress.

With the Irvine Company taking the organizational lead, other allies like Pacific Mutual, a large regional life and health insurance company, plus C.J. Segerstrom and Sons—owners of South Coast Plaza, office buildings, and apartments—formed the nucleus of the Coalition. Soon, Disneyland, Rockwell, Beckman Instruments, and Lucky Stores (which owned a giant distribution warehouse in Irvine) joined. So did the two big South County ranches that had become planned communities—Mission Viejo and Rancho Mission Viejo.

The Coalition hired a lobbyist in Sacramento and set out to conduct a sales tax election in Orange County.

Rather than attack the tax issue directly, the Coalition first sought legislation that would create a new, separate Caltrans district in Orange County. At the time, the county was lumped together with giant Los Angeles County in a single district. The result was a set of administrative formulas, plus the sheer political power of LA's large legislative delegation, that gave priority to Los Angeles transportation projects.

The strategy was to ease into the politically sensitive issue of raising taxes.

The Republican Party dominated Orange County politics, and within it, there was a tradition of resistance to tax increases and government interference in business and personal issues. And while the county's changing demography was making it lean more to the center of the political spectrum, the anti-tax mood remained strong.

However, seeking a separate Caltrans district provided common ground across the political spectrum, and a bill allowing the split was passed and signed by Governor Jerry Brown. He had been lobbied on the matter by Richard J. O'Neill, whose family owned the cattle ranch that became the Mission Viejo Company as well as lands being developed further south under the banner of Rancho Mission Viejo.

O'Neill was a maverick in Orange County politics, a wealthy landowner who was not just a Democrat but had risen to become chairman of the state party.

With the Caltrans district victory notched, the Coalition began to lay the groundwork for a bill that would allow a vote on a sales tax for transportation. It was the next step in a process that led to a multi-election effort that would take more than a decade to settle.

***

Compare a road and highway map of the Irvine Ranch from, say, 1970 to one dated 2000, and an immediate change is apparent. One that in many ways altered the future of Irvine Company properties as well as communities in two neighboring counties.

Three bold lines stand out on the 2000 map that were absent in 1970: toll roads—multilaned, designed to accommodate more than 175,000 cars per day. The three were built by a unique public agency after a prolonged, bitter fight on battlefields that extended from Orange County to Sacramento, Washington, D.C., and Wall Street. The complete story would require a separate book; here we offer an outline of the highlights and reveal some of the behind-the-scenes machinations by the Irvine Company and other developers, attorneys, and elected officials who played a role in the outcome.

The idea of building a toll road anywhere in California was beyond radical.

While the bridges over the San Francisco Bay had been tolled for decades, in southern California, where the car was king, local authorities had designed and built hundreds of miles of freeways, with emphasis on "free."

However, in Orange County in the late 1970s and early 1980s, worsening traffic conditions were forcing business and government leaders to begin to think the unthinkable.

The need for additional highways in the county had been identified in the late 1970s in two studies of the northern and southern sectors of the county. In the north, already highly urbanized and, in some cities, densely populated, the potential existed to widen the 55 Freeway but not much else. From time to time, the concept of running a traffic corridor down the middle of the Santa Ana River, which was concrete from end to end, courtesy of the Army Corps of Engineers and the county flood control district, was floated. Preliminary engineering studies showed the idea to be impossible not to mention financially infeasible.

In the south, the potential for new highways was greater, as there was substantial open property, much of it owned by a handful of large landowners like the Irvine Company, Mission Viejo Company, Rancho Mission Viejo, and a few others.

Informal discussions began between the Irvine Company and county supervisor Thomas Riley, who represented nearly every acre of land owned by the company and the other large developers in the southern part of the county. County traffic engineers struggled to grasp such an unusual undertaking—the lead engineer so frustrated Irvine Company transportation guru Hugh Fitzpatrick that Riley was quietly persuaded to have him replaced.

It became apparent that the county could not muster the resources or single focus necessary to accomplish the goal of creating a new transportation system, which by then everyone agreed had to be funded with tolls. A joint powers authority (JPA) was created, with three county supervisors as members plus a representative from each city that the new system would benefit. Two JPAs were formed, one to build the San Joaquin Hills Corridor and one to build the Foothill Eastern Corridor. Each was named a Transportation Corridor Agency (TCA).[9]

To jumpstart the project and to assure critics that no tax dollars would be used, an initial funding system was created that levied a fee on every new home, apartment, and commercial or industrial project built within an "area of benefit" that generally overlaid the boundaries of cities and the county supervisorial district in each JPA. The fees would fund TCA operations until bonds could be sold and tolls collected.

The first section of the toll road system was small, 3.2 miles of the Foothill Corridor at the base of the Lomas de Santiago hills in what became the city of Lake Forest and the community of Foothill Ranch. Because it was little more than a symbolic start that piggybacked on a previous route pioneered by the county, there was little opposition when it opened on October 17, 1993.[10]

The most virulent resistance would be focused on a 14-mile toll road on the western side of Irvine, the San Joaquin Hills Transportation Corridor.

A large area west of the I-5 freeway, directly across from Mission Viejo, was destined to become the city of Aliso Viejo. Running through the middle of the acreage, which was nearly empty at the time, was a planned roadway, known by various names, including the Aliso Corridor or Aliso Parkway.

The Aliso Corridor started at Laguna Canyon Road and ended at Interstate 5, just above San Juan Capistrano. By the early 1980s, the county was so convinced that at least that portion would ultimately be built, the roadway had been rough graded; several areas had been carved out from substantial hillsides, and abutments for future bridges were compacted.

At Laguna Canyon Road, an extension of the Aliso Parkway alignment to Newport Beach and UC Irvine was logically a straight line passing north through thousands of acres of undeveloped open space owned by the Irvine Company. And while substantial grading and filling would be required along several sections of the right of way, much of it would run along a long ridgeline.

Environmental activists in Laguna Beach, joined by national, regional, and state anti-growth, anti-road, and wildlife protection advocates, formed a coalition of opposition to the TCA and its plans for the toll roads, metaphorically staking their claim to the right of way for the San Joaquin Corridor and vowing a fight to the death on the project.

Over more than ten years, political, legal, and public relations warfare played out in city halls, courtrooms, backrooms, and the halls of the state legislature and in Washington, D.C. There were four main stages where the toll road drama played out that will be discussed next.

Environmental

Federal law forbade highways from being built in parklands; the Irvine Company acreage had been dedicated to the county as open space. Obtaining an exemption to the law (Section 4F), was accomplished by deft political lobbying in Washington, D.C., led by the Irvine Company.

A tiny bird, the gnatcatcher, inhabited the open space along the corridor right-of-way. Along with the cactus wren, it was listed as an endangered species. Impacting it required an exemption from the U.S. Fish and Wildlife Service. TCA staff and its star environmental attorney, Rob Thornton, labored for years to obtain the exemption.

(The gnatcatcher never went away. It would play a critical role in the creation of an entirely new and untried conservation strategy, see Chapter 14)

The project was, of course, so substantial (and had federal funding) it required both state and federal environmental approval. This meant litigation, with the Natural Resources Defense Council (NRDC) filing a series of lawsuits alleging various destructive outcomes if the road were built. Thornton and his firm, Nossaman, beat back the lawsuits.

California Coastal Commission

Two small areas of the right-of-way were in the California Coastal Zone, giving the commission jurisdiction over the entire project.

The commission was generally sympathetic to environmental arguments affecting housing, commercial, or other projects within the coastal zone. But the toll road's limited impact near the Upper Newport Bay allowed TCA staff to propose several mitigation strategies—including building a three-acre artificial wetland—that the commission staff agreed to.[11] It also helped the TCA's cause that the road's crossing point at Laguna Canyon—the focus of Laguna Beach objections—was miles outside the coastal zone.

The other factor that pushed the commission to approve the project was the state of the economy, which was grim. Arguments from labor representatives, citing the hundreds of jobs the construction of the toll road would create, resonated with the swing vote commissioners.

Financing/Construction

The cost of the San Joaquin Corridor would touch $1 billion. TCA staff struggled for months to convince bond rating agencies on Wall Street that Orange County drivers would pay to use the toll road. Joining the argument were the men and women from investment banks Smith Barney and First Boston who had signed on years before to promote the project. In the end, only one of the major bond agencies, Fitch, gave a rating of BBB-, one tiny notch above what would have been considered a junk bond.

Toll road critics sniped repeatedly at the traffic studies supporting the road.

Only a handful of construction/engineering firms had the capacity and experience to build the project. TCA staff and board members desperately wanted to attract as many bidders as possible for the tremendous job to demonstrate its viability. They were disappointed when only two firms made an offer. The winner was a partnership of Kewit, an Omaha-based construction giant, and Granite Construction, which did most of its work in California.

Political/Demonstrations

Toll road opponents in some of the cities that had joined the TCA attempted to inject the toll roads into city council elections, without much success. In Irvine, the city council held an advisory election on the San Joaquin toll road. Still, the vote was in favor, quieting officials both in Irvine and Newport Beach, where a similar advisory vote had been discussed.

As the toll road passed through the City of Laguna Beach, it was eligible to join the TCA as a voting member. However, to the surprise of TCA staff and others, the city took a principled stand and refused to do so. It was a glaring strategic error—had the city joined the TCA board, it would have been the proverbial fox in the project's henhouse, allowing attendance in closed sessions on legal matters as well as access to sensitive financial documents.

After the coastal commission approval for the project, more than $1 billion in a series of bonds were issued by the TCA. And when last-minute legal objections were swept aside, it became apparent to opponents that the last arrow of resistance left in their quiver was to place their bodies in front of the giant D9s that would begin to shape the contours of the toll road.

On June 14, 1994, along the right-of-way in Laguna Canyon, the TCA started to clear, dozens of individuals showed up with signs and cries of outrage in an attempt to impede the bulldozers, whose drivers either moved around them or found additional right of way to clear. Ultimately, a handful of arrests were made. Bitterly, the toll road opponents waved the white flag and tearfully retreated from their impassioned last stand. On November 21, 1996, the San Joaquin Corridor opened to traffic.

The success of the San Joaquin Corridor paved the way for the next major toll road to proceed with little opposition. Connecting both Interstate 5 in Tustin as well as an extension of Laguna Canyon Road, State Route 133, the Foothill Eastern Transportation Corridor was also more than a billion-dollar project. It made a huge impact when completed in 1999 by allowing drivers from Riverside and San Bernardino Counties the ability to avoid the hated 55 Freeway, which until then had been the only viable route into job-rich Orange County.

***

After conceiving and nurturing the Orange County Transportation Coalition, the Irvine Company joined with the OCTA and other county developers and business owners in 1984 to place a transportation sales tax measure on the county ballot.

The plan asked for one cent of additional sales tax revenue for 20 years to be spent entirely on transportation issues. However, included in the list of projects eligible for funding were the toll roads and heavy rail transit. When opponents focused on the two elements, they found that voters reacted negatively; Measure A went down by a landslide, 30 percent in favor, 70 percent opposed.

The Coalition licked its wounds and began to fashion a transportation plan they hoped voters would find more palatable, reducing the cost to a half cent and eliminating funding for toll roads or transit. They also dumped the local political consultant who ran Measure A in favor of a Sacramento firm that was building a record of success with sales tax efforts. Again, their efforts failed.

Finally, in 1990, a half-cent sales tax for transportation, labeled Measure M for mobility, found favor with the electorate, promising about $1 billion in funding.

Transportation planners at the OCTA had carefully crafted Measure M with something for every part of the county. Several large north/south arterials like Beach Boulevard would be redesigned and improved as "super streets." Other projects identified critical intersections where the campaign consultants could target mailing pieces to neighborhoods close by, motivating a positive vote. The intersection of the 5, 22, and 57 freeways, known as the Orange Crush, would be upgraded to improve flow.

But the main event was a series of improvements to Interstate 5, which touched nearly every city in the county, with the *pièce de résistance* focused on the notorious blend of the 5 and 405 Freeways, known to every Orange County driver as the El Toro Y.[12]

Heavily congested almost every hour of daylight, the Y would receive a $166 million makeover, featuring new lanes and flyovers, a separate lane for truck

Figure 12.1 The El Toro Y where I-5 and I-405 merge has the most lanes of any freeway interchange in Orange County, CA

Source: Courtesy of OC Drone Shoots

traffic, and other traffic adjustments that would ultimately result in 26 lanes of traffic. The Y may be one of the few road interchanges with its own Wikipedia entry.[13]

Measure M proved to be a continuing hit with Orange County voters. It was reauthorized in 2006, when voters approved an extension until 2041. The last funding vehicle is expected to raise more than $13.2 billion.[14]

Especially for the big land developers, the twin successes of building the toll roads and midwifing the sales tax exclusively for transportation were more than a financial victory. In many ways, the two efforts subdued the county's anti-growth attitudes, which would never again achieve the level of power it had demonstrated in the 1970s and early 1980s.

***

As a young planner just out of college, Ray Watson worked for an architectural firm that, among other things, designed schools.

To assist school districts obtain funds to build new schools, Watson explained in his oral history, the architects and contractors in the area would help district parents organize and fund a bond election.[15]

When Watson arrived at the Irvine Ranch in 1960, there were hundreds of decisions that needed to be addressed. Yet there is but a single mention in his oral history about schools; he thought it would be best for the new community the company was creating to have the same boundaries as the school district serving it.

In 1960, there were five school districts on Irvine property: Newport Beach Elementary, Newport Union High School, Tustin Elementary, Tustin High School, and San Joaquin Elementary.

The only elementary school in the San Joaquin District was located near what is now the intersection of Sand Canyon Avenue and the Santa Ana Freeway. It was built by James Irvine II in 1899; by 1911, more than 100 students were in attendance, even though it had only 1 room when built.

The students were children of local farm workers and tenant farmers. Linda Irvine Smith, daughter of Myford Irvine, recalled in an interview with the Balboa Island Museum in 2024 that when she moved to the family mansion and company headquarters in the early 1950s, she attended Tustin High School, about three miles away.[16]

In 1967, shortly after homes began selling in University Park, the first new school in modern Irvine Ranch history opened as University Elementary. In 1970, to serve the second new community, Turtle Rock Elementary opened.

Former Irvine councilwoman and mayor Sally Anne Sheridan recalled in a 2024 interview that many Irvine residents who had moved to the new community were unsatisfied with the San Joaquin Elementary District's level of educational competence.[17] A movement to establish a district that would serve the new community began, and in 1973, the Irvine Unified School District was created.[18] So, too, was the Tustin Unified School District, which shared a border with Irvine.

It is unclear how much—if any—coordination took place when the lines were drawn. The City of Irvine had set its western boundary coincident with Tustin's, but when the school district was created, its western boundary ran along Culver Drive in the northwestern part of the city.

And on the eastern edge of the district the Saddleback Unified School District also came into existence in 1973. When the Saddleback district drew its boundaries, it took the unusual step of sending an arm of land into the El Toro Marine Corps Air Station. The apparent motive was to pick up some of the industrial property in the Irvine Industrial Complex-East along the edge of the base.[19] Property tax revenue at the time was very valuable for schools, something that changed dramatically in 1978 when Proposition 13 passed.[20] As was the case in Tustin, it would be decades before the odd boundary decisions affected students and their parents.

School overcrowding was a compelling issue in the 1950s and 1960s throughout California. And there was also a racial element at work in nearby Los Angeles County, where the potential for busing minority students from one district to the other had many parents either upset or organizing to resist it.

Race wasn't a factor in Irvine, as the overwhelming majority of families moving to the isolated new community were white. But concern about availability of seats in the new schools was top of mind for parents, dreading the possibility that their children might be forced to attend a half day of school in double sessions.

It became apparent to Irvine Company officials that the quality of schools, both in terms of curriculum and space, would be a critical factor in the success of the New Town they were planning and building.

Funding for construction of the first few schools in the new district was assured in 1972, when a $50 million school bond was approved by voters, followed by voter approval of a loan of $44 million from state school funds.[21] The district went on a building spree with the funding, and by 1978, there were two new high schools, two new middle schools, and seven new elementary schools. (One of the high schools, University, had been completed in 1971).

Then came Proposition 13, which in addition to dramatically impacting city finances, threatened the Irvine district's future funding to build new schools. A solution came four years later in Sacramento with passage of the Mello-Roos Community Facilities Act.[22]

The new law allowed public agencies, including school districts, to create areas where current and future development was expected to take place and levy a fee to raise funds.

The fund rate would be set for 20 to 25 years and collected on top of the general property tax rate, which because of Proposition 13 was capped at a limit of 2 percent annually. Moreover, the new Mello-Roos district could be established on uninhabited land by the landowner. Only when more than 12 people resided in a district would an election be required.

For the Irvine Company and the Irvine Unified School District, it was a lifesaving law, which was no surprise to Sacramento insiders, as Irvine Company lobbyists and executives were deeply involved in crafting the legislation.

With the new law in place, the company carefully drew the first community facilities district for schools in the Irvine Unified School District, excluding any inhabited parcels of land, giving the company the power to establish the district. With future funding assured, the district issued bonds to fund more school construction.[23]

By this time, the company had determined that land for schools would be sold to the district at fair market value.[24] There was also the desire that the appearance of the schools reflect both the architectural and planning theme of the community where they were built as well as a high level of quality in building materials and landscaping.

With such high demand for new schools in the rapidly growing Irvine district, it became critical to find state funds to supplement the bond funds the Mello-Roos program would generate. And, of course, the more money that came from the state, the less would come out of Irvine Company income and profits. Jeanette Justus had joined the company in late 1985 after working as the director of housing in the city of Garden Grove when she found herself with an unexpected assignment. "I got back from a conference, and I found there's a couple of binders on my desk, and they are all about schools," Justus recalled in a 2025 interview.[24]

She had been assigned a job that hadn't existed until then: develop plans with school districts—initially just in Irvine, but later in Tustin and Orange—that would justify receipt of state funds for school construction.

> So, I started doing these assessments where you look at the student generation rates . . . having eligibility (for state funds) means your schools would be full. So we would build for 700 students and plan to put one thousand in there, with portables, And at the time there were about one thousand kids there with the portables, the next school would open. I spent a lot of time in Sacramento and a lot of time understanding how to make sure we got those state funds. It was like a math algorithm.[25]

Justus received support internally as she was learning a new task because, she said, it was clear that without adequate school capacity, development would be slowed or halted. She also credits Ray Watson, who had returned to the company, with supporting efforts to connect with students and parents in ways that transcended funding for construction of new schools. Those programs, Justus said, "were getting the community to know who the Irvine Company was."[26]

Justus and others at the company developed vehicles aimed at student recognition for academic achievement, including scholarships given to top performers in every high school where the company did business.[27] School district trustees, staff and parent groups were recruited to evaluate the students who competed, giving Justus and other company the executives a vehicle to form personal relations with decision makers.

Justus recalls Irvine school district elected officials and staff as highly cooperative and supportive of joint efforts to obtain school financing. In the older established districts like Tustin and Orange, the reception to company overtures to assist with funding efforts was cooler.[28]

Tustin was an especially critical assignment for Justus; the new community (and nearby development) required a high school, middle school and three elementary schools. Recalling that the school superintendent at the time was not a fan of the Irvine Company, Justus and other company staff built good relationships with a few school board members to build a foundation of trust for the eventual agreement. The details, however, were complex.[29]

> What were the most important elements? Was it the money? The size of the school? The timing? These were the real key points that the company had to focus on to get it done. . . . Everthing comes down to dealing with architects and consultants (retained by the school district). And those guys,,the longer the agreement goes on, the more money they make . . . Student generation rates were a major point of negotiation because they determined when the school needed to open and also impacted things like getting your state funding match.[30]

Justus worked on increasingly complex school issues until 1993, when she was laid off. A week after boxing up her personal items and walking out of her office at 550 Newport Center Drive, she signed a consulting contract with the company and continued her efforts. The knowledge she accumulated during her decade at the company, as well as in following years, served as a foundation for her to form her own consulting company. For the next 20 years, Justus assisted land developers and builders across California with school finance issues.[31]

***

Unlike the city's often contentious city council elections, school board governance in Irvine has been non-controversial to the point of boredom since its inception. The only point of real contention came in 2022 when, like the city council, districts were established for school trustees.[32] Previously, trustees were elected at large.

Since its founding in 1972, the district has had just four superintendents, including the first, who served 15 years, and the third, who served 30 before retiring. Candidates for school board almost without exception seemed genuinely concerned about providing quality education for the district's children (Gordon Getchel, who worked as a planner for the Irvine Company, won two terms on the school board. He told an associate that occasionally, a parent would run for the board to address a single issue affecting their child. When the issue was solved, he observed, their interest in broader educational matters faded and they usually chose not to run for another term).[33]

The new district was able to recruit many high-quality teachers–the classrooms were new, clean and safe, and students were largely from middle to upper-middle class families that valued education. But new school funding formulas from the state changed after Proposition 13 passed, and because at the time the Irvine district had a low population, funding for operations became a persistent problem.

In 1991, the district proposed to tax property owners $35 per parcel to raise extra funds for school operations. The biggest impact would have been on the Irvine Company, but it raised no objection to the plan. To the surprise of many people, residents rejected the parcel tax.

The district would try the same strategy again in 1999 proposing a tax of $95 per parcel. The election was held in February, and again voters rejected it.[34] Hoping to demonstrate greater need for the revenue, the school board cut the district budget by $5 million and put the measure on theballot again in April 2000.[35] The Irvine Company supported the effort and assigned an employee to help organize the campaign.

Six years later, Irvine Company owner Donald Bren stepped up with a gift of $20 million, spread over ten years, to assist the district with funding classes in art, music, and science. In 2010, largely at the urging of Councilman Larry Agran, the city placed a ballot measure before voters asking that $1 million in city funds be allocated to the school district for three years. It passed easily, and in 2012, it was extended again by a vote of the electorate. Since then, the city council has routinely allocated similar funding on an annual basis for the school district. And in 2016, Bren extended his $20 million gift to 2026.[36]

It was also in 2016 the district tried a new tactic to convince voters that additional funds were needed for Irvine schools. Measure E argued that more than $300 million in bond funds were needed not for teacher salaries or school operations, but to modernize classrooms and equipt them with computers and other state-of-the-art technology. It passed by a 60-40 percent margin.[37]

In 2025, for the ninth consecutive year, the Irvine Unified School District ranked first in student performance for California public school districts with more than 35,000 students in sciene, mathematics and English language arts/literacy testing.[38]

Notes

1 "Pacific Coast Highway," *California Department of Transportation*, accessed September 11, 2025, https://dot.ca.gov/caltrans-near-me/district-12/district-12-projects/pch.
2 "Orange County Airport History," *John Wayne Airport*, accessed September 11, 2025, https://www.ocair.com/about/history/.
3 "Airport Noise and Curfew Regulations," *City of Newport Beach*, accessed September 11, 2025, https://www.newportbeachca.gov/government/departments/public-works/airport-issues.
4 "John Wayne Airport," *Los Angeles Times*, February 5, 1982.
5 "Campus Drive Bridge Approved," *Los Angeles Times*, December 21, 1972.
6 "Campus Drive Bridge Approved," *Los Angeles Times*.
7 E. F. Schumacher, *Small Is Beautiful: Economics as if People Mattered* (Harper & Row, 1973).
8 "San Joaquin Hills Transportation Corridor," *Transportation Corridor Agencies*, accessed September 11, 2025, https://thetollroads.com/about/projects/san-joaquin-hills/.
9 "Foothill/Eastern Transportation Corridor," *Transportation Corridor Agencies*, accessed September 11, 2025, https://thetollroads.com/about/projects/foothill-eastern/.
10 "California Coastal Commission Decisions," *California Coastal Commission*, accessed September 11, 2025, https://coastal.ca.gov/meetings/.
11 "El Toro Y Freeway Expansion," *Los Angeles Times*, November 22, 1996.
12 "Orange Crush Interchange," *Wikipedia*, accessed September 11, 2025, https://en.wikipedia.org/wiki/Orange_Crush_interchange.
13 Orange County Transportation Authority, "Renewed Measure M (2011–2041)," accessed January 14, 2026. https://www.octa.net/programs-projects/programs/measure-m/about-measure-m/renewed-measure-m-2011-2041/

14 Ray Watson, interview by Samuel C. McCulloch, May 24, 1989, UCI Oral History Project.
15 Linda Irvine Smith, interview with Balboa Island Museum, 2024.
16 Sally Anne Sheridan, C. Michael Stockstill, Irvine, CA, 2024.
17 City of Irvine, "Irvine Unified School District Formation Records," 1973, City Clerk Archives.
18 California Department of Education, "School District Boundary Adjustments: Saddleback and Tustin," 1973.
19 California Legislative Analyst's Office, "Proposition 13: Impact on Local School Funding," 1978.
20 Irvine Unified School District, "School Bond Election Results," 1972–1978.
21 California State Legislature, "Mello-Roos Community Facilities Act," Statutes of 1982, Chapter 82.
22 Irvine Unified School District, "Community Facilities District Bond Issuance Records," 1983.
23 Irvine Company internal planning memorandum, "School Site Design and Land Valuation Policy," 1985.
24 Jeanette Justus, interview by C.Michael Stockstill, Irvine, CA, 2025.
25 Justus interview.
26 Justus interview.
27 Irvine Company, "Education Partnerships and Scholarship Program Overview," internal report, 1990s.
28 Justus interview.
29 Justus interview.
30 Justus interview.
31 Justus interview.
32 Irvine Unified School District, Board of Education Meeting Agenda and Minutes, February 16, 2022. https://iusd.org/about/board-education/board-meetings/2021-22-school-year/february-16-2022-board-meeting
33 Author's recollection.
34 Oange County Registrar of Voters, Statement of Vote: November 2, 1999 Consolidated Election, Irvine Unified School District Measure B, accessed January 26, 2026, https://ocvote.gov/data-and-results/election-results/1999/1999-consolidated-election
35 Irvine Unified School District, "Parcel Tax Election Campaign Records," 2000.
36 Donald Bren Foundation, "$20 Million Gift to Irvine Unified School District," *Press Release*, 2006.
37 Irvine Unified School District, "Measure E Bond Initiative," 2016.
38 California Department of Education, "California Assessment of Student Performance and Progress (CAASPP) Rankings," 2025.

13 The battle of Laguna Canyon

For many residents of Laguna Beach, Laguna Canyon was an inspiration.

There were few opportunities to camp, picnic, or stroll on the acreage, so it was the visual beauty of the canyon landscape that appealed to those who commuted from their homes on the coast to jobs in central Orange County. Many people said that gazing onto the canyon's rolling hills and oak-dotted slopes helped them decompress and relax.

For the Irvine Company, Laguna Canyon was mostly forgotten, a finger of open land where the only company employees who cared much about it were on horseback, tending to cattle.

That changed in early 1984, a few months after Donald Bren took control of the company. He began to study the thousands of acres he owned that were not yet zoned for development.

In the northern section of the ranch, he pondered the potential of Gypsum Canyon, which was more a large sloping parcel dotted with scrub oak and sagebrush than an actual canyon, immediately adjacent to the busy SR 91 freeway. There was a similar, much larger parcel adjacent to Irvine Lake in the foothills east of the Lomas de Santiago, with the potential to locate an entire new village.

Gypsum Canyon was within the sphere of influence of the City of Anaheim; the Irvine Lake area within the sphere of the City of Orange. Both cities had reputations as much more development-friendly than Irvine or Newport Beach.

And then there was Laguna Canyon.

A seasonal stream bisected the canyon. Laguna Canyon Road, which roughly followed the path of the watercourse, was not much more than a rough-graded path until around 1914, when it was improved into a road for autos and trucks. In 1926, Pacific Coast Highway was extended through downtown Laguna Beach, allowing access from Newport Beach to the north and San Juan Capistrano to the south.

By virtue of the Irvine Company's ownership of the vast Rancho San Joaquin, all of Laguna Canyon, roughly from the streambed to the Pacific Ocean, was part of the Irvine Ranch. To the east, a small portion of Rancho Niguel, eventually owned by Louis Moulton and renamed the Moulton Ranch, bordered the Irvine Company holdings.

But the boundaries of the two ranchos were not coincident. When drawn by the king of Spain, the rancho boundaries diverged at what are now the hillsides of

DOI: 10.1201/9781003683087-13

Laguna Beach, forming a rough triangle of land that was not included in any of the rancho grants. The area was reportedly labeled on some historic maps as *sobrante*, translated from the Spanish as "leftover."

Thus, the town was destined for multiple owners, which occurred as individuals taking advantage of the Homestead Act began acquiring properties. James Irvine II helped the process along when he sold off most of his land along the coastal hillsides west of Laguna Canyon Road, likely to raise capital to invest in agricultural operations.

Semi-isolated, far from urban Orange County, Laguna Beach was a small town of just 6,600 souls in 1950, growing to 9,200 ten years later. The rapid suburbanization of the county didn't make much of an impact until the San Diego Freeway punched through Irvine to connect with the Santa Ana Freeway in 1968, allowing faster access via Laguna Canyon Road to north county jobs; the city population grew to 14,500.

During its early years, in addition to a solid core of middle-class residents, Laguna Beach had become a tourist destination, attracting visitors who packed the tiny coves and few large beaches to enjoy a dip in the temperate Pacific Ocean. But the laid-back small town vibe also beckoned to artists, writers, and entertainers. The atmosphere morphed upon itself over time, prompting local activist Mark Chamberlain to describe the town as "a place with character, that over time drew more characters."[1]

The collection of Laguna characters was colorful, outspoken, and fervent when they embraced a cause or a controversy. Liberal, far-Left, and political/social positions on the margins of anarchy all found fertile intellectual ground in homes along the shoreline and in the hidden, leafy enclaves of residential pockets in the inhabited sections of Laguna Canyon.

By the mid-1970s, Laguna Beach was a year-round carnival of colorful human activity. A band of Hare Krishna adherents had purchased a former church where they served free vegetarian meals; sidewalks along Forest Avenue and Pacific Coast Highway rang out with cymbals and bells as the orange-robed Krishnas chanted their way from one end of the small downtown to the other. An offshoot group promoting the vegan lifestyle—Love Animals Don't Eat Them—paraded a similar route with a small camel in tow.

And these were the organized elements of the fringe. Joining throughout the year, thousands of teens and young adults poured into Laguna Beach to soak up the sunshine, frolic in the surf, and cruise Pacific Coast Highway day and night seeking fun, adventure, and probably some weed or, on occasion, a tab of LSD. During the summer, hundreds of art lovers attended the Festival of Arts and the Sawdust Festival.

Bren may or may not have been familiar with the human kaleidoscope of Laguna Beach as he considered his vision for a golf course community in Laguna Canyon. But, in retrospect, had he inquired about the history of Laguna Canyon Road and the decades of conflict engendered by efforts to widen it, he might have asked himself a question: if a mere roadway could create such a pitched battle that lasted for decades, might the prospect of 3,200 houses in beloved open space stimulate even more opposition?

***

In its early life, Laguna Canyon Road was a two-lane rural thoroughfare, 24 feet wide, with 8-foot shoulders and no lighting. Like other roadways in the county, such as Beach Boulevard, Imperial Highway, and MacArthur Boulevard, Laguna Canyon Road was owned by the state and maintained by Caltrans.

Many newspaper stories published in the mid-1940s were painfully similar when the subject was Laguna Canyon Road; speeding autos regularly ran off it and careened into the brushy sagebrush. More often than not, the driver and passengers were young Marines from El Toro Marine Corps Air Station in eastern Irvine. Most escaped with cuts and bruises and an occasional broken bone. But the unforgiving symmetry of the roadway, combined with speed, recklessness, or both, also brought death.

Tracking efforts to widen the road mirrors the ebb and flow of the political complexion of the Laguna Beach City Council. Despite the city's well-deserved reputation as an outpost of tolerance, there was still an active and organized echelon of business-oriented moderates that elected representatives to the council. For example, in 1970, the council asked Caltrans to widen Laguna Canyon Road from two lanes to four; six years later, the council reversed itself and wanted only two lanes.

The engineers at Caltrans had their own ideas about widening the road, pointing out that from 1973 to 1977, there had been 322 accidents, and in 1976, 8 fatalities. But, bending to the reality of local control and watching the seesawing of council positions on the matter, Caltrans simply stopped placing money in its budget for any improvements to Laguna Canyon Road.

The climax over the road widening Donald Bren should (and may) have considered came in early 1988, when yet another Caltrans proposal to widen and improve the road was first opposed by the city council before landing in front of the California Coastal Commission for consideration (the coastal zone was roughly 1,000 feet from the shoreline but extended inland along rivers and streams. Thus a small part of Laguna Canyon Road fell under Coastal Commission jurisdiction).

After packing city council chambers to oppose the widening, dozens of opponents, citing potential environmental damage (while apparently ignoring the rising death toll) were bused to the Coastal Commission hearing in Long Beach to rail against the project. In addition to playing the conservation card, a few people voiced their opposition, saying simply they didn't want more people and cars pouring into town, period.

Adding to the comedy/drama of the opposition campaign was the nocturnal activity of a person who came to be known as the "Phantom of the Canyon." He or she began planting homemade, Burma Shave–style signs along the canyon road in the deep darkness, attacking the widening plans. Newspapers loved the quirky spectacle and played it up in print.

On a 6–5 vote, the commission rejected the Caltrans plan. It would be 20 years before the deadly road would be widened into a four-lane, divided highway. And the highway fight proved to be but the skirmish in what would become the Battle of Laguna Canyon a few months later.

***

The development plan for the Laguna Laurel project was a revision, relocation, and reinterpretation of a similar proposal the Irvine Company had taken to the County of Orange in 1979. And while the county had granted preliminary approval, there were enough outstanding issues and initial opposition related to the plan that sent it into limbo. It was also likely that the 1981 blowup over land leases in Newport Beach and Irvine convinced the Irvine Company to put the project on hold.

But nine years later, Bren was eager to move forward with development all over the ranch. He tasked planners and traffic engineers within the company to write a plan to transform Laguna Canyon into a golf course community with up to 3,200 homes, schools, and a retail center.

By then, the concept of creating a greenbelt of protected open land around Laguna Beach was well in place, both in theory and in practice.

The greenbelt was the dream of an eccentric but persuasive local bookstore owner named James Dilley. Inspired by a visit to areas of the United Kingdom where greenbelts had been implemented, Dilley asked, "Why not here in Laguna Beach?"

Buoyed by the growing national environmental movement and its eager acceptance in Orange County, Dilley became a relentless advocate, winning endorsement for the idea not only from the Laguna Beach City Council but, surprisingly, from the Orange County Board of Supervisors, which, at the time, was considered by some as little more than an extension of the powerful real estate development lobby.

The city made a portion of the greenbelt a reality in 1968, when it purchased a 521-acre parcel of land between Laguna Canyon Road and El Toro Road known as Sycamore Hills. A low ridgeline ran the length of the property, which was dotted in sagebrush and oaks. Still, there was also a large, flat area adjacent to Laguna Canyon Road suitable for parking and public gatherings (a notorious music festival was held there in 1970, concluding with rogue event organizers dropping 2,500 postcards dotted with LSD tabs from a helicopter into the crowd).

The cost of the purchase was $6.7 million, which the city partially covered by selling 62 acres of the property facing El Toro Road to KB Homes for $5.4 million. KB built dozens of condominiums on the property. The city spent another decade cobbling together funds to complete the purchase.

Dilley created the Laguna Greenbelt Inc. as the city was completing the Sycamore Hills purchase. The non-profit was an immediate success, attracting funding and intellectual firepower from a community populated by a large number of intelligent, creative and affluent residents eager to protect their unique lifestyle. By 1988, with twenty years of advocacy and environmental victories under its belt, the organization was more than ready for a fight.

***

Business was good all over California in 1988, and Orange County was enjoying success as well. The go-go years of the early 1980s had paid big returns for nearly every segment of industry and commerce. However, a closer examination of the local economy revealed that the fat times—now six years long—wouldn't last forever. This reality, combined with Bren's mantra to his employees to build more and build faster, underscored the desire to obtain zoning for Laguna Laurel.

The property was entirely within the jurisdiction of the County of Orange, located in the fifth supervisorial district, represented by retired Marine Corps general Thomas Riley. He had been appointed to the post 14 years earlier after then-supervisor Ron Caspers was lost at sea in a boating accident off Baja California. Riley was supportive of land developers and builders but had gradually become more sympathetic to environmental issues.

In the early months of 1988, Irvine Company employees gamed their strategies for the upcoming hearings at the county planning commission and board of supervisors. At the same time, forces in Laguna Beach began to plot plans for opposition.

A new group broke off from the Laguna Greenbelt Inc. as the year began. The Laguna Canyon Conservancy was established to focus exclusively on preserving Laguna Canyon, as by that time, the Greenbelt was taking positions on both local and regional environmental issues. A teacher named Lida Lenney would soon become the face of the conservancy. She had parlayed her focus on environmental issues into a seat on the city council in 1986.

There was also another facet of opposition taking shape, led by a pair of photographers. Mark Chamberlain and Jerry Burchfield had migrated to Laguna Beach and were besotted with the people and the geography, especially Laguna Canyon. In 1980, they began to photograph parts of Laguna Canyon Road in daylight and later at night, creating an archive of photos they used for display as well as casting their actual photographic journey in what Chamberlain described as "performance art," telling the story of their concern for potential development in the canyon. He also noted that their efforts won them the title of "Lagunatics of the Week" in the local newspaper as well as favorable coverage in the *Los Angeles Times* and *Orange County Register*.

Chamberlain and Burchfield were about a year away from the creation of what would become their signature piece of propaganda/art installation when hearings began at the Orange County Planning Commission and Orange County Board of Supervisors on the Laguna Laurel project.

The planning commission meeting was relatively subdued; Laguna Laurel was one of four major land use proposals before the board. But when the scene shifted to the board of supervisors two weeks later, the atmosphere turned ominous.

Presenting the project for the Irvine Company was Carol McDermott. Her position as project manager was a reflection of her combination of intelligence, fortitude, and charm that made her one of just a handful of women at the most senior level of the Irvine Company. Tall, poised, and always fashionably dressed, McDermott was unflappable before audiences that ranged from mildly concerned about a project to those that were overtly hostile.

There were perhaps 200 seats in the meeting chambers, and none were empty. By design, dozens of company employees and consultants had been recruited to arrive early and secure seats, thereby denying opponents a majority presence. But there were plenty of passionate individuals who had claimed a place, and a few of them were seething with barely contained anger.

The tension in the room unnerved even some veteran Irvine Company executives, who over their careers had faced down residents unhappy with plans for new development in their backyards. But this crowd was different; scattered among the suit-and-tie city elected officials and some of the more affluent and sophisticated opponents, there were a handful of canyon denizens dressed more for a rugged hike in the underbrush than a public hearing.

"It wasn't the men in that group that scared me," one company executive recalled. "A young woman was lurking in the back who looked like a lioness stalking its prey. I was afraid she had a knife and was going to stick it in me." Unbeknownst to the audience that day, the Orange County Sheriff's Department had assigned a group of plainclothes officers to attend and be prepared for any trouble.[2]

After hours of passionate argument, the board voted 3–2 in favor of the project, with supervisors from the western and central districts casting no votes. Supervisor Riley, whose district included Laguna Laurel and Laguna Beach, voted yes. On her way out of the meeting, McDermott recalled that she linked arms with Keith Greer and Brad Olson, two of the largest men in the company contingent, to walk to her car.

Shortly after the County approved of the project and just before the issue submerged for several months in the public eye, the concept of stopping the project by using dollars instead of demonstrations surfaced in the press.

Four months after its formation, the Laguna Canyon Conservancy held a press conference at Sycamore Hills proposing to purchase the land from the Irvine Company with a $10 million grant from an anonymous benefactor. Details from Conservancy chairman Richard Harris were vague at best, and the Irvine Company pooh-poohed the idea, telling the *Los Angeles Times*, "The land is not for sale. We've been pursuing this investment in both the land and development, (spokesman Larry) Thomas said. "We did not invite them to come back with an offer."[3]

Neither the benefactor nor the money appeared in the days following the announcement, but a seed had been planted in the canyon soil that would ultimately prove fertile.

***

In late spring 1989, a wooden structure began to rise up from the floor of the city-owned open space preserve in Sycamore Hills.

The final burst of performance art/photo mural/opposition symbol, *The Tell*, was a pastiche of two-by-fours and plywood forming the rough profile of a recumbent female, plastered with hundreds of photographs, written messages, and other mementos collected and attached to *The Tell*.

At 634 feet in length and 34 feet at its tallest point, *The Tell* raised a giant middle finger at the Laguna Laurel project. *The Tell* provided a daily reminder to drivers in Laguna Canyon of the impending development and the various efforts underway to stop it. But it also fostered a sense of community involvement, as hundreds of individuals from Laguna Beach and beyond had contributed to its construction and adornment.

The Tell proved to be a rallying point for two events soon thereafter that, in retrospect, marked the critical turning point in the Battle of Laguna Canyon.

***

Trailing the administrative evaluation process for the Laguna Laurel project was the environmental impact report (EIR), hundreds of pages of data detailing how it would be built and what would happen to the land as a result.

Local hearings on the EIR were scheduled for late September, providing opponents with an opportunity to focus their objections in a public setting likely to attract media attention. But as discussions were held to prepare for the hearings, another, more direct and radical strategy eventually came to fruition, a plan to channel the frustration and anger the project had generated directly at the man who owned Laguna Canyon, Donald Bren.

Bren owned a home on Linda Island in Newport Beach, an enclave of wealth and privilege that had a single access point—a small bridge over an even smaller channel of water on Newport Bay. In early September, at the entrance to the island along Bayside Drive, about two dozen sign-bearing protestors, including a city councilwoman, picketed up and down the street. In addition to targeting Bren, Supervisor Thomas Riley was also shamed for his support of Laguna Laurel.

The protest lasted just a couple of hours, and as the opponents hoped, it received ample press coverage. The Irvine Company said Bren was aware of the protest and had no comment. But according to Paul Freeman, who would play a crucial role in the eventual settlement of the Laguna Laurel conflict, Bren "complained bitterly" about what he saw as an "uncivil" invasion of his privacy.[4]

The picketing of Bren's personal space set the stage for the event many people believe sealed the fate of the Laguna Laurel project, altering its course from development to preservation: the Walk in the Canyon.

It took place on Veterans Day, November 11, 1989. Thousands of development opponents, most from Laguna Beach, assembled at the Festival of Arts grounds a few blocks from Laguna Beach City Hall. Led by city officials and organizers from Laguna Greenbelt and the Laguna Canyon Conservancy, an estimated 6,000 to 8,000 people made the trek up the canyon to Sycamore Hills.

Momentum for the march had been building for days, including radio ads that highlighted Bette Midler's opposition, who had a home in Laguna Beach. Meandering with the crowd, largely unrecognized without makeup, was Robert Englund, "Freddy" from the horror film *A Nightmare on Elm Street*, who had recently moved to Laguna as well. Englund got off a line even Irvine Company public relations staff privately admired when he told the crowd, "People ask me what scares me, and I think it's Donald Bren."[5]

The march, accompanied by aerial photos to emphasize the event's magnitude, was featured on the front pages of the *Times*, *Register,* and *Daily Pilot*. Readers throughout the county, who just two days earlier had read the sensational news that

Figure 13.1 Thousands marched in opposition to the Laguna Laurel project

the Berlin Wall had come down, absorbed the images and were left to ponder how it might impact the years-long conflict.

The next official step in the path to development of Laguna Laurel was a hearing in late November at the county planning commission. A few days before it was scheduled to take place, the Irvine Company unexpectedly asked for a postponement. Laguna Beach officials and anti-development leaders took note; it appeared that the solid wall of Irvine Company advocacy for Laguna Laurel had developed a crack.

The request to put off the meeting came a few weeks after the company had lobbed what many saw as a less-than-serious offer to sell 270 acres of Laguna Canyon for $38 million, conditioned on the city accepting a downsized project (no golf course, 500 fewer homes) as well as support for a new toll road that would cross the canyon. The city had quickly rejected the offer.

But now, the unexpected delay in a critical path to the development's success caught Laguna Laurel opponents off guard. It was as if a football team that was powering toward a touchdown had stopped play with a timeout.

The Irvine Company had decided to talk.

Another factor that likely moved the company to negotiations was a last-minute ploy to have the City of Irvine extend its boundaries into upper Laguna Canyon and approve the development. The city quickly rejected that overture.

The initial meetings on a possible compromise had been held between city officials and Gary Hunt and Carol McDermott of the Irvine Company. Soon thereafter, a new player arrived on the scene to assume the role of lead negotiator for the city, the Laguna Greenbelt, Laguna Canyon Conservancy, and what he described as "stakeholders and representatives from national and state environmental groups."

Paul Freeman had moved to Laguna Beach in 1988 to establish a think tank for a local developer who had supported Gary Hart's unsuccessful presidential campaign. Freeman had worked in Hart's senate office. But the think tank failed to find adequate funding, and Freeman was left to his own devices. He was doing some freelance public affairs work (including managing the visit of the Dalai Lama to Southern California) when a friend recommended that he talk to the Laguna Laurel opponents.

In a 2024 interview, Freeman recalled that he was hired to develop a negotiating strategy for the disparate groups because "No one knew me. Different interest groups were vetoing the choice of the others. I was the candidate with no baggage, plus I interviewed well."

After a couple of initial meetings with the Irvine Company, Freeman understood that the foundation of a deal was the sale of the property. At the same time, the company was willing to discount the value, money had to change hands—from the opponents into the coffers of the Irvine Company. And it would not come cheap.

Freeman developed three principles to guide the negotiations. "First was fair market value, which was critical to the Irvine Company, and that was tricky. For the environmentalists, a discount was the most important element."[6]

Second was to define "what is the canyon? We came up with a definition based on where the water flows, which kept some areas at the northeastern end of Laguna Canyon Road (which was suitable for development) outside of the definition."

Third was "a principle that there would be some sort of risk/reward on how the land was set aside and then paid for. This led to . . . the environmentalists seeing if they sued or did not pay for the parcels, they would not get the open space."

Thus, began months of discussions between Freeman and the Irvine Company. Freeman said the process involved a great deal of "subtle diplomacy" as well as a particularly intense session with company vice chairman Ray Watson, who came on so strong that Freeman was asking himself if the normally affable Watson had been cast as the Bad Cop in a Good Cop/Bad Cop negotiation role usually reserved for crime drama.[7]

McDermott remembers the negotiations as long and complex, with Hunt being "very, very effective" in shaping the parameters of the agreement. McDermott also recalls that Bren kept in touch with her and Hunt to monitor the points being discussed.[8]

Freeman got a taste of the Irvine Company's almost limitless resources when he called Hunt to report that, after relaying the most recent deal points to his clients, a serious impasse had arisen, requiring an immediate face-to-face meeting. Hunt and McDermott were at a board meeting in Arizona, so Hunt ordered a jet from Bren's Skybird Aviation to fly Freeman to Arizona. The meeting took place, a compromise was reached, and negotiations resumed.

As the discussions neared a conclusion, it became evident that a significant payment could most quickly be achieved if the city authorized a bond. The timing was good—it was mid-summer in 1990 and the November statewide elections were coming up. Freeman had already done polling to test the willingness of voters in the highly liberal city to increase their property taxes in return for saving their beloved canyon. The poll said they were.

The months of closed-door negotiations concluded in early October with the announcement that the City of Laguna Beach would purchase 2,150 acres of Laguna Canyon for $78 million.

The city would be required to pay at least $33 million by June 30, 1991, giving it 85 percent of the most environmentally sensitive property. The remaining land would be divided up into four parcels with separate prices and deadlines for payment stretching into the next decade. (Five months later, the deal was amended slightly, redefining some payment amounts and deadlines, but the foundation of the agreement remained the same).

Hunt and Freeman took a victory lap in the press, telling the *Los Angeles Times* on October 8, 1991, "There is no question in my mind this is a milestone (Hunt)," and "it is a milestone in terms of dispute resolution involving historically contentious interests (Freeman)."[9]

Freeman monitored the bond campaign, which resulted in an overwhelming endorsement of the bond measure, far exceeding the two-thirds vote required by law. Four years later, he won the first of two terms on the Laguna Beach City Council and became the senior public affairs officer for C.J. Segerstrom and Sons.

Hunt and McDermott spent a significant part of their subsequent careers at the Irvine Company helping to identify and capture funds to assist the city and the environmental interests in purchasing the remaining open space parcels. Bit by bit, finding money from state bond issues, county open space funds, and a collection of other sources, the entire price was paid.

Why then did Donald Bren decide to sell?

Bren himself provided an answer when he made a rare public appearance in Laguna Beach a few days after the agreement was announced.

At a reception at the Hotel Laguna, attended by hundreds of people, Bren accepted a "Save the Canyon" T-shirt and wore a button that read "Don Bren Says Save the Canyon." The *Los Angeles Times* reported that

> during an interview following his brief speech, Bren said he never intended to sell the property—independently appraised at $105 million—below fair market value. But I believe that once we understood and were educated by the Laguna community, and they understood us, we were able to reach an agreement.

Bren also conceded that he was persuaded to meet the other side at the negotiating table after a march through the canyon last November drew an estimated 8,000 people:[10] "I was struck with that—the number of people who turned out. I believe the number and the sincerity impressed me, and we moved forward with our discussions and our negotiations."

McDermott and others believe Bren saw an economic downturn coming and was eager to put some money in the bank before it arrived. McDermott also believes that the company could have realized a tax benefit by selling the land at a discount.

And McDermott remembers the day when the first $33 million payment was scheduled to be made:

> He (Bren) called my office probably five times. He had never rung my extension in my life. He calls, "Has it come yet, has it come yet?" When it came, I got a big canvas bag, filled it with play money, and walked down to Gary's office. He (Bren) came in, and I handed him the bag with the play money. He reached down and took some of the money, and he put it in his pocket like a pocket square, and he reached over and kissed me on the cheek and said, "Thank you."[11]

Notes

1 Mark Chamberlain, *The Laguna Canyon Project: Refining Activism* (Laguna Beach: Laguna Wilderness Press, 2018), 10.
2 Author's recollection.
3 Larry Thomas, quoted in "Laguna Canyon Conservancy Offers $10 Million for Land," *Los Angeles Times*, April 26, 1988.
4 Interview with Paul Freeman by C.M. Stockstill, April 17, 2024.
5 "Thousands March to Save Laguna Canyon," *Los Angeles Times*, November 12, 1989.
6 Interview with Paul Freeman.
7 Interview with Paul Freeman.
8 Interview with Carol McDermott, by C.M. Stockstill and H.P. Oliver March 20, 2024.
9 Gary Hunt and Paul Freeman, quoted in "Laguna Canyon Deal Reached," *Los Angeles Times*, October 9, 1990.
10 Interview with Paul Freeman.
11 Interview with Carol McDermott.

14 This land is your land

The desire to preserve land in its natural state became something of a subplot in the larger drama of controlling growth and development in Orange County in the late 1970s and early 1980s.

Grassroots movements to preserve natural or undeveloped areas had succeeded in Newport Beach and Huntington Beach. But each was an island of open space bordered by neighborhoods of homes and businesses. The big prizes that environmental groups aspired to capture were found in Irvine and the communities south of there, where thousands of acres of hillsides, canyons, and cattle pasture were still untouched.

Irvine had been largely immune to any significant open space preservation advocacy since incorporation, but that changed when a flock of Canadian geese foraging on a small hillside became the inspiration for what became a landmark preservation agreement between the city and the Irvine Company.

At the time, no one realized that the initial engagement over 360 acres of land would become the first installment in a public and ultimately private process that would result in a new federal policy trading habit for permission to develop and, finally, the decision of Donald Bren that would define his legacy as the owner of the Irvine Ranch.

***

Quail Hill ran parallel to the San Diego Freeway from University Drive to Sand Canyon Avenue, about half a mile away. Few people knew its name, so when the city council in 1984 began to cast about for locations to build a permanent city hall, some council members saw the hilltop ridge as a possibility for the new structure.

But for city councilman Larry Agran, Quail Hill offered an opportunity.

Agran was always alert for causes that could be championed and turned into political capital. In the 1980s, the mood of the entire county, as well as Irvine, was shifting toward increasing concern about the pace of development and the resulting loss of open space. Agran correctly sensed that the captivating image of geese on a grassy green slope immediately adjacent to his political stronghold of University Park could be a winning vehicle to exploit. During much of 1984, Agran rallied followers and new recruits in University Park and Turtle Rock to "Save Quail Hill," with flyers, events, and newspaper ads.

DOI: 10.1201/9781003683087-14

Agran kept the pressure on until early 1985 when the council rejected Quail Hill as a city hall site. But that was just the warm-up for the June 1986 election, when an Agran-backed initiative created the new post of directly elected mayor in Irvine (previously, as with most other Orange County cities, the mayor's post was largely ceremonial and rotated among council members.)

For more than a decade, a pro-growth majority controlled the Irvine City Council, with Agran and a series of his allies consistently losing 3–2 votes. That changed in September 1985 when council member David Sills was appointed to the Orange County Superior Court. Facing a council vacancy, the two sides searched for a compromise candidate, settling on a UC Irvine professor named Ray Catalano.

Catalano had served on the city's planning commission and was admired for his intellect and academic background, which included urban planning. But it quickly became evident he would become a swing vote on the divided council. Agran was quoted after backing Catalano's appointment as saying, "Ray better not be in the middle," while Irvine Company president Thomas Nielsen praised Catalano intelligence, adding, "He will be in a position to play a strong role." (Catalano knew he would be the swing vote on a variety of issues. After playing the role for the remainder of Sills' term, he decided not to run for election and returned to academia).[1]

Catalano's appointment came as the initiative effort to preserve Quail Hill ramped up. At the same time, the city was starting its state-mandated update of the general plan. It became evident to the Irvine Company that the votes were present to adopt a comprehensive open space preservation policy that would go well beyond Quail Hill. The word was quietly passed from company headquarters in Newport Beach to Agran that the potential for a major agreement was ripe. Talks began in private, with Catalano and councilwoman Sally Anne Sheridan in the lead for the city, Michael LeBlanc for the company.

The result was what became known as the 88–1 initiative. The language plainly stated the intent:

> to assure that the conservation of open space areas designated . . . for public ownership will be obtained through the provision of a phased dedication and compensating development opportunities . . . which transfers development opportunities from conservation and open space areas and consolidates them in appropriate development areas.[2]

Accompanying the language was a map of the areas to be conserved, most of them east of Jeffrey Road and north of Portola in the foothills above the city. Also shown was a rectangle of land with the phrase "golf course overlay." That designation would turn out to be a ticking time bomb that exploded 37 years later (see Chapter 26).

Voters approved the open space initiative by a vast majority.[3]

But few, if any, of the voters saw the document accompanying the initiative, the Memorandum of Understanding Implementing Initiative Resolution 88–1 (MOU), which was approved three months after the June vote. In excruciating detail, the MOU described how open space would be transferred into public ownership by linking each gift of land to specific zoning actions and issuance of building permits.[4]

It was a watershed moment in the city and the company's history. While the overall development framework of the town remained as defined in the general plan, the MOU made wholesale alterations to the intensity of residential, retail, and commercial development never contemplated either in the original 1960s-era master plans conceived by William Pereira or in subsequent general plans developed by the Irvine Company and the city (which were very similar).

Ultimately, what did the Open Space Initiative deliver?

For Agran and his allies, it was a major political victory with tangible results. He and others could stand before places like Bommer Canyon and Quail Hill and rightfully proclaim that they would remain in their natural state forever. After years spent sniping at the Irvine Company and pro-development council members, Agran's reputation shifted from obstructionist to achiever.

For the company, it was an agreement providing a significant degree of immediate certainty plus the guarantee that in future years, much more development, both in numbers and intensity, would take place that otherwise would have been the subject of a long, parcel by parcel slog potentially impacted by changing council political views.

There was also the question—never raised publicly—of just how much open space had really been given up. Quail Hill could have doubtlessly been developed but would have required extensive grading, and the homes built there would have had a less-than-inspiring view, looking down on the San Diego Freeway.

Bommer Canyon, the site of the former cattle roundup facility, had such a sentimental reputation it is doubtful the company would have ever considered significant development there, if any. And the foothills, while able to be graded into homesites, were remote; development there was likely decades away. The land with the greatest value was the large, flat acreage next to the intersection of Jeffrey Road and Irvine Center Drive. A golf course was built there, and a legal restriction known as a conservation easement was scheduled to be recorded on the property, but the land was never transferred to the city.

Agran's moment of glory proved brief. Two years later, he narrowly lost the mayor's race to Sally Anne Sheridan in a bitter, caustic election.[5] Agran would endure political exile until 1998, when he reclaimed a council seat, followed by the mayor's chair in 2000, 22 years after his first election to the Irvine City Council.

There is also another interpretation of the open space election and subsequent dialogue between the city and the company. It may have sparked, either directly or indirectly, Donald Bren's thinking about the subject and how it would impact the future of the Irvine Ranch. He was 56 at the time, five years into complete control of Irvine Company operations. As he often said to trusted employees regularly, and to the media on rare occasions, he had a long-term vision for the Irvine Ranch and was willing to be patient. Was it just a coincidence, then, that in 1990, it was reported that Bren began conversations with the Nature Conservancy about a potential collaboration?[6]

***

Very large birds—geese—ignited the initial open space issue in Irvine. It was a tiny songbird—the gnatcatcher—that precipitated the second, creating an impact that reached all the way to Washington, D.C.

The gnatcatcher was one of the least well-known birds in the diverse avian life of Southern California. Ask a local resident to name a favorite species, those close to the coast might cite any number of gulls, ospreys, or grebes; further inland, it was not unusual to see a blue heron or an egret in the flood control channels. Watching intently from high-voltage electric towers or perched at the top of a eucalyptus tree, one of several varieties of hawks scanned the landscape for prey. And in backyards throughout the county, the hummingbird was ubiquitous.

The gnatcatcher lived almost exclusively among the thousands of acres of hills and mountainsides covered in a variety of plant life that naturalists labeled coastal sage scrub. There, the tiny bird flitted about in relatively small spaces.

It was the Endangered Species Act that hurled the gnatcatcher into an arena of battle in Orange County, eventually pitting developers, homebuilders, business interests, and a host of elected officials against environmental organizations, lawyers, and activists—and, eventually, the Irvine Company.

Bruce Babbitt, the former attorney general and governor of Arizona, was serving as secretary of the interior in the Clinton administration when a petition arrived at his office seeking to list the gnatcatcher as an endangered species. In his book *Cities in the Wilderness*, Babbitt recalled that he approved the application in March 1993 without realizing a political firestorm would follow when the designation was made.[7]

After all, he knew, in 1980, when the listing of a rare butterfly nearly shut down development near the Silicon Valley in northern California, Congress had created a process called a habitat conservation plan to balance preservation of land necessary for the survival of the affected species with legal guarantees to landowners that they could use their land for farming, other business, or development. In 1991, in California, the legislature had created a legal methodology to accomplish similar goals. It was called the Natural Communities Conservation Act, shorthanded to NCCP.

When the gnatcatcher listing became official, "the storm broke," he wrote:

> Sage scrub habitat extended across much of the best (and highest priced) land remaining between Los Angeles and San Diego . . . by placing the gnatcatcher on the endangered species list, we had . . . dropped a blanket moratorium on much of the remaining developable land in the fastest growing real estate market in California.[8]

The action did not go unnoticed. Soon, Babbitt and his staff were under pressure from business and development interests, as well as the White House, reminding him that Clinton won California's 54 electoral college votes and wanted to do so again. Babbitt learned that California governor Pete Wilson had declined to list the gnatcatcher on the state's endangered species list, which was similar to the federal version. This caused the state program, in Babbitt's words, "to begin to fall apart" when the federal listing took place. Babbitt believed, however, that a solution could be crafted given enough time:

> We began casting about for a place to begin. Coastal Orange County . . . seemed the logical place. We soon discovered, however, there was just one

> landowner in Orange County who really mattered: the Irvine Company which was actually just one person, Donald Bren . . . In conservative Orange County, he was accustomed to having things done his way and to public officials who understood who was really in charge. [This] did not bode well for a federal wildlife agency about to become a partner in making his land use decisions.[9]

While much of the Irvine Company's land was free of coastal sage scrub, some of the most valuable—along the Newport Coast—was just being developed. An area above Turtle Rock called Shady Canyon was also planned for luxury homes set above a links-style golf course that was almost entirely covered in coastal sage habitat. In the rolling hills near Irvine Lake, virtually every acre was potential habitat, not for the gnatcatcher, but for other species that were now being aggressively proposed for listing by environmental forces. Worse, the planned new toll road, the San Joaquin Hills Corridor, would likely never be built if the gnatcatcher was listed.

Babbitt decided to engage. "I arranged a discussion with (Bren's) representatives, agreeing to meet in Phoenix, where we would attract less attention."[10] Attending the meeting were John Cox, at the time chairman of the San Joaquin Toll Road Authority, Gary Hunt, Monica Florian, and the toll road's attorney Rob Thornton.

Thornton had become the toll agency's point man for environmental issues and was widely recognized as one of the most knowledgeable legal experts on endangered species litigation. Thornton had focused many of his arguments on the gnatcatcher listing by challenging the findings of biologists claiming the gnatcatcher was a unique species. The Orange County gnatcatcher population was estimated at fewer than 2,000 pairs, but Thornton noted that birds that were biologically near-identical to the local version that existed in the tens of thousands just across the border in Mexico.

For Thornton, the tiny difference would become a lifelong professional obsession, one that was argued about again and again in legal documents and private conversations. Thirty years later, the distinction remains in law, and the gnatcatcher continues to enjoy federal protection.

Babbitt wrote in his book that he insisted any potential solution involving habitat preservation would mean a lot of it—thousands of acres and no golf courses masquerading as places where birds could thrive:

> The company representatives did not disagree, and I began to sense they knew more about these issues than I was giving them credit for. By the end of our lunch . . . we agreed to work to include in the plan not just the gnatcatcher, but all other potentially threatened or endangered species so that we would not need to repeat the process every time another plant or animal appeared on the list.[11]

The responsibility to manage the experiment in environmental governance from the company's perspective was given to Florian, who had risen in the decade she had worked there from a planning manager to senior vice president, the highest-ranking woman in a management structure filled almost exclusively by men.

Florian had come from the city of Huntington Beach, having previously gained planning experience in Riverside County, where she had grown up. After joining the company, her relentless work ethic and shrewd approach to problem-solving were recognized by Hunt and others, and she was promoted, eventually overseeing entitlement and community relations staff. Interviewed in 2024, she recalled the NCCP process as one of the most difficult and draining assignments she had ever managed.

The company's approach to what initially seemed to be an intractable problem was the same as what Florian considered to be its standard approach to dealing with issues: "what is the big picture, what is the long-term solution, and then how do we deal with it. And that's pretty much what happened."[12]

She continued:

> We brought in consultants and we started talking with environmental groups, credible, national ones, Environmental Defense Fund, some others that were obviously supportive of the Endangered Species Act, but not so much focused on (using) it to stop a particular development.[13]

Those discussions, as well as with Bren, led Florian to conclude that

> maybe we need to not just say no, maybe there's a way to deal with this in a way that would allow life to go on, maybe not as we knew it, but life to go on with changes.
>
> And that probably was a unique approach at the time because everybody was opposed to it or wanted it to stop everything . . . and to defeat it, it would just mean one more time another species would come up and (cause) chaos through all the process, so how do we get order here?[14]

Florian said that after working with experts from all over the country, "credible ones and scientists," absorbing their view that it was less the individual birds or animals than the places where they lived,

> the more we realized, that makes a lot of sense . . . and we have a lot of habitat, a lot of open space, and we can try to deal with this, and if we can get it all wrapped up at once, that would be desirable.[15]

As Florian noted, the thousands of acres of land owned by the company—much of it that would never be developed—was a powerful weapon in a new game where the ability to protect habitat was emerging as the key element to success. But, she said, to get over the finish line, the plan had to include others.

> We engaged environmental groups; we were building coalitions. The Irvine Ranch, I think was a perfect laboratory to test this idea because local governments, private landowners and government agencies had a history of working together, of doing big things. We got landowners, the developers, utility

companies into one group, and we were trying to marshal our resources in favor of this idea.[16]

And while portraying the effort as a cooperative undertaking motivated by sound science and the spirit of the local building community's can-do attitude, Florian also remembered what she considered to be an equally compelling rationale: fear.

> Fear on the environmentalist side that this is going to blow up because once they have stopped the economy of Southern California, they were afraid the Endangered Species Act would be thrown out the window. The government agencies and private sector were afraid that the process would be so brutal and disruptive that nobody would be able to do anything, and there would be chaos.[17]

The Irvine Company likely had the most to lose in the battle, but it also had a reputation within the county that dated back to the days of Ray Watson's tenure as the logical place to look for leadership on significant issues. Bren had continued the tradition not by personally assuming a public role but by empowering people like Hunt and Florian with the prestige and resources to take the mantle of leadership and wield it.

Three years after the gnatcatcher listing that instigated a process Babbitt admitted he gave "a thirty percent chance" of success, a ceremony was held in Shady Canyon marking the signing of an agreement to create the first NCCP preserve in Orange County, a total of 37,000 acres, 21,000 from the Irvine Company.[18]

Remembering the day, Babbitt wrote:

> I turned to Doug Wheeler (secretary of the California Resources Agency) and exclaimed, "How did this happen?" He reflected a moment and replied, "public support, and good press, the fact we worked together . . . and Donald Bren. You know that he is a friend and big supporter of the governor, and I expect that had a lot to do with it."[19]

Babbit also wrote:

> The settlement at Irvine Ranch was a defining moment in the emergence of the Endangered Species Act as a land use planning statute, capable of forging a balance between development and the preservation of large ecosystems. Fortunately, Bren was out to solve a problem, and in the process, he had given credibility to a statute that might not have survived much longer without some demonstrable evidence that it could be made to work on private property.[20]

Florian's recollections primarily focus on the company's interactions with business and development interests.[21] An insight into how the environmental community viewed the NCCP process is found in a white paper commissioned by the County of Orange, containing interviews with county, state, and federal environmental officials who would ultimately manage the NCCP as well as two well-known and respected leaders of environmental groups, Dan Silver of the Endangered Habitats League and Peter DeSimone, manager of the Starr Ranch wildlife sanctuary and a representative of the Audubon Society. All except the county representative were members of what became known as the Working Group.[22]

Silver noted in his comments that the initial reaction from the environmental group to the potential for land restrictions by the Irvine Company was cynical. "Some interpreted reserved designs that used previously set-aside lands for mitigation as 'double dipping' . . . and considered only conservation benefits to be a result of a paltry 3,000 acres of additional Irvine Company lands." Silver disagreed: "We had to live with a lot of fixed constraints. Considering this and local politics, we got a lot in terms of the 15% of lands not there before, the connection, restoration, and the coordinated management."

As the NCCP process continued, Silver said the Working Group sessions

> allowed us time to understand how the plan developed and evolved . . . we had an opportunity to satisfy ourselves that we couldn't do certain things like connect the two reserves directly. So while it didn't allow us to influence the ultimate decisions much, it was valuable in our understanding.[23]

DeSimone criticized an initial decision to include, then eliminate, three potential species from the program, which was done in his view "with nothing more than a gut feeling and with no baseline information." Of the overall Working Group effort, his opinion was "you have to look at (the Working Group) from the context of what would have happened without it. The plan . . . would have been much worse for sure."

Currently, in California, there are 17 approved NCCPs which together cover more than 8 million acres of protected habitat.[24] In Orange County, the county has the responsibility to manage the 38,000 acres but also coordinates with groups that oversee acreage donated by utilities, the Toll Roads agency, and others.[25]

Florian believes the NCCP experience was the preamble to what would become Bren's final chapter in his personal history of open space preservation.[26]

***

An appreciation of the natural features of the Irvine Ranch is most accurately defined as five major landforms: the coast, encompassing shoreline and foothills; the western foothills; the central plain; the eastern foothills; and the eastern mountains and canyons. There are also two major wetlands on the property, the San Joaquin Marsh and Upper Newport Bay, the former largely freshwater, the latter a saltwater estuary.

In 1983, when Donald Bren became the controlling owner of the Irvine Company, the fate of the central plains was firmly set—it would be completely developed. The only areas left in open space would be designated as parks, with almost all of them improved and landscaped, and just a long sliver of a creek remaining in

Figure 14.1 The Sinks in Limestone Canyon

Source: Image courtesy of Orange County Parks Department

its natural form along Campus Drive, at what had become Mason Regional Park. In his first years as majority owner, Bren had infused his staff and consultants with an urgency to speed up development of residential and commercial properties to make up for what he considered lost time during 1977 to 1983 when he was outvoted on company policy by his partners.

However, by 1988, Bren was facing an open space battle in Irvine, followed a few months later by a similar, but more intense, conflagration over Laguna Canyon.

In both instances, Bren ultimately chose to negotiate rather than fight, achieving, in the first case, deferred intensity of development for phased donations of land and, in the second, receiving millions of dollars in cash.

And as described previously, in 1993, Bren was confronted by what could have been an existential crisis with the listing of the gnatcatcher as an endangered species.

Bren was known to his senior employees as someone who spent a great deal of time reading and thinking. He preferred the solitude of his ninth-floor office to long strategy sessions or briefings. He assembled a board of directors not for corporate governance but to seek expert opinions. Mostly, he kept his own counsel, turning over in his mind possibilities large and small that could impact his vast land empire.

When he thought about the northern, mountainous section of the ranch—by size just under a third of his total holding—he knew that the vast majority of the land would never be developed. The three major canyons—Weir, Limestone, and

Fremont (which had been called by some the "Yosemite of the Irvine Ranch)—were deep and barely accessible, even to cowboys on horseback who had herded cattle there for decades. The canyons were bordered by rugged hills, on their verdant floors sycamore, California oak, and other flora thrived. One particular section of a limestone-rich area was known as the Sinks—though small, it bore an uncanny resemblance to the orange rock spires of Utah's Bryce Canyon.

Two relatively flat areas in the mountainous region could be developed; one just west of Irvine Lake, which had been built by James Irvine II for irrigation in the 1930s, and along the SR 91 freeway at Gypsum Canyon. The rolling hills near the lake would require little grading but were miles from established water, sewer, and electric lines. Gypsum Canyon was steep, but homes could be built there with some creative engineering.

The hillside of the Lomas de Santiago overlooking the center of the ranch was gentle compared to the craggy ridges further north. Here, the potential for development was high, and there were fewer scenic canyons; nearly all of the land had been planted in citrus for the past century, leaving little in a natural state.

On the other side of the great center, the San Joaquin foothills were even more gentle than the Lomas de Santiago, rising to a ridge that formed a spine at the top of the Newport Coast. Bren had already given up Bommer Canyon and Quail Hill as part of the 1988 city agreement. The substantial open space lay in the thousands of acres of hills along the Newport Coast, which flowed down into Laguna Canyon, much of which had been sold to the city of Laguna Beach in return for the elimination of development there. Moreover, some of the land in the Newport Coast was being preserved in its natural state as part of the NCCP: the shoreline plus one major canyon had been sold to the state for Crystal Cove State Park. As for the NCCP, the legal timeframe for preservation was 75 years, but the political reality meant it might as well have been 750 years—once given over for habitat protection, it would have been almost impossible to reclaim.

Monica Florian explained how Bren reached the seminal conclusion that would define his legacy as the sole master of the Irvine Ranch:

> There then came a period where we had all of this land that was kind of off the table for development, but nothing was happening to it. I think that Bren focused on why the land was just sitting there. I would say that is kind of the beginning of the next period of the open space evolution, and it was him saying, "The land is already gone, how do we make it useful?"
>
> He was quite adamant that we need to give this as much attention to the open space lands as we do the development land, and his goal was that if these lands are set aside, they ought to be producing value to the community.[27]

Bren assigned Florian the responsibility of developing a plan to ensure that whatever open space was eventually given away would have a robust public access element. Looking back 23 years later, she said the assignment "would change my life."[28]

She continued:

> Bren's question about what can we really do with the land started a search . . . all over the country, talking to different people, environmental groups, land management groups, and conservation organizations. That's when we arrived at the idea that a stewardship plan would be a good idea, and (we thought it didn't make sense for a development organization to do a plan by ourselves, so looking for a partner became the next step).[29]

The concept of being a steward of the land as opposed to just owning it was not new at the time. Years earlier, Tom Wilck had developed the beginnings of a public relations campaign emphasizing the company's objectives as something beyond the significant development underway, but it never gained much traction.

Bren's thinking, Florian believes, focused on the fact that even after the 21,000-acre open space dedication to make the NCCP a reality, a look at the ranch map showed gaps between those specific lands and other acreage that either had been protected or likely would be as time passed.

She said,

> I think he was concerned about how it was still a piecemeal approach, because the land was being donated piece by piece as projects came forward, but then there were these conservation easements . . . which limited what anybody could do with (the land).[30]

Florian's efforts culminated in a carefully planned series of announcements unveiling the news that Bren was taking two unprecedented and intertwining actions. First, 11,000 acres of land in two significant segments of the northern ranch would be donated for eventual public ownership. Second, Bren had retained the Nature Conservancy to manage the 11,000 acres. That was the news on November 28, 2001, as delivered at the Four Seasons Hotel in Newport Beach to a crowd of more than 200 "conservationists and public officials."[31]

Bren made a rare public appearance at the event and said that the land gift fulfilled his goals: "my dream is that the Irvine Ranch will be known and celebrated as much for what is not developed here . . . as it is for the outstanding communities we have planned and built." The *Times* also reported that Bren was adding $10 million to a $20 million fund "established last year to support management and restoration of natural habitats on the land and improve public access."[32]

Reaction in the tastefully pink-shaded meeting room at the company-owned hotel was overwhelmingly positive. One well-known and often crusty environmental advocate said she was "moved to tears" at the news, while another confessed that it "was a very emotional moment."[33] A month later, the same person was writing a glowing opinion piece in the *Times* praising the open space gift.

In March, the open space news blitz continued, this time emphasizing how the company's stewardship plans would shape the future of the massive gift of acreage.

Newspaper stories on March 10, 2002, said the Nature Conservancy had a ten-year management contract valued at $20 million. "We should be clear that one of the key points that the Nature Conservancy and the Irvine Company have talked about is being able to provide opportunities for public access," said Graham Chisholm, director of the California Nature Conservancy office.[34]

Five months later at a ceremony noting the donation, the *Times* reported that the management deal was for two years, not ten, that the land being managed had grown from 11,000 acres to 17,000 acres (the additional land in Emerald Canyon and Boat Canyon in Laguna Beach) and would now be known as the Irvine Company Open Space Reserve, and that the lands would be dedicated to the public over "the next 20 to 25 years."[35]

Bren explained the management plan:

> Our goal is to protect these natural resources in perpetuity, enhance the wildlife habitat areas found here, and provide for early public use of this special land. I believe this program offers us the opportunity to make a lasting contribution to California's rich natural resources.[36]

Media praise for the land gift continued in the *Times*. On May 29, 2002, writer John Balzar reported on correspondence he had received from Bren on the matter. He began with his own question: "I'm curious. Just how does a land developer, a man whose father was also a developer, come to express the idea that quality of life is measured just by what was built up, but what was not? I asked him."[37]

Replying in writing, Bren said,

> I believe open legroom means different things to different people. Not only does it mean freedom for those who want to use it, gain access to it, but I also believe it also provides a sense of comfort and security as a defined boundary that won't be developed, as a buffer, as elbowroom.[38]

Bren closed his thoughts on a personal note, telling Balzar, "I love the sheer joy, the sense of freedom, even escape that I had in the outdoors as a child."[39]

The media coverage of the open space donations and the reaction to them was almost uniformly positive.

Almost.

As might have been expected, reporters and others knew that a gift of land meant it was undoubtedly worth something and, as such, potentially eligible for a tax break.

Land values across the entire Irvine Ranch had come up in 1987 during the legal proceedings in Michigan that determined how much money Joan Irvine Smith and her mother would be paid for the shares in the company they had sold in 1983. Testifying before a court-appointed referee in the case, Thomas H. Nielsen, then company president, said that 42,000 acres of land could not be developed because of topography and proximity to the El Toro Marine Corps Air Station (those lands became *very* valuable after El Toro closed).[40]

In his 2005 oral history, Ray Watson also addressed the issue. Asked about the tax implications of the open space dedication, he said,

> That isn't the reason he did it. Yes, he was influenced by the fact you can get a tax deduction for it . . . but the fact is, he could have just sat on it and not given it, and at some time in the future, when it becomes even feasible, maybe to develop it (which in that case would mean) it wouldn't have been open space.[41]

Times reporter Janet Wilson dove into the possible tax implications of the open space deal in a long story accompanying the straight news on the open space dedication. She focused on the potential write-off available for placing a conservation easement on the land. Citing a variety of tax experts, she said the land value of the dedication could range from $88 million to up to $500 million.[42]

Irvine Company officials at the time deflected the story with their usual refusal to address company finances. Monica Florian recalled in her 2024 interview discussing the issue with the Internal Revenue Service. She said the IRS had rarely seen anything of the scope of the dedication and was having a tough time figuring out how to value it.

***

The Nature Conservancy's management of the open space lands concluded in 2004. A year later, a new organization was created that became the Irvine Ranch Conservancy. Michael O'Connell, a veteran environmental leader, was named to

Table 14.1 Irvine Ranch Land Preservation—Management Breakdown[44]

Managing Entity	*Approximate Acreage*	*Description*
Irvine Ranch Conservancy (IRC)	Approximately 30,000 acres	Lands managed in partnership with OC Parks, the City of Irvine, and the City of Newport Beach. IRC provides ecological stewardship, public access programs, and habitat restoration.
California State Parks (Crystal Cove State Park)	Approximately 2,400 acres	Coastal and backcountry parkland originally part of the Irvine Ranch. Managed independently by California State Parks.
OC Parks (County of Orange)	Approximately 20,000 acres	Includes Limestone Canyon, Weir Canyon, and other regional preserves. IRC co-manages many of these areas under long-term agreements.
City of Irvine	Approximately 5,000 acres	Includes Bommer Canyon, Quail Hill, and other city-owned preserves. IRC provides stewardship and public programming.
City of Newport Beach	Approximately 150 acres	Includes Buck Gully Reserve and adjacent coastal parcels. IRC completed infrastructure and trail restoration projects.

Source: Note: Acreage figures are approximate and based on agency reports, land transfer records, and published coverage between 2001 and 2025. Microsoft Copilot created the table from an author-provided prompt.

lead the Conservancy, and Bren promised to donate $50 million over time towards its stewardship efforts.

Records of the Conservancy for 2022, 2023, and 2024 show a consistent annual donation of just over $5 million, presumably from the Irvine Company, as records of Bren's foundation do not show any recent donations.[43]

Figure 14.2 Irvine Ranch open space

Source: City of Irvine

The Conservancy now manages approximately 30,000 acres of the 57,500 acres of open space lands sold or gifted by the Irvine Company, the most recent parcel coming in 2014.

Table 14.1 summarizes the approximate acreage and stewardship responsibilities for preserved lands originating from the historic Irvine Ranch. It reflects publicly available data from the Irvine Ranch Conservancy, California State Parks, OC Parks, and the Cities of Irvine and Newport Beach.

The open space story has become one of the foundational elements of the company's and Bren's continuing marketing and public relations efforts. Bren's website features open space as one of his four principal charitable and business focuses. In the monthly tabloid the company publishes, *The Irvine Standard*, rarely more than two issues pass without a new description of the open space donation benefits.

When text descriptions run short, there is no shortage of photos of beautiful vistas, rolling hills, quiet canyons, and stunning coastal shorelines. Supporters and doubters together must acknowledge the incredible landscape, as has the U.S. Department of the Interior, which in 2006 declared the Irvine Ranch a United States National Landmark, one of 38 in California and 606 in the continental United States. California has a similar program and included the ranch on its list in 2008.[45,46]

It is worth noting that Bren almost never attended groundbreaking or opening ceremonies for villages, buildings, or commercial centers. Still, he always made it a point to appear in person for an event related to open space.

Notes

1 "Catalano Appointed to Irvine Council," *Los Angeles Times*, June 23, 1986.
2 City of Irvine. Initiative Resolution 88–1: An Initiative Resolution of the City of Irvine Directing the Amendment of the Conservation and Open Space Element and the Land Use Element of the Irvine General Plan. Irvine, CA: City of Irvine, 1988. https://alfresco.cityofirvine.org/alfresco/guestDownload/direct?path=/Company%20Home/Shared/CS/Environmental%20Programs/Environmental%20Policies/Open_Space_Ballot_Initiative_Resolution_88-1.pdf
3 "Open Space Initiative Passes," *Los Angeles Times*, June 8, 1988.
4 Memorandum of Understanding Implementing Initiative Resolution 88–1, City of Irvine, 1988.
5 "Irvine Mayor's Race Turns Bitter," *Los Angeles Times*, November 6, 1988.
6 "Bren Begins Talks with Nature Conservancy," *Los Angeles Times*, October 12, 1990.
7 Bruce Babbitt, *Cities in the Wilderness: A New Vision of Land Use in America* (Washington, DC: Island Press, 2005).
8 Babbitt, *Cities in the Wilderness*, 63.
9 Babbitt, *Cities in the Wilderness*, 63.
10 Babbitt, *Cities in the Wilderness*, 67.
11 Babbitt, *Cities in the Wilderness*, 69.
12 Interview with Monica Florian by H.P. Oliver and C.M. Stockstill, May 28, 2024.
13 Interview with Monica Florian.
14 Interview with Monica Florian.
15 Interview with Monica Florian.
16 Interview with Monica Florian.
17 Interview with Monica Florian.
18 "NCCP Agreement Signed in Shady Canyon," *Los Angeles Times*, July 18, 1996.

19 Babbitt, *Cities in the Wilderness*, 73.
20 Babbitt, *Cities in the Wilderness*, 73.
21 Interview with Monica Florian.
22 County of Orange, *Orange County Central-Coastal NCCP/HCP White Paper*, undated.
23 *Orange County Central-Coastal NCCP/HCP White Paper.*
24 California Department of Fish & Game, "Natural Community Conseration Planning," *California Department of Fish and Game*, accessed on January 17, 2026, https://wildlife.ca.gov/Conservation/Planning/NCCP.
25 Natural Communities Coalition, "About NCC," accessed January 17, 2026, https://occonservation.org/about-ncc/.
26 Interview with Monica Florian.
27 Forian interview.
28 Florian interview.
29 Florian interview.
30 Florian interview.
31 "Bren Donates 11,000 Acres for Open Space," *Los Angeles Times*, November 29, 2001.
32 "Bren Donates 11,000 Acres for Open Space.
33 Bren Donates 11,000 Acres for Open Space.
34 "Nature Conservancy to Manage Irvine Lands," *Los Angeles Times*, March 10, 2002.
35 "Irvine Company Expands Open Space Reserve," *Los Angeles Times*, August 7, 2002.
36 John Balzar, "Bren's Vision for Open Space," *Los Angeles Times*, May 29, 2002.
37 "Bren's Vision for Open Space."
38 "Bren's Vision for Open Space."
39 "Bren's Vision for Open Space."
40 "Testimony by Thomas H. Nielsen," *Los Angeles Times*, August 6, 1987.
41 Raymond L. Watson and Ann Lage, *Planning and Developing the New Town of Irvine, California, 1960–2003* (Berkeley: Regional Oral History Office, Bancroft Library, University of California, 2005).
42 Janet Wilson, "Tax Implications of Bren's Land Gift," *Los Angeles Times*, March 10, 2002.
43 Irvine Ranch Conservancy, Annual Reports, 2022–2024.
44 Irvine Ranch Land Preservation—Management Breakdown table was compiled using publicly available data from the Irvine Ranch Conservancy, California State Parks, OC Parks, and the Cities of Irvine and Newport Beach. Figures reflect agency websites, annual reports (2022–2024), land transfer announcements, and coverage in the *Los Angeles Times* between 2001 and 2025. While Crystal Cove State Park is included in the total preserved acreage attributed to the historic Irvine Ranch, it is managed independently by California State Parks. The segmented breakdown clarifies jurisdictional boundaries and highlights the collaborative stewardship model that defines Irvine Ranch conservation.
45 U.S. Department of the Interior, "National Natural Landmark Designation: Irvine Ranch," 2006.
46 California Department of Parks and Recreation, "California Natural Landmark Registry," 2008.

15 The Spectrum

On a bright spring day in Sacramento in the early 1960s, a group of engineers at the California Division of Highways ordered up a topographical map of Orange County from a shelf deep in the department's records warehouse.

Their task was to plot a line for the extension of the new San Diego Freeway, Interstate 405, which had been built to the point where Los Angeles and Orange Counties met.

Examining the maps, the engineers calculated the cost of bridges over the Los Angeles River channel as it passed through Long Beach, then drew the right-of-way to steer east of the Seal Beach Naval Weapons Station and its low bunkers filled with ammunition for the nation's warships.

From there, heading south, there was little in the way of the giant Caterpillar D9s that would scrape out a patch to accommodate four to six lanes of concrete: almost all the land from Seal Beach to the San Diego County line was being farmed or grazed.

Planning for the freeway had started years earlier and was being monitored with more than casual interest by an aspiring land developer.

Henry Segerstrom was the son of a family of Swedes who had come to the fertile fields of the south-central area of Orange County to farm. Like James Irvine II, they planted lima beans on acreage between what would become Harbor Boulevard and Bristol Street.

Segerstrom was in college when World War II began. He enlisted, was commissioned a lieutenant in an artillery unit, and had fared well until the final months of the war in Europe. Wounded at the Battle of the Bulge, he was shipped home to heal and then graduated with both a bachelor's and a master's degree from Stanford. After recovering, Segerstrom convinced the family to take a big gamble—bid on buildings declared surplus at the nearby Orange County Air Base. The Segerstroms won the bid, and within a few months, Henry had leased every one.

As the family began planning to turn a large section of their farm into a giant retail center, Henry knew that it would benefit immensely if the San Diego Freeway alignment ran nearby. Segerstrom

> exercised his already considerable political influence to ensure that the exact location of the freeway would be advantageous to South Coast Plaza.

DOI: 10.1201/9781003683087-15

> The company archives include documents revealing Henry's acute attention directed toward the location and design of the freeway, down to the nuances of the entrance and exit ramps near South Coast Plaza.[1]

Segerstrom's prescience and willingness to massage the state road building bureaucracy would make him and his family multimillionaires as South Coast Plaza grew into perhaps the most successful retail mall in California and its surrounding acreage was developed into office buildings and apartments.

Further south, there are no historical records to indicate that the owners or managers of the Irvine Company were undertaking efforts similar to those of Segerstrom. Company planner Al Trevino said that the original freeway alignment sliced into his plans for the Village of University Park and that company president William Mason's offer of a fill dirt site in Laguna Canyon for freeway construction persuaded the state planners to tweak the alignment away from the planned new homes.

As the drafting continued, Division of Highways engineers considered just where to bring the San Diego Freeway into a junction with the existing Santa Ana Freeway, now known as Interstate 5. Cradling a protractor in hand, an engineer likely traced a relatively straight line from University Park to the I-5, hugging the land at the base of some low foothills to the west through a gap adjacent to a small knoll known as Barton's Mound.

In drawing the final line, the engineer may have noticed how the intersection of the two freeways became the point of a triangle, with the north/south alignment of Laguna Canyon Road forming the base of the pyramid-shaped parcel of land that in the future would become known as the "Golden Triangle."

The Golden Triangle became a legendary designation, but it would remain empty for decades. In reality, the triangle was only a small element in a development that was planned to spread out over both sides of the coming San Diego Freeway, with much of it between there and the El Toro Marine Corps Air Station.

The new employment center was known as Irvine Industrial Complex-East (IIC-East), a bookend commercial and industrial development that would mirror the rapidly expanding Irvine Industrial Complex (IIC) taking shape around Orange County Airport, about six miles to the north, also next to the San Diego Freeway.

The IIC had defied real estate and economic norms on its way to becoming a huge success for the Irvine Company, which had planned it in much the same fashion as it had its residential villages. The IIC required new owners of properties to conform to strict design and landscape guidelines when building a new factory or warehouse. There was ample distance between structures, and landscaping was required. No barbed wire fencing was allowed. Known as an industrial park, the focus and vibe in the IIC was clearly on "park."

While the IIC expanded, plans for the IIC-East languished; infrastructure costs were going up, and environmental regulations were increasing. And while an economic downturn was showing signs of easing, a lawsuit challenging the development had been filed by a group of Irvine residents to pressure the Irvine Company on the issue of low-cost housing.

Irvine Company planners promoted IIC-East as something quite different than just another collection of manufacturing plants and generic buildings. It would become "a group of industrial neighborhoods (where) commercial services will form the core of a series of rings, with offices and small industries in the center rings, and larger industries surrounding them."[2]

And, in the continuing (but ultimately futile) quest of Irvine Company planners, starting with William Pereira a prediction was made that the IIC-East neighborhoods would be "designed to encourage short walks . . . and to eliminate, as much as possible, the need for automobiles."[3]

After new owners took control of the Irvine Company in late 1977, the lawsuit was settled and development of the IIC-East began with the installation of a large auto center and a few small tilt-up industrial buildings. It would be seven years later before a new vision for the area would take shape under a revised plan and a fresh name: the Irvine Spectrum.[4]

***

As Irvine Company executives considered how to plan, develop, and market the Irvine Spectrum, they relied on economic research the company had commissioned. It showed one element of the county's still-burgeoning economy that was likely to grow—and needed room to do so—was related to biotechnology.

And unlike Silicon Valley, which some marketers wanted the Spectrum to mirror both in name and composition, biotech was comprised of a more varied group of companies with an extensive and diverse array of technological interests.

The county already had a nucleus of biotech players, including the major eye care company Allergan, which had established its headquarters and labs in the business area adjacent to Orange County Airport. Edwards Life Science, which specialized in heart valves, was just across the freeway. But the economic research showed that there were potentially dozens more biotech firms in the early stages of development that needed a place to grow.

In fall 1984, the Irvine Company announced the creation of the Irvine Industrial Research and Development Company, focusing exclusively on the initial development of the Spectrum as a destination for the biotech industry. Selected to lead the new subsidiary was Richard Sim, who had arrived at the company three years earlier and had begun a 20-year career that would take him to the highest levels of company decision-making.

A Pennsylvania native, Sim grew up helping his parents run a small hotel. After graduating from Villanova, Sim found his way to California; he was CEO of Kaiser Development Company when he joined the Irvine Company. Sim's management presence was soon reverberating through the company, which, in his view, was caught up in endless meetings and needless bureaucracy. A self-proclaimed "clean desk" executive, Sim proudly declared that any paper that ended up on his desk would stay there only once, destined to be delegated or read and discarded. Colleagues enjoyed ribbing Sim about his habit, but his take-charge, make-a-decision attitude was welcomed by Donald Bren, who would promote Sim once he gained control of the company.

Sim's initial management of the Spectrum development was undertaken in what was for the Irvine Company an untried business arrangement—partnering with other developers to speed up construction of infrastructure and buildings. Three firms—the Koll Company, O'Donnell Brigham, and Birtcher—signed on to kick development into high gear. In most cases, the Irvine Company would maintain ownership of the land and the buildings, with the three partners receiving a fee to do the work. On a few deals, the partners were granted actual ownership.[5]

Development of the Spectrum also initiated a renewed partnership with UC Irvine to promote increased and expanded science curriculum at the university to provide graduates with the expertise to staff the growing biotech businesses locating in the Spectrum.

The Irvine Company–UC Irvine relationship had been strained; plans to build a hospital on campus were thwarted months earlier when a competing drive led by Irvine residents and political leaders championed an independently-owned hospital in the center of the city. The company had sided with the residents, leaving some university faculty and staff with lingering resentment.

The Spectrum development provided common ground for the university and company to find reasons for cooperation, resulting in a number of initiatives at the university to connect with the emerging biotech presence.

Marketing the Spectrum was enhanced with the creation of the Irvine Exhibit. Located at the company headquarters in Newport Beach, the exhibit was two-fold: a scale model of the coastal and central sections of the Irvine Ranch that measured 80 feet by 40 feet and an inspirational short film projected onto a giant movie screen. The film was titled *Roots and Wings*, featuring gauzy, uplifting images of happy families living productive lives in Irvine Company communities. The message: putting down roots was necessary to allow dreams to take wing.

Typically, company executives would seat guests in the Exhibit (it cost $1 million to build, according to Sim, which would be equivalent to about $3 million in 2025 dollars), show *Roots and Wings*, and then unveil the giant map/model, where visitors could walk around and see a spot in the Spectrum waiting for them to make a purchase. The Exhibit was also used regularly to impress other guests, including politicians and finance professionals.[6]

To lay the foundation for the Spectrum, the Irvine Company spent a reported $227 million for initial infrastructure, including paying for freeway interchanges. Part of the money also went into what had become Bren's signature landscaping element, giant palm trees planted streetside and in medians to form towering pathways through the new development.[7]

Using its own funds and those of its partners, development slowly ramped up, then accelerated. In May 1997, the business section of the *Los Angeles Times* looked back at the decade of Spectrum growth and reported that "the 3,600-acre Irvine Spectrum emerged relatively unscathed from the California recession . . . and has been growing at breakneck speed ever since. All told, 2,200 companies employing more than 36,000 people now make their home there.[8]

The Spectrum was being compared to national tech hubs like Boston Route 128 and the Research Triangle in North Carolina as well as the standard for

California—Silicon Valley. "The Spectrum is a uniquely Southern California phenomenon," the *Times* story continued. "The concentration of unassuming beige two-story buildings . . . is designed to collect the 'gazelles,' fast-growing entrepreneurs flowing out of the urban core of the LA basin."[9]

The story adds:

> Many of the young companies in the Spectrum start out leasing 1,000 square feet in one of the center's small office buildings, then graduate to a larger building. Eventually, sometimes within just a few years, they build their own headquarters. These small, fast-growing firms have become the backbone of the Spectrum and have fueled further spin-offs. A whopping 77 per cent of the Spectrum's companies employ 24 employees or fewer, according to a company study.[10]

The story identified three companies that had located and grown in the Spectrum, their CEOs citing the fact it was close to their homes and they liked the area's "no-frills, sterile environment that sends a message to investors that money is not being squandered."[11]

The one complaint voiced by early company owners in the Spectrum was the relative remoteness of support businesses, especially a place to eat. It took a decade to solve that problem, which the Irvine Company accomplished in spectacular fashion.

***

The appeal of the Golden Triangle as a retail center was evident to both professional planners and amateur observers. Two major freeways—one an interstate running from Canada to Mexico, the other a new corridor just over the hill from affluent coastal communities, from Newport Beach and Laguna Beach—would bring hundreds of thousands of drivers past the location every day.

In the early 1970s, Irvine Company executives sketched out conceptual plans for a major shopping center at the Golden Triangle, but there was little follow-through. It was the news from Mission Viejo, about ten miles south, that spurred the company to take a more detailed look at the development potential there.

The Mission Viejo Company had been purchased in 1972 by cash-rich Phillip Morris, enabling it to ramp up development of homes and boost its image by building a new professional tennis facility to host the Virginia Slims tournament. The female athletes might not have widely used the cigarette aimed at women, but the big-money purses offered at the contest attracted famous names like Billie Jean King and a young phenom from Czechoslovakia, Martina Navratilova.[12]

In 1976, the Mission Viejo Company announced plans to build a new regional shopping mall at the southern end of its property. It would be the first major shopping center south of Fashion Island and the extraordinarily profitable South Coast Plaza. Developer Edward DeBartolo, whose family at the time owned the San Francisco 49ers football team, was slated to build and own the center.[13]

The Irvine Company's plan for a retail center in the triangle was standard fare—a large mall with department stores as anchors and an entertainment area called

Tivoli Terrace. Not much had been done with the plan when the company was sold in 1977 to new owners, including Al Taubman, who had made a fortune building and operating retail malls. Taubman ordered the plans for the triangle set aside until he could put his own touch on them. Five years later, when Taubman and his other partners sold their shares to Donald Bren, nothing had been done at the triangle.

In a speech to the Urban Land Institute, Dick Sim described the genesis and evolution of how the Golden Triangle became first the Irvine Entertainment Center and, finally, the Spectrum Entertainment Center:

> We began by considering a site to locate perhaps four restaurants . . . for Spectrum employees. However, Rick Evans, our retail division president, realized that while the restaurants would have built-in lunchtime business . . . attracting diners in the evening would be more difficult . . . We began to consider an anchor that would attract customers in the evenings and on weekends . . . our first thought was a multi-screen cinema.[14]

Enter movie theater mogul James Edwards Sr., the owner of a chain of theaters all over Southern California, he had started in the 1930s. A diminutive dynamo, he had retired in his 80s but got bored and was back in the office every day, checking the receipts from the previous night's films.

Edwards lived in Newport Beach and had located his headquarters there next to two giant theaters across from Fashion Island ("Big Newport"), so he was a natural for the Irvine Company to seek out as a potential partner. According to Sim, he leapt at the chance. "Jim is ever the dreamer. As we evolved, he asked me, 'How can I open a 20-screen cinema, so near to the 21st century. So he decided on 21 screens including an IMAX theater.'"[15]

"Our demographic study of the Irvine Entertainment Center excited not only Jim Edwards, but quickly caught the attention of four outstanding restaurant operators as well." (Interviewed in 2024, Sim recalled that in its first year of operation, the 21-screen theaters generated $3 million in profit split equally between Edwards and the Irvine Company).[16]

With Edwards (age 89 at the time) on board, Sim listed six characteristics the company factored into consideration as to how the entertainment center would be designed and operated:

- Location—high visibility and traffic, good access.
- Demographics—high population numbers, accompanied by high disposable income.
- Major attraction—to draw people to the center.
- Universal entertainment—a variety of attractions and food offerings for all ages.
- Safety and security—Irvine was a safe city, but private security would be added.
- Viable tenants—seasoned merchants with quality reputations.

"Donald Bren envisioned the entertainment center as an escape, somewhere where you left your car behind and put a smile on your face."[17]

Bren had traveled to Spain and visited the Alhambra, the magnificent fortress/palace complex in Granada, built during the centuries when Islamic rulers occupied the country. Located on a plateau above the city, the Alhambra is a mystical collection of Islamic and Spanish Renaissance architecture, manifested in fountains, gardens, and soaring spaces richly decorated with tile and intricate carvings.[18]

Bren determined that the spirit of the Alhambra, combined with similar Moroccan-style architecture, would become the theme of the entertainment center. The cinema building was envisioned as a Moroccan palace, with a five-building restaurant/retail complex as the marketplace. Uniting the two elements is a grand plaza paved with intense colors recalling waves of wind-blown sand. Adding to the sense of North African adventure are palm trees, olive trees, banners, and portable Moroccan bazaar-like kiosks.[19]

Irvine Spectrum opened in 1995. Four years later, the entertainment center was expanded, adding more retail and restaurants. A giant Ferris wheel went up next to a large open area where an ice skating rink was installed during the Christmas holidays.[20] In a nod to the needs of thousands of apartment dwellers who were residing in new Irvine Company apartments across the street from the center, a Target

Figure 15.1 Alhambra Palace in Granada, Spain

Source: Author's image

department store was opened. And after years of complaints about parking from visitors who sometimes had to trek hundreds of yards to reach the center, parking structures immediately adjacent to shops and restaurants were built.

Lit up at night by neon and a giant Jumbotron-type sign, in the early 2020, the Spectrum has become a Southern California phenomenon. Visitors from all over

Figure 15.2 Influence of the Alhambra Palace at the Spectrum Entertainment Center

Source: Author's image

the southland make it a destination trip for the day, taking in a movie or two, strolling along the walkways to select lunch and dinner, then enjoying an ice cream and people watching into the evening.

Teens and preteens are deposited by parents with cash or credit cards midday and picked up as closing time approaches. In summer, the entertainment center is jammed with young people from every corner of a four-county area. And while there have been a few high-profile property robberies at Fashion Island and South Coast Plaza, the Spectrum has remained largely crime-free.

An estimated 17 million people visit the entertainment center annually.[21] As it has grown, there has been substantial turnover of restaurants and shops; the Edwards Theaters chain was sold to Regal Cinemas a few years after James Edwards Sr. died at 90.[22] Retail offerings largely reflect a youthful market, and only one major department store remains after the Macy's outpost closed.

Many people consider the Spectrum Entertainment Center Irvine's downtown. And virtually all who visit it likely fulfill Donald Bren's vision that they leave with a smile.

Notes

1 "Henry Segerstrom Biography," *HenrySegerstrom.com*, accessed September 12, 2025, https://www.henrysegerstrom.com.
2 *New Worlds Magazine*, "Industrial Neighborhoods and Walkable Design," May 1976.
3 "Industrial Neighborhoods and Walkable Design."
4 Former Irvine Company executive Robert Osbrink claims credit for naming the Spectrum.
5 Interview with Richard Sim, conducted by H.P. Oliver and C.M. Stockstill, February 29, 2024.
6 "Roots and Wings Exhibit," Irvine Company. 1985. (no longer in existence).
7 "Spectrum Infrastructure Investment Tops $227 Million," Orange County Business Journal, March 12, 1996.
8 "Irvine Spectrum Emerges as Tech Hub," *Los Angeles Times*, May 18, 1997.
9 "Irvine Spectrum Emerges as Tech Hub."
10 "Irvine Spectrum Emerges as Tech Hub."
11 "Irvine Spectrum Emerges as Tech Hub."
12 "Virginia Slims Tournament Draws Stars to Mission Viejo," *Los Angeles Times*, April 3, 1976.
13 "Mission Viejo Mall Proposal," *Los Angeles Times*, October 12, 1976.
14 "Spectrum Entertainment Center Opens," *Los Angeles Times*, November 22, 1995.
15 Richard Sim interview.
16 Richard Sim interview.
17 Richard Sim interview.
18 "Donald Bren's Alhambra Inspiration," *Orange Coast Magazine*, April 16, 1998.
19 "Donald Bren's Alhambra Inspiration."
20 "Ferris Wheel and Ice Rink Additions," *Los Angeles Times*, December 10, 1999.
21 "Spectrum Draws 17 Million Visitors Annually," Orange County Register, July 7, 2023.
22 "Regal Acquires Edwards Theaters," *Variety*, August 15, 2002.

16 The last coast

The four miles of coastline between Irvine Cove and Corona del Mar was the last major landmass on the Irvine Ranch to be planned and developed.

For most of modern Orange County history, it had been relatively remote, despite its location between the outskirts of Newport Beach and the tiny coastal enclave of Laguna Beach.

Historic photos show the Irvine family gamboling along the rugged shoreline. They had picked their way to the beach from one of the few spots where the 60-foot cliffs gave way to trails alongside El Morro or Los Trancos Canyons. Prior to the construction of Pacific Coast Highway (PCH) in 1926, they had likely arrived in a horse-drawn wagon from far inland, trekking down Laguna Canyon from the family home at the base of the Lomas de Santiago foothills, more than a dozen miles distant.

Around the time PCH opened, a tiny hamlet of tents had formed at the beach where Los Trancos Canyon met the Pacific Ocean. Dubbed Crystal Cove by a frequent visitor, weekend structures and campsites matured into more permanent abodes under the benign tolerance of James Irvine II. He had watched the area accumulate a beachside population, and as most people who arrived were either ranch employees or friends, he collected a modest fee for the properties and instructed one of his employees to oversee the area.

Crystal Cove and its residents were destined to play a large role in the future of the coast, but in the 1930s through the 1950s, news from the area fell into two major categories: automobile accidents and drownings.

Pacific Coast Highway was relatively flat and straight along the coastal plateau until it reached Los Trancos Canyon. There, it curved and dipped sharply, creating—for speeding vehicles—a potentially dangerous and sometimes lethal driving environment.

Newspaper stories began cropping up that described how cars had skidded, lurched, or flown off the highway and down the embankments into the tiny Crystal Cove community. Worse, cars going in both directions on the highway sometimes slammed into each other in murderous head-on crashes. In one particularly gruesome incident, three people were killed in a head-on crash.[1]

The ocean off Crystal Cove was abundant with marine life, attracting fishermen and divers. The waters were also filled with kelp forests, and newspapers began to

DOI: 10.1201/9781003683087-16

Figure 16.1 The beach at Crystal Cove State Park
Source: Author's image

report drownings due to entanglement in the kelp as well as swimmers who misjudged the power of the surf that crashed onto the beach.

James Irvine II apparently saw little value in his coastal properties. Upon inheriting the massive landholding at the turn of the 20th century, he sold off the northern section of Laguna Beach, including Emerald Bay (for unknown or undocumented reasons, he retained ownership of Irvine Cove, which was developed after his death). Historical speculation suggests that Irvine used the proceeds from coastal property sales to fund farm infrastructure in the rich fields and orchards of the central part of the ranch.

However, in the late 1920s, Irvine's oldest son, James Jr., gazed out at the empty coast and saw the potential for development.

In July 1927, the *Los Angeles Times* announced that Irvine Jr. and a number of business partners joined forces to create the Country Club Estates on "624 acres of the vast Irvine Ranch . . . situated south of Corona del Mar." The plans included a clubhouse, golf course, and polo field at an initial cost of $450,000 (almost $8 million in 2025 dollars).[2]

However, less than a year later, the partners were suing one another in a dispute over the potential sale of the company and land. Nothing ever came of the planned development.

More than 30 years later, it was Irvine's other son, Myford, then president of the Irvine Company, who announced that the company would team up with bandleader Freddie Martin to build a $5 million hotel at Crystal Cove. And again, after news of the planned development was reported in the *Long Beach Independent*, nothing happened.[3]

Cottages at Crystal Cove, a trailer park at Morro Canyon and a primitive horse stable near Corona del Mar remained the only three signs of civilization on the

Irvine Coast as the Irvine Company geared up for residential and commercial land development in the 1960s.[4] A few years later, company executives were travelling to Spain, Portugal, France, and Italy to see how the beach communities there might be translated into an Orange County version of St. Tropez.

The trip marked the beginning of a process that would take decades to transform the vacant coast and would not be completed until a single individual once again owned the Irvine Company.

***

After returning from the European trip in 1969, Ray Watson began to sketch out a strategy for a development plan along the coast. It combined large sections of open space along the narrow bluffs at the shoreline, overlooked by thousands of homes stair-stepping up the hillsides. And in the midpoint of the coast at Crystal Cove, he proposed a village inspired by what he had seen at the small Italian fishing village of Portofino.

Watson and company planners drew up maps showing a potential population of up to 50,000 people on the coast. As he explained in his oral history, the number was merely a placeholder, a means to estimate the amount of water, sewer, and transportation infrastructure required to serve the final number of homes and businesses.

Watson had convinced company managers that the land along the coastal strip had to be held in public ownership. Feelers were sent out to the California state park officials about a potential purchase. After an appraisal set a price of $8 million, the company discounted it and began to discuss how a sale might be concluded.

Joan Irvine Smith saw the proposed price and threw a fit, pointing out that beach acreage in nearby Huntington Beach had commanded upwards of $250,000 per acre, while the proposed Irvine Coast price was almost 90 percent lower. As was usually the case when Joan objected to a company action, she went to court to stop it.

As planning continued, events were underway that would recast the future not only of the company-owned land but also every mile of the California coastline.

The California Coastal Act was enacted by initiative in November 1972, establishing regional commissions as well as a statewide commission to regulate development and access along a roughly 1,000-foot strip from where the tides touched the shoreline to a point inland (but farther to the first ridgeline where there was hilly topography such as along the Irvine Coast.)

Creation of the state and regional coastal commissions and subsequent passage of implementing legislation upended what would have otherwise been a traditional zoning effort at the Orange County Board of Supervisors (the coast was still under county jurisdiction). Moreover, the fledgling environmental movement in the county had matured quickly, bringing local groups like the Laguna Greenbelt into the fray. Also, because the property was virtually the last major open coastal space in Southern California, the big guns of the environmental movement—including the Sierra Club—took notice.

The Irvine Company had opposed the coastal initiative but acknowledged that it would be the critical factor in coastal planning. To accommodate the new reality, Watson and other company executives decided on a risky strategy of bringing known and potential opponents of their plan to the table to hash out a compromise. Dubbed TICMAP (The Irvine Company Multi-Agency Planning Process), the result was disappointing. For two years, the process dragged on with little agreement. Finally, with the sale of the company looming due to the change in federal tax law, the plan was tossed up to the Orange County Board of Supervisors, who approved it in 1976.

It was a Pyrrhic victory: with the plan dead on arrival at the coast commission, it went on a shelf for the new owners to deal with after purchasing the company assets in late 1977.

***

It was not lost on company president Peter Kremer how much potential value the coast presented. A few months after he was hired, the existing coastal plans were dusted off and re-examined in the context of just what could be approved by the coastal commission. An internal team led by a crew of planners began to revise the previous Watson-era documents, significantly scaling down the concept of building more than 30,000 homes on the property. A lobbyist who had achieved success in gaining approval for other coastal developments was retained as was noted Canadian architect Arthur Erickson. Members of the team flew to Vancouver to see Erikson-designed homes and buildings, catching a glimpse of the aftermath of the Mt. St. Helens eruption on the way.

As the outlines of a new coastal plan began to emerge from meetings, white papers, and discussions among planners and policy advisors, another major initiative began to mature in Sacramento that would dramatically reshape the coastal planning process and eventual outcome.

State park officials had long coveted the Irvine Coast. Money had been piling up in the state treasury for park acquisition, not only for coastal property, but for other open space suitable for recreation and to protect wildlife. Nor had the allure of the last stretch of undeveloped shoreline gone unnoticed by the newly empowered California Coastal Commission and its young staff.

While there is no historical or legal evidence that ties approval of an Irvine Company coastal plan to an open space dedication or transfer, it was assumed within the company that letting go of prime coastal acreage would be the *quid* in an unwritten *quid pro quo,* allowing development to take place.

Private discussions with state officials became public on December 5, 1979, with the announcement that the California State Parks Department would purchase 1,896 acres of the coast for $32.6 million, at the time the largest amount ever paid for a park. The sale gave the state a strip of land from the shoreline to (roughly) Pacific Coast Highway encompassing the bluffs and the beaches along the four miles, extending from the homes in Corona del Mar to the rugged cliffs at Irvine Cove. The land purchase also included Morro Canyon at the southern end of the property, extending from the beach to the ridgeline nearly two miles inland.

The final element of the deal was an option for the state to purchase another 393 acres of a canyon for $3.25 million. (In 1980, the Irvine Company donated 500 acres of coastal canyon to the state, bringing the total park acreage to 2,791.)

A few months before the park purchase was announced, rumblings were heard in the county's environmental community about a plan to create a national park on the Irvine Coast.

It remains unclear 46 years later exactly where the idea originated, but history shows that two county members of Congress—Jerry Patterson, a Democrat, and Robert Badham, a Republican—joined forces to bring the concept to reality by introducing legislation in Washington, D.C. to fund federal acquisition of coastal property. Patterson and Badham, with Patterson's staff in the lead, believed the state purchase would boost their efforts. The two had plugged a $38 million number into their legislation but did not have specifics as to exactly what the money would buy. The Irvine Company, Patterson said, would not discuss the matter with them until the state sale had been consummated.

Hopes were high for eventual success, but Patterson and Badham soon found little legislative appetite for a massive allocation of funds for a park with such a nebulous definition. A $250,000 feasibility study was proposed, with the legislative vehicle for funding identified in a bill being crafted by Rep. Phillip Burton, chairman of the National Parks Subcommittee of the House Interior Committee.

Burton's interest in parks and recreation belied his persona. Tall, overweight, a heavy smoker and drinker, Burton wore a perpetual scowl on his face, was notoriously profane, and rarely spoke below a bellow, intimidating colleagues, staff, and visitors alike. In his decades in the House of Representatives, he had developed an intricate knowledge of procedural rules as well as an uncanny sense of how to create coalitions to achieve his unabashedly liberal goals.

If Burton had taken a hike in a forest or enjoyed a picnic on a beach anytime in his adult life, no one was aware of it. But he was a champion of parks and open space for his working-class constituents in San Francisco and had achieved success in building, piece by piece, a series of land acquisitions that were creating the Golden Gate National Seashore in Marin County on the north side of the Golden Gate Bridge.

Greg Sanders was serving as Patterson's chief of staff at the time. Interviewed in 2024, he recalled how a meeting with Burton set the stage for the demise of the national park.

> Burton wanted some due diligence on the proposed park, so he called the Irvine Company and representatives of the County of Orange to a meeting in his office in the Capitol. The attendees were Burton, Patterson, Peter Herman (an aide to Supervisor Thomas Riley), Robert Shelton, and Peter Kremer from the company, and me.
>
> It was 6 pm. Burton had his shoes off and was drinking vodka. He would point to various elements of a map that was spread on the floor of his office. Kremer was on his knees, explaining what was there. Burton saw a large red section of the map and pointed to it with his toes. "What is that?" he asked. Kremer explained that it was a hotel and golf course complex.
>
> "No fucking way, not in my national park," Burton said. Kremer replied that the hotel was not negotiable and rolled up the map, ending the meeting.[5]

Patterson was stunned. He managed to salvage $50,000 of the original request in the final House version of the bill, but it was eliminated in the Senate version. The coastal national park boomlet faded away and disappeared.

***

With the remnants of plans for a coastal national park left on a vodka-stained floor of Phillip Burton's Washington, D.C., office, the Irvine Company ramped up plans to file an application for development with the California Coastal Commission.

Gone were the hopes of multiple thousands of homes cascading down the ridgelines of the property as well as a resort complex at the heart of Crystal Cove. What made it into the final language of the application were 2,600 homes, many of them on large lots, a 1,500-room hotel above Pacific Coast Highway in the center of the property, a golf course, and up to 200,000 square feet of offices in buildings as high as ten stories.

In addition to building a major access road to the coast beginning near UC Irvine, a second major roadway was proposed at the southern end of the project area, an extension of Sand Canyon Avenue in Irvine, which at the time ended at the 405 Freeway. The company proposed a four-lane road, but the commission reduced it to two lanes. At the junction of the proposed Sand Canyon extension and Pacific Coast Highway, the company planned to build 25,000 square feet of commercial development as well as a 225-room hotel (Sand Canyon was never built).

In late July 1981, on an 8–1 vote, the coastal commission approved the project. Two months later, Friends of the Irvine Coast filed a lawsuit against the commission, the county, and the Irvine Company challenging the ruling.

Friends of the Irvine Coast had been formed five years earlier by Fern Pirkle. She had arrived in Corona del Mar in 1970 when her husband took a job at UC Irvine. Pirkle had a long history of activism in Louisville, Kentucky, where she marched against the Vietnam War and led the Louisville League of Women Voters and the Kentucky Civil Liberties Union. She continued her civic engagement in Orange County by joining the local chapter of the League of Women Voters.

While Pirkle is credited with providing much of the organizing and fundraising energy that made the Friends credible and influential, it also had something of a secret weapon in an aspiring young planner named Terry Watt.

Watt was the daughter of Jean Watt, who, along with Frank and Frances Robinson, was one of the most influential environmental leaders in Newport Beach history. Watt was the key founder of Stop Polluting Our Newport (SPON), organized initially to fight noise from the Orange County Airport. Ironically, the Watt family lived on Harbor Island, where Irvine Company owner Donald Bren also had a home.

Terry Watt was working as a planner when she sat in on a meeting with her mother and attorney Clem Shute. Jean Watt hoped to recruit Shute, whose firm specialized in environmental litigation, to assist with the airport battle. Terry Watt was "mesmerized by the idea of working for the firm and going to law school." Before long, Terry was volunteering for SPON and serving as the executive director of Friends of the Irvine Coast (she never went to law school).[6]

Terry Watt entered the contentious realm of environmental activism with an attitude that was at odds with her mother's. Jean Watt and her allies at SPON, in her daughter's estimation, viewed any development of any kind as entirely worthless. Alan Beek, who, along with Jean Watt, became one of the best known SPON advocates, referred to "any kind of development as quote, the destruction area," she said.

Terry Watt described her approach as more practical. "My bias was always, who do we need to get in a room with and have a conversation about this and sort it out?"[7] She got the opportunity to test her personal theories of how best to resolve environmental battles over the next decade as she worked for the Friends and their lawyers in San Francisco at Shute, Mihaly, and Weinberger.

It took more than two years for the Friends lawsuit to work its way through Orange County Superior Court, where it was rejected. Another two years passed as it was considered by an appellate court. But during the time the legal process continued, Donald Bren had taken control of the Irvine Company.

Bren asked former company president Ray Watson to return as a vice chairman, filling a role as something of an elder statesman and keeper of institutional history. Watson was pleased with his assignment but also wanted some hands-on responsibility—he asked Bren if he could take control of the process to create yet another coastal plan, one that could once and for all allow construction on the still-empty coastal acreage to proceed. Bren agreed—he had already concluded that a new coastal plan had to be developed.

After Bren gave the go-ahead, Watson recruited a planner, Jon Worden, and the two of them began brainstorming new approaches to the coastal strategy. Interviewed in 2024, Worden recalled that he had been hired by then-planning director Chuck Trevasian to "stir the pot, be an outside voice" in reviewing work by other planners. He had fulfilled that role earlier, he said, by revising a plan for a large lot coastal development.

Working in something of an "undercover" environment, Worden developed more comprehensive coastal plans for Watson, who then presented them to Bren.

Forty years later, precisely what Watson and Worden came up with remains unclear. One indication of the results of their work was that some of the fiscal assumptions on which previous plans were based were flawed.

Testifying in 1987 during the lawsuit to determine the company's value (see Chapter 11), Watson said he found that original cost estimates to build the coastal development were too low. "Watson said studies showed the initial (coastal) plan would have yielded the company a pretax profit of at least $40 million over a period of 20 years on an initial investment of about $240 million."[8]

On another floor of the company, development executives were plotting how to move forward to a final plan approval when and if the lawsuit was concluded. Carol McDermott had returned to the company to work on major entitlement issues just as Bren had decided to seek a rapprochement with Friends of the Irvine Coast. With Gary Hunt assigned to manage the high-level strategy, McDermott became the ground soldier and began meeting with Terry Watt to see if a compromise could be achieved.

In separate interviews, McDermott and Watt described how their personalities and attitudes about seeking an agreeable outcome built mutual trust. Both recalled meetings in comfortable, informal settings, including the occasional tea party. McDermott made sure to bring baked goods to early morning meetings. As time went on, the two women gravitated to each other, working out deals that each could take back to their principals. "Terry had a real good ability to focus on what they needed and trying to figure out a way to get what they wanted in the context of what we (at the Irvine Company) were doing," McDermott said.[9]

Fern Pirkle, McDermott said, was "closed, and not very friendly at all during the process," but "trusted Terry to carry out their objectives of retaining as much open space as possible, to make sure they didn't see development from Pacific Coast Highway, and limiting, of course, the number of units. So the business side of the company had developed the sense that we could live with 2,150 hotel rooms and 2,150 residential units."[10]

Achieving an agreement in 1987 was critical for the company: a coalition of anti-growth forces was gathering signatures to place a countywide growth limitation measure on the June primary ballot in 1988.

The settlement between the Friends and the company was signed on September 29, 1987. While the number of allowed homes was set at 2,600, the settlement required that additional open space be dedicated, resulting in 7,234 of the project's 9,432 acres remaining undeveloped. Also, the height of office buildings that had so infuriated the Friends was reduced from ten stories to seven (no office buildings were ever built in the project).[11]

A copy of the agreement is contained in the Friends papers in Special Collections at the UC Irvine Langson Library. It offers a fascinating glimpse into what the Irvine Company was willing to concede to make peace.

The price was small, $195,000. Of that total, Shute, Mihaly, and Weinberger received $85,000, about two-thirds of what they calculated would have normally been billed. But the company also agreed to pay the Friends $45,000 to compensate Terry Watt to continually monitor the implementation of the coastal plan during permitting, construction, and beyond. Balancing out that element of the deal, the company required the Friends to obtain legally binding promises from other major county environmental groups, such as SPON and the Laguna Greenbelt, that those groups would not sue to void the agreement.

As was often the case in dealings with opponents, if forced to concede money, the company tied payment to deadlines, ensuring, to the degree possible, there would be no backsliding or buyer's remorse that could squelch the agreement. Thus, the language noted that as of September 29, 1987,

> the sum of $195,000 was placed into escrow . . . from that deposit $15,000 will be released to the Friends on approximately January 29, 1988. All the remaining funds, including one-half of the accrued interest, will be released to the Friends 120 days following the LCP certification *and* approval of a development agreement for the Irvine Coast.[12]

The company didn't give Terry Watt a blank check. She invoiced the Friends, who, in turn, billed the company for her services, which then doled out compensation incrementally until the $45,000 total was reached. Thus, after receiving a total of $195,000, paying the law firm, and retaining money to enable Terry Watt to observe and comment on future company work, the Friends banked approximately $65,000 to use for ongoing activities and advocacy. And that is what they did the next decade.

***

With years of planning, analysis, revisions, public hearings, and lawsuits over, it was time to create a new community.

The landscape where it was to take place was daunting.

The Irvine Coast has two distinct geographies. The first is the shoreline and bluffs bordered by Pacific Coast Highway. The second is a series of canyons and ridges between them that meet at the crest of a hill stretching from the upper neighborhoods of Corona del Mar to a point about two miles south.

Analyzing the myriad of environmental and physical issues before them, Irvine Company planners could largely write off the first section; it now belonged to the State of California, which would have the responsibility of creating Crystal Cove State Park. (Controlling runoff into the ocean also had to be addressed, a matter that would lead to serious consequences in the future). Of the four major canyons, Morro Canyon was also part of the state park, off-limits to development, leaving Muddy and Los Trancos in the middle of the project, with Buck Gully at the northern boundary of coastal Corona del Mar.

To prepare the thousands of acres of land for new homes, resorts, a golf course, and a small commercial center, grading would likely be the most expensive and potentially sensitive task. Initial planning had identified, in rough fashion, determined largely by geography, four large sections on the land that would be transformed into neighborhoods.

The first was a small rectangular parcel immediately adjacent to PCH and Corona del Mar that would stand alone on the bluff facing the Pacific Ocean. It would be entirely residential, surrounding four holes of one of the two 18-hole golf courses destined to be built above the coastal road. A tunnel had to be built under PCH to accommodate construction and provide access to the golf course in the long run.

Two hillside communities would be built on the ridges adjacent to Los Trancos Canyon and Muddy Canyon, with the lower reaches of Los Trancos shown on maps as Pelican Hill.

Below Muddy Canyon's ridgeline, another community, reflecting the name Wishbone Hills, was planned. It would later be named Crystal Cove. On the crest of the coastal hills, an extension of San Joaquin Hills Road was built, and above it, just outside the coastal zone and the jurisdiction of the Coastal Commission, a separate community called Newport Ridge would ultimately be built.

Interviews with former Irvine Company employees and consultants who supervised the initial grading operations as well as data from Sukut Construction, whose bulldozers did much of the actual work, result in a variety of recollections about the magnitude and cost of the operation.

Joe Davis had been placed in charge of overall coastal development, with Bernard Maniscalco operating as the day-to-day project manager. Recalling the start of construction, Davis said in a 2025 interview that company leaders were weighing how to fund the huge and complex amount of grading without raising the ire of environmental groups.[13]

Davis said that had the project been

> graded incrementally, somebody at some point midway could file a challenge of some sort, and if it stopped the project, the adverse financial impacts on the company would be huge. So there was a deep gulp taking on the 9th floor of the building (the location of Irvine Company executive offices, including Donald Bren's), and the decision was made to do all the rough grading all at once to mitigate the potential danger of a shutdown.[14]

As grading for the residential and commercial areas began, the company discovered landslides, which necessitated the construction of retaining walls, thereby adding to the cost. The Sukut Construction website, which lists dozens of past projects, noted that on the two Irvine Coast projects it completed, a total of 32.5 million cubic yards of dirt was moved; for just one of the projects, Sukut was paid $53.2 million.[15]

Another element of the coast project requiring substantial grading was the construction of Newport Coast Road (called initially Pelican Hill Road). Winding from Pacific Coast Highway to Bonita Canyon Drive in Irvine, the road climbs along the edge of Los Trancos Canyon, where, as construction took place, an old landfill was discovered. To allow construction to continue, contractors built a fortified sleeve with a metal weight inside and crammed down the decaying garbage in the landfill until it was flat, then covered it with dirt. Jim Lorman, who worked with Davis and Maniscalco on the project, said that the average depth of asphalt and concrete on the road was 4 to 8 inches, rising to 15 inches at some points.[16] Estimates of the cost to build Newport Coast Drive vary from $75 to $90 million (equivalent to $160 to $190 million in 2025 dollars).

As the rough grading got underway, archeological sites were discovered, yielding fossilized scallop shells, evidence that the 1,200-foot-high ridgelines were once part of a prehistoric seabed. But when human remains were uncovered, work stopped.

By law, the county coroner had to examine any body in any condition found in such circumstances. After archeologists determined that the remains were probably 1,500 years old, the county official's visit was perfunctory. But after his analysis was complete, the law also required that living descendants of the Native Americans identify the remains and provide them with a dignified burial.

David Belardes and Jim Velasques presented the Irvine Company with government documents proving their lineage to the ancient tribes that had inhabited Orange County. Belardes had what the company considered the more compelling connection; he was the tribal chief of the Juaneno tribe based in San Juan Capistrano. Velasques traced his ancestry to the Navajo nation and was often at odds with Belardes and the Juanenos about the authenticity of their heritage claims. Velasques was adamant on the point, but as he was equally adamant about his claim that aliens had abducted him from outer space at one point in his life, it didn't help his credibility.

Under pressure to keep to a tight construction schedule, Maniscalco made it a priority to inform Belardes or Velasques immediately whenever remains were discovered. Quoted in a *Los Angeles Times* story about Native American burial sites, he said, "All human remains and funeral-related materials found on the Newport Coast site have been returned to Mother Earth."[17]

His description was accurate: when remains were found, either Belardes or Velasques were summoned to the location—if they were not already there. They clambered into the dirt, dropped to their knees with a trowel, and carefully removed the remains, placing them in a container for reburial in a remote location on the coastal property or somewhere else in the county.

In the first three years of coastal construction, the Irvine Company spent more than $6 million for archeologists and paleontologists to evaluate ancient geological, animal, and human remains.[18]

While it was an enormous relief for the Irvine Company to finally start work on the Irvine Coast, it came at a tenuous time.

The latter part of the decade of the 1980s was, for most real estate developers, a journey through a land of milk and honey after surviving exceptionally high interest rates of the earlier part of the decade. And for the Irvine Company, it had been especially sweet.

The national economy, fueled by lower interest rates, was churning out an abundance of jobs, mainly in the service and retail sectors. Statistics showed far more people being hired in healthcare and financial services than in manufacturing, which meant that the new office buildings the company was opening would soon be full.

In 1984, the company opened the Four Seasons hotel in Newport Center, bringing a heightened level of elegance and sophistication to Newport Beach. Across the street in Fashion Island, which was undergoing a multimillion dollar upgrade, the ultra chic (with prices to match) Amen Wardy boutique was crowded with the wives and girlfriends of real estate investors and developers eager to upscale their wardrobes.

The residential housing market exploded in the second half of the decade. In 1986, home prices in the county went up 9 percent, and a year later 13 percent. By 1988, the annual increase was an incredible 25 percent, followed by an 18 percent rise in 1989.[19] For the Irvine Company, it was a double bonus. Not only was the company reaping huge revenues from the sales price of land, the company was also taking a percentage of the price of each home when it was sold. Builders complained about the double dip into their revenues but were willing to pay up to build in an Irvine Company village.

And then, the party was over.

The nation slipped into a recession as the new decade arrived. Land sales, the lifeblood of Irvine Company revenues, slowed dramatically. The company reduced expenses and began raising funds by other means (see Chapter 25). But to fund the massive cost of grading and installing infrastructure on the coast, the company turned to the county and a series of public debt obligations.

Issuing tax-exempt bonds was nothing new; almost all such issues in the state were based on legislation dating to 1915. But in the aftermath of Proposition 13 in 1978, bond financing for large projects increasingly turned to assessment districts as well as the newly invented Mello-Roos Community Facilities Districts (CFDs).

Both financing vehicles could be established without a vote if there were no residents within the district boundaries (or, in the case of a CFD, fewer than 12). After the company established the Irvine Ranch Water District in the early 1960s, the first order of business was to sell bonds. And in the 1980s, the company created huge CFDs within the city of Irvine to provide funds for school construction.

Assessment districts were typically used to fund initial community improvements: water and sewer, utility systems, streets, curbs, and gutters. Given a choice, the Irvine Company preferred assessment district financing. So in 1988, a giant district, labeled AD 88–1 by county finance officials, was drawn up to cover the coastal property, authorizing $165,705,000 in bonds. A second assessment district for the coast was created in 1992, AD 92–1, and bonds totaling $51,845,000 were issued, bringing the grand total of publicly issued debt for the coast to $219,345,000 (equivalent to $545 million in 2025 inflation adjusted dollars).

The 88–1 and 92–1 districts reflected the total debt that could be issued, but the actual debt was allocated incrementally to individual assessment districts drawn up to reflect specific communities of homes. Over the life of 88–1, for example, more than 17 assessment districts were created under its authority.

When an assessment district was established, bonds were sold and the landowner began paying back the principal and interest. As residents moved into the assessment district, they, too, began paying off the remaining amount due on the bonds. Thus, the faster the development took place, the sooner the burden of repayment on the landowner was reduced or eliminated. (In 2001, the Irvine Company asked the county for permission to pay off early the debt on at least one assessment district, making it more attractive to sell homes in it).[20]

The $219 million paid for a great deal of the cost to build the coastal project. But none of it went to the design or construction of two 18-hole golf courses, or the 1,200-room resort, or the retail center, or the huge entry monument towers at the intersection of PCH and Newport Coast Drive. Nor was any of it spent on the lush landscaping and hardscape of the high-quality Donald Bren was known to favor or for the thousands of trees that were planted on the property (a number were planted, only to be removed after local regulators said they would negatively impact wildlife habitat).

Ask the people who worked on the coast plan and its implementation just how much in total was spent to create it, from assessment districts as well as directly by the Irvine Company, estimates run from $400 million to $600 million and perhaps even more.

***

While the market for residential housing was weak, one builder in early 1992 was ready to take a chance.

Bramelea of California had the honor of building and selling the first homes on the Irvine Coast. The homes were large, the smallest model just under 5,000 square feet. All the homes had three-car garages. The lot sizes started at just over a quarter of an acre and went up from there. Prices ranged from $1.4 to $2.2 million.[21]

Unfortunately for Bramelea, things didn't go well. Roberta Marshall, who had started work as a project manager on the coast a few months earlier, recalled in a 2025 interview that after building 30 of the planned 58 homes, the homebuilder went bankrupt. The Irvine Company, she said, built out the streets and landscaping, then sectioned off the remaining homesites and converted them to custom lots.[22]

Later that year, RecreActions Group of Companies (RCG) took the plunge on 111 semi-custom homes just up the street from the Bramelea project. Sales were strong, and by late 1994, Irvine Company real estate ads showed sales opportunities at three custom home locations, with lot prices ranging from $400,000 to $3 million; four neighborhoods of single-family homes in the heart of the coast, and six communities in Newport Ridge (outside the coastal zone) with prices there beginning at just $190,000 for the smallest condominium. Earlier that year, the company proudly announced that 500 homes had been sold on what was now being marketed as the Newport Coast.

(Before home sales began, company marketing executives tasked the government relations staff with obtaining a separate zip code for the coast. A meeting was arranged with Orange County postmaster Hector Godinez, paperwork was filled out, and in 1990, the zip code for what would become the most expensive neighborhood in Orange County debuted as 92657.)

Rudy Svrcek was in charge of marketing the custom lots on the Newport Coast. He built a large sales pavilion at Fashion Island to show off the stunning views that almost all the lots provided for the fortunate few who could afford them.

As coastal planning began, real estate gurus in Newport Beach speculated that many, if not most, of the buyers of the custom lots would come from outside of California and the United States. Visions of Middle Eastern royalty, Russian oligarchs, and Wall Street executives danced in the minds of those who considered such matters.

Initial sales proved otherwise. In September 1994, Svrcek reported that 161 lots had been sold. The majority of the buyers were local (Orange County or Southern California) business owners, 23 percent were physicians, 20 percent senior executives, and 6 percent were attorneys. More than 60 percent of the buyers paid cash for their lots.

Interviewed in 2024, Svrcek recalled one memorable lot sale.

Apparel designer Mossimo Giannulli approached Svrcek in 1996 with an offer to purchase six custom lots; he wanted to combine them and build a family compound. Svrcek quoted a price, but Giannulli countered with a request for a discount on such a large sale. Svrcek had to check with Donald Bren.

Svrcek said he brought a map of the properties to Bren's office and explained Giannulli's request for a discount. Bren asked the price, and Svrcek told him. Bren mused for a moment, Svrcek remembers, then said, "Maybe we should be asking more."[23]

(*The Los Angeles Times* reported on February 25, 1998, that after business reversals, Giannulli put the six lots he had purchased in 1996 for $10 million back on the market, asking $12 million.)[24]

***

It was 13 years after grading and initial construction began on the Irvine Coast that the Irvine Company announced plans to build a resort and country club in the middle of two 18-hole golf courses.

Famed golf designer Ted Robinson had been sought out to envision the courses, sited just above Pacific Coast Highway. The location would offer golfers spectacular, sweeping views of the Pacific Ocean, with the potential to create fairways leading to greens with iconic vistas rivaling those on two famous Monterey coast golf courses, Pebble Beach and Spanish Bay (company planners visited both).

Robinson roughed out an approach, but it was not to the liking of Donald Bren. "The Robinson plan filled in too many canyons for Bren's taste," recalled Norm Witt, a project manager on the coast development plan. This reflected the green side of Bren "that few people appreciated."[25]

Bren then connected with Tom Fazio. Witt said that Fazio was reluctant to take on a public course on the West Coast, far from his North Carolina headquarters. Interviewed June 1, 2008 by the American Society of Golf Course Architects, Fazio said he was "forced" into the job. "I really didn't want to design those Pelican Hill courses, but Ueberroth would not take no for an answer."[26] (Peter Ueberroth, at the time, was a member of the Irvine Company board of directors).

Fazio came to appreciate the challenge of the unique geography the two courses presented and came up with a solution. The two courses—Ocean North and Ocean South—are consistently acclaimed in travel and golf publications as nothing less than world-class.

(Interviewed in 2024, Richard Sim said the cost of the first course was $18 million, a price that Bren considered too high. Sim said he took charge of the second course construction and brought it in for far less, in large part by reducing the number of trees. Sim also claims credit for renaming the courses, when the Ocean course became overcrowded because everyone wanted to play on it to the detriment of the Links course. Sim's solution: rename the courses Ocean North and Ocean South.[27]

Similar praise has been heaped upon the Resort at Pelican Hill, the 504-acre hotel, spa, and golf center located just off Newport Coast Drive. When plans were first announced in mid-2003, the architectural theme was envisioned as Spanish Revival, something a traveler might see in Santa Barbara. But when the time came for Donald Bren to describe the resort that many would call his masterpiece, he turned to a Renaissance architect he had become increasingly enamored with: Andrea Palladio.

"Applying Palladio's distinct style, the resort's design goals focused on the simplicity of form and geometry, as well as balanced landscape architecture," reads the resort's promotional description. Trees became a signature element of the landscape design, with 750 mature olive trees planted throughout the resort, creating an image that "recalls the manicured maturity of the countryside of Tuscany or Umbria."[28]

Other trees in the landscaping palette include huge Canary Island palms, as well as Italian cypress, Italian stone pines, Moreton figs, citrus, and Aleppo pines. The hardscape elements of the resort are described as

> courtyards, or auto courts, featuring custom paving tiles . . . laid in intricate Italian patterns. The main and largest auto court is modeled after the Piazza del Campo in Siena, Italy . . . conceived as the main square, the Coliseum Pool is the resort's heart, hub and alluring pleasure forum.[29]

(The Coliseum pool has achieved legendary status in the telling and retelling of the Pelican Hill Resort story, first because at 136 feet across, it is one of the largest—if not the largest—circular pools anywhere in the world. The second reason is the tale repeated by former Irvine Company employees that when all or nearly all of the pool's 1.1 million hand-cut glass mosaic tiles were installed, Bren gazed on his creation and gave it a Caesar-like thumbs down. All the tiles were removed and replaced to meet the owner's critical eye.)

A walk through the Pelican Hill Resort offers an immersion in luxury set in the architectural context of a man born in 1508, channeled through the eyes and exacting standards of a man born in 1932. A visitor might be forgiven if, after waking in one of the well-appointed villas and strolling to breakfast along a terrace paneled with limestone and hand-carved fountains, he or she believed they had been transported to a villa in the Italian countryside.

And Bren's scrupulous attention to detail is reflected in the artworks, rugs, and tapestries in the grand lobby interior of the resort he personally selected. On the night the resort opened, Carol McDermott remembers celebrating there with other company employees and consultants. One of them told her that Bren had been spotted there the day before, inspecting electrical outlets; he wanted to be sure the screws affixing them to the walls were all precisely lined up parallel to the floor.

***

As coastal development continued, Irvine Company executives began to take a closer look at what would likely be the final section of the giant property to be developed.

Adjacent to the Morro Canyon section of Crystal Cove State Park, a large bluff overlooking the beach had originally been planned for custom lots. However, market projections indicated that demand for houses on the coast was so strong that revising the project to focus on individual home sales would generate greater profits at an accelerated pace.

Around the same time, residents of Crystal Cove, as well as members of environmental groups that monitored ocean issues, began to report concerns about the quality of the water running off the coastal properties, focusing intently on the amount of fertilizer that might be mixing with golf course runoff.

The 1998–99 rainy season was heavier than average, almost 20 inches. A substantial storm would produce a robust gusher of runoff, turning the waters off the state park muddy. There was also suspicion that runoff from the Pacific Coast Highway was contributing to what critics saw as a toxic brew that could harm sea life.

The Irvine Company public relations staff was soon fending off charges that the coast project was polluting the sea. It didn't help their cause when a professor at nearby Orange Coast College claimed to have documented a dolphin birthing site a few dozen yards off the Crystal Cove beach.

When the company project managers applied for permits from the county to build 635 homes in the project, it now labeled Crystal Cove, they discovered that the county's authority to process the application was no longer valid because of what was essentially a paperwork error. Seizing the opportunity to assert jurisdiction for a project many coastal staff members and commissioners never liked anyway, the commission ruled that the company would be required to bring its plans to them. Also joining the fray was the Environmental Protection Agency (EPA), alerted to the water quality issue by press reports.

For almost a year, the Irvine Company replanned the drainage collection and filtration system, not only for the 635 new homes, but for the two golf courses as well. Next to Pacific Coast Highway at the base of El Morro Canyon, a site that had been planned for a hotel was selected to accommodate a huge settling basin to collect runoff. The basin is more than 600 feet by 400 feet in size, and more than 40 feet deep.

Joe Davis believes it cost $25 million to, as he put it, "replumb the water and runoff system" to meet the water quality standards set by the California Coastal Commission and the EPA.

***

When the state purchased Irvine Coast beaches and canyons for the new Crystal Cove State Park, it also acquired two residential enclaves.

The first was at Crystal Cove, where, as previously noted, over the decades, a small group of families had quietly transformed 43 modestly built shacks into a community of relatively sturdy homes. The second settlement was located at the southern end of the coast, at Morro Cove, where approximately 258 mobile homes were arranged along the shoreline, as well as on bluffs above PCH (access to the beach was provided by a tunnel). The owners of the trailers were, in large part, newer to the community and more affluent than those in Crystal Cove. When the state purchased the land for the park, knowing it would take years to plan and build it, the residents of what was known as the El Morro Community Association were offered a 20-year lease for their individual units.

A similar leasing arrangement was brokered at Crystal Cove. Residents there had resigned themselves to their eventual eviction (it took place in 2001), when state officials announced a plan to lease the cottages to a developer.

Michael Freed, who had built a similar project in the Big Sur coast, won a 60-year lease on the property, agreeing to spend some $35 million to repair and rebuild the sewer system, upgrade the condition of the cottages and add a pool and restaurant. In return, Freed would rent out the cottages for a predicted $375 per night. But before Freed could gain traction for the planned improvements, opponents raised a host of objections, lobbying state and local officials to cancel or modify the agreement.

And then Joan Irvine Smith joined the fray.

In the latter years of her life, Smith had become an environmental advocate, speaking out on a variety of issues and devoting long elements of her biography to subjects like climate change, open space preservation, and her perennial favorite, water supply.

Smith's historical connection to the coastal property, her star power, and willingness to appear at public meetings to heap criticism on the Freed deal soon reversed public opinion and motivated state park officials to reconsider their entire approach to Crystal Cove preservation. By early 2001, the state paid Freed $2 million to break his lease.

Five years later, the state put out a bid to manage rental and upkeep of the Crystal Cove cottages. Only one bid was received from the non-profit Crystal Cove Alliance, which is organized and led by Laura Davick, whose family had owned one of the cottages. The Alliance raised its own funds to pay for cottage restoration, but the state—largely using bond monies—has steadily allocated millions of dollars to repair the site and rebuild the cottages. In 2025, the price of an overnight stay at one of the cottages ranges from $150 to $375.

Residents at El Morro did not go quietly when their leases ran out.

The first deadline was in 1999. What followed was, according to the *Los Angeles Times*, "lawsuits, legislative hearings, hot tempers and pleas for more time," all of which combined to buy the holdouts another seven years of coastal residency. It ended for them on March 1, 2006, when the state began removing the remaining trailers and turning the area into a campground for tents and recreational vehicles.[30]

***

The City of Newport Beach annexed the Newport Coast on the first day of 2002. Water and sewer service transferred to the Irvine Ranch Water District, which agreed to pay the city $25 million over more than a decade to build out the necessary water supply network.[31]

Celebrities began locating on the coast, including author Dean Koontz, who combined several lots to create a 30,000-square-foot home. (Koontz sold it for $50 million in 2020, relocating over the hill to Irvine's Shady Canyon).[32] Kobe Bryant was perhaps the best-known celebrity to reside there until his accidental death in early 2020. There are various reports, all unconfirmed, that other famous and near-famous individuals purchased homes there, including Tiger Woods, former KISS leader Gene Simmons, and self-help guru Tony Robbins.

Crystal Cove State Park notched 1,092,000 visitors in the 2022–23 fiscal year.[33] The cottages at Crystal Cove are sold out year-round, and the Beachcomber Cafe at the center of the community is always busy. The Crystal Cove Alliance has a small bookstore on site and organizes exhibits about the unique settlement's history as well as a host of educational and artistic programs.

The continuing wow factor is, of course, the elegance and extravagance of the homes that have been built there and their astronomical prices. Custom homes

Figure 16.2 Restored cottages at Crystal Cove

Source: Author's image

with all the over-the-top architectural, design, and landscaping accouterments are usually priced in the $40 million and up range. In 2025, the highest price ever paid for a home on the Newport Coast was $61 million for 46 Deep Sea, six bedrooms, 14,500 square feet in the Crystal Cove development, overlooking the Pacific Ocean. Ironically, the home built on the largest lot on the coast—Villa de Lago—faces inland, with views of a deep, verdant canyon. The 17,000-square-foot home features a small lake, tennis courts, an organic garden, a 14-car garage, and a conservatory on its 12.4 acre parcel. It was originally built as a speculative investment and was last listed for sale at a price of $55 million.

The Pelican Hill Resort remains a serene island of luxury and exclusiveness, although anyone can play golf there ($425 a round for the public, $320 a round for guests). Guest suites start at $1,500 per night and rise to $6,000 per night for a four-bedroom villa located a few yards above the main resort complex.

***

Who better to have the last word on the coast than Ray Watson?

The diary he kept during his first year as a junior planner at the Irvine Company mentions a trip up to the rugged high point on the western edge of the ranch, Signal Peak. It would have afforded Ray and his colleagues a sweeping vista of the empty coastal acreage touching the vast Pacific Ocean. Was his inspiration for an eventual community there sparked on that trip?

Or was it a few years later, during the adventure along the southern coast of Europe, motoring through dozens of picturesque towns and villages with white-washed walls and rust-colored tile roofs, crammed tightly together on steep slopes that looked down on a tiny harbor of sparkling blue Mediterranean waters?

As president of the company, Watson had struggled to define a workable plan for the coast but failed. He received a second chance, developed an approach, and then handed it off to a new generation of employees who ultimately pushed the monumental effort over the regulatory goal line.

Asked to sum up all that had preceded the coastal effort, and considering the outcome, his view was succinct: "In the end," he said, "the coast is Don's baby."[34]

Notes

1 "Three Persons Killed in Head-on Crash," *Anaheim Bulletin*, January 9, 1950, p. 1.
2 *Los Angeles Times*, July 1927.
3 "Report on Proposed $5 Million Crystal Cove Hotel," *Long Beach Independent*, June 26, 1956.
4 The spelling of Morro Canyon is often Moro Canyon on maps of various origins.
5 Greg Sanders, interview by C. Michael Stockstill, 2024.
6 Terry Watt, interview by H. Pike Oliver and C. Michael Stockstill, March 26, 2024.
7 Watt interview.
8 *Los Angeles Times*, August 12, 1987.
9 Carol McDermott, interview, H.P. Oliver and C.M. Stockstill, March 20, 2024.
10 Carol McDermott, interview.
11 Friends of the Irvine Coast Papers, "Settlement Agreement Between Friends of the Irvine Coast and The Irvine Company," September 29, 1987, Special Collections, UC Irvine Libraries, Langson Library.
12 "Settlement Agreement Between Friends of the Irvine Coast and The Irvine Company," September 29, 1987.
13 Joe Davis, interview, H.P. Oliver and C.M. Stockstill, April 10, 2025.
14 Joe Davis, interview.
15 Sukut Construction, "Project Summaries (Irvine Coast)," accessed September 13, 2025, https://www.sukut.com/completed-projects/.
16 Jim Lorman, interview, C.M. Stockstill, October 29, 2024.
17 *Los Angeles Times*, September 23, 1990.
18 *Los Angeles Times*, September 23, 1990.
19 California Association of Realtors, "Historical Home Price Data," accessed September 13, 2025, https://www.car.org/marketdata/data/.
20 Author's recollection.
21 *Los Angeles Times*, November 26, 1991.
22 Roberta Marsall, interview, H.P. Oliver and C.M. Stockstill, April 14, 2024.
23 Rudy Svrcek interview, H.P. Oliver and C.M. Stockstill, June 6, 2023.
24 *Los Angeles Times*, November 26, 1991.
25 Norm Witt interview, H.P. Oliver and C.M. Stockstill, January 16, 2025.
26 American Society of Golf Course Architects, "No One Likes to Be Turned Down—Tom Fazio," June 1, 2008, accessed September 13, 2025, https://asgca.org/no-one-likes-to-be-turned-down-tom-fazio/.
27 Richard Sim, interview, H.P. Oliver and C.M. Stockstill, February 29, 2024.
28 *Los Angeles Times*, June 21, 2003.
29 The Irvine Company, The Resort at Pelican Hill, accessed January 17, 2026, https://pelicanhill.com/.

30 *Los Angeles Times*, February 27, 2006.
31 *Daily Pilot*, January 29, 2000.
32 Orange County Register, June 18, 2020.
33 Crystal Cove Conservancy, "Financial Information," accessed January 17, 2026, https://crystalcove.org/financial-information/.
34 *OC Metropolitan*, December 1, 1992.

17 Villages and villagers

The residential enclaves planned, designed, and built by the Irvine Company evolved over the six decades of Irvine's existence.

As first envisioned by Ray Watson and his colleagues in the early 1960s, the village served as the foundation for a comprehensive plan for an entirely new city. Drawing on concepts developed by planning experts and philosophers, the Irvine Company created a set of guidelines to follow, among them:

- Each village will have an identity.
- The physical nature of the area will be reflected in the siting and development of villages.
- Innovative designs will be incorporated into the village.
- Villages will have a range of housing types, at different prices, to encourage diversity.

Additionally, in almost all cases, distinctive edges defined villages, some of them heavily landscaped, reinforcing the identity of one from another.

As this book is written, Irvine has 21 villages designed and built entirely by the Irvine Company. There is also one that existed either before or at the time the company's general plan for the area was approved by the County of Orange, a few years before Irvine incorporated as a city.

While two of the communities built on land owned by others are fairly large, in a city whose population now exceeds 300,000, the Irvine Company villages define the city's architectural, design, and sociological identity.

Each village is in its own way unique, but four villages exemplify how their creation reflects the evolution of a city of villages. They are:

- University Park
- Woodbridge
- Westpark
- Woodbury

Additionally, there are three areas with one-of-a-kind elements that merit briefer attention: Tustin Ranch, Shady Canyon, and residential development in the Irvine Business Complex (IBC).

DOI: 10.1201/9781003683087-17

Inherent in the story of the villages is the passage of time, especially a few critical points where economic, political, financial, and personal circumstances combined to alter the village concept conceived by Watson and his early-day planning compatriots.

Figure 17.1 Location of selected Irvine Ranch villages and the Irvine Business Complex (IBC)

Source: City of Irvine with village names added by the authors

University Park

Driving into University Park today presents the visitor with a vision dominated by beige walls, red tile roofs, and swaths of greenery. Leave a vehicle behind to walk or bike, and the greenery expands into acres of man-made grasslands that meander between homes.

The greenbelt defines University Park. It was imported in large part from Eastbluff in Newport Beach, the first village the Irvine Company built in that city. As in Eastbluff, University Park also featured homes built as condominiums, with attached walls, a concept almost unheard of in Southern California in the 1960s.

For some University Park residents, the greenbelt became their backyard. And while there was typically room for a couple of patio chairs next to the back wall of their house, gone was the large backyard had they bought in Huntington Beach or Fullerton. And for some new owners, the yard on each side of the home was gone. They had purchased a home with a "zero lot line," which placed the boundary of their lot perpendicular to the wall of their home on one side. Doing so increased the size of the side yard on the *other* side of the house (written into the deed was an agreement that neighbors would allow access to each other's property for maintenance).

Common areas and zero lot lines, along with other design innovations, meant confusion for county planners and zoning administrators. Issuing permits became difficult, prompting the Irvine Company to cooperate with county officials in creating a new set of building guidelines that accommodated the changes. Planned Unit Development (PUD) guidelines would become standard for nearly every succeeding development in the city. After Irvine incorporated, a second iteration of the concept was enshrined in a set of design standards hammered out between the company and city planners—the Urban Design Implementation Plan. The new rules recognized the unique nature of Irvine Company-created communities, all of which feature landscaping and amenities far more generous than those in surrounding older cities.

An equally radical idea came with the attached homes of University Park—land that was leased, not sold.

A reflection of the almost biblical desire of Irvine Ranch owner James Irvine II that his property remain forever in his ownership, the Irvine Company began offering homes for sale with the underlying lot available for purchase or lease. Many customers chose to lease because it dramatically lowered their mortgage payments. Left to deal with 25 years later was the potential for a higher annual payment when the property was re-appraised. (Had Irvine been alive, the purchase option would have likely not been offered, but by the mid-1960s, the increasing desire of homebuyers to own their land forced the company's hand).

Streets in University Park meandered, creating a landscape of neighboring homes at angles to one another. A driver might see a backyard fence facing the street, curving away onto a greenbelt, instead of row upon row of front yards, neatly arrayed on a straight line typical of tract homes in virtually every other community in Orange County at the time.

A key element of the University Park design was a community center with a large pool and tennis courts owned by the residents as members of the mandatory homeowners association (HOA). This Irvine Company commitment to a generous shared recreational amenity would become a signature element in every village that would follow, as would the insistence on a resident-run community association to maintain the greenbelts and community centers. While the concept of an HOA is now accepted in almost all new planned communities, in the 1960s, it was an idea as radical as shared open space.

The density of University Park is low, about five units to the acre. Moreover, there is only one apartment complex within the village boundaries, Parkwood Apartments, with 296 units, located at the far southern end of the community. Directly across Culver Drive from the first section of University Park, the Irvine Company built a large apartment complex, Park West. Even though it bordered the new community, it was never marketed as part of the village or integrated into its design.

(Built in 1977, Park West is the oldest apartment complex in the city, and at 880 units, until recently, the largest. The complex's architectural design had been awarded to a small firm in Los Angeles owned by Frank Gehry, who had served in the U.S. Army with company planner Al Trevino. Gehry had designed the home-finding center for the New Town of Columbia. William Watt, who was in charge of building Park West, had seen that structure and was impressed with its appearance, he recalled in a 2025 interview. In an inkling of the future innovative—even radical—designs Gehry would become famous for, he had come up with a concept of installing movable storage units for the interiors of the apartments. Watt said the idea was too unusual for Irvine Company executives, and it went nowhere.)

University Park and Turtle Rock were the first two neighborhoods built in concert with the opening of UC Irvine in 1965. Turtle Rock's geography and design made it the more expensive of the two, and many people who were either affiliated with the university or wanted to live in a university-oriented community decided to live in University Park.

And with a university came professors and administrators who were Jewish.

In the mid-1960s, Orange County was just coming out of a period of strong conservative political attitudes. The John Birch Society's popularity was fading, and the sometimes shrill anti-communism espoused by some elected officials had muted. Nevertheless, academics organizing the new faculty and curriculum at UC Irvine heard murmurs of concern about "radicals and Jews" coming to work at the new campus.

For its proximity to the university as well as other factors, University Park began to attract Jewish families, initially a group of Hasidic Jews from South Africa. As the years passed, two synagogues and a Chabad center were located in University Park. Turtle Rock began to see more Jews move in as well, and a 2003 study conducted to locate the Samueli Jewish Campus about a mile from the University Park/Turtle Rock boundary found that "eighty percent of Orange County's Jews were within fifteen to twenty minutes of the campus!"[1]

All of University Park and the neighboring Rancho San Joaquin community are enclosed in an *eruv chatzerot,* a symbolic ritual enclosure that allows activities to take place during Shabbat that would otherwise be forbidden by religious law. The *eruv* is marked with a thin monofilament line strung along signs and trees and is maintained by Temple Beth Jacob.

Figure 17.2 University Park greenbelt and townhomes

Source: Author's image

Home prices in University Park hover just above $1 million in all neighborhoods, with the most recent built more than a decade after the initial enclave along Michelson Drive at Culver. There are two small shopping centers at either end of the community. National chains originally owned the grocery stores at both locations, but today the shops are ethnically oriented.

Combining greenbelts, street medians, parkways, and parks (public and private), fully 25 percent of the total acreage of University Park is open space.

Woodbridge

At just over 1,700 acres in size, Woodbridge is the largest village in Irvine. But beyond its size, Woodbridge has, for decades, been widely acknowledged as the physical manifestation of the most creative and lasting elements of Irvine Company planning.

Located in the center of the city and built on flatlands that were once the heart of the giant Irvine Ranch agricultural empire, Woodbridge united the southern section of the city—UC Irvine, University Town Center, Turtle Rock, and University Park—with the pre- and incorporation-era communities of Deerfield, The Ranch, El Camino Village, and Culverdale.

The site of the village is an immense set of fields that slopes imperceptibly to the San Diego Creek, a natural drainage that bisects central Irvine. From an engineering standpoint, it was a gift—grading would be minimal, streets could be laid out with few impediments, water and sewer pipes would fall neatly into place.

Figure 17.3 Woodbridge north lake and the wooden footpath bridge

Source: Author's image

For the planners, the large, featureless landscape presented a significant marketing challenge: how to create a compelling image for an entire community that would lure buyers to a city less than a decade old?

The solution became iconic—two man-made lakes on either side of the San Diego Creek, with a stylish wooden bridge crossing each one to name the new community.

Recreation became the operative theme for Woodbridge, with dozens of community pools and small parks connected by trails—not greenbelts—scattered throughout the community. Each lake had a beach club and tennis courts available to all residents who belonged to the community-wide homeowners association. The village has two large athletic complexes, which combine baseball diamonds and soccer fields at opposite ends of the community.

Planners created a design theme of river rock and timber walls for the entire village, giving a rustic cohesion to every neighborhood. Setbacks along Yale Loop, the circular roadway that united the entire village, were generous and heavily landscaped. Many of the entry streets connecting to Yale Loop wind slightly as they curve into the interior. Hidden there, the homes are lined up on a traditional grid. The effect is a softened urban landscape where structures, houses, plants, and trees blend together.

Along Yale Loop and through the middle of the community are two design creations that were never duplicated in any future Irvine Company village.

The first was a series of homes created by Ken Agid, at the time the Irvine Company's chief advertising and marketing guru. Designed to present an image of style and affluence, homes along Yale Loop (the Estates) were built on a mound elevated about ten feet above the street. They were also grouped as duplexes or triplexes, each unit individually owned but with a common facade and architectural detailing. The result was the illusion of large, stately mini-mansions one would have expected to see in Pasadena or Beverly Hills. And to maintain the unbroken landscaping spreading out from each home, garages were set behind, accessed from a small alley. While most of the Estates were built in the first of four quadrants that would make up the village, the concept was continued to a lesser degree in the other three.

At the other end of the economic spectrum from the Estates were the first and only for-sale units built to satisfy a new city policy for inclusion of moderate-income homes. Scattered throughout Woodbridge, dozens of townhomes were constructed and sold to individuals or families that met a strict income/assets formula. After Woodbridge was completed, the company and the city reached an agreement to satisfy the lower cost obligation with units built in the apartment complexes the company was increasingly locating in new developments.

For all the planning, design, and landscaping successes embodied in Woodbridge, one element that comes closest to being labeled a failure extends from one end of the village to another.

The Activity Corridor was located between Alton and Barranca Parkways, which crossed the village on an east/west alignment on either side of San Diego Creek (which had been improved into the first major flood control channel in the city). The idea was simple: locate most of the retail, small commercial, educational, and social services into the very middle of the village, in theory just a short walk away for much of the population.

The Irvine Company kickstarted the Activity Corridor by building the Woodbridge Village Center overlooking the South Lake at Barranca Parkway. Anchored by a large grocery store, other uses included a bank, restaurants, a bakery, and a stationery store. Offices suitable for dentists, lawyers, or other professionals were created on the second story of the center. In 1980, a five-screen movie theater—the city's first—would be located there.

While initial business activity at the center was brisk, in time it began to level off. The mom-and-pop stores went out of business, and a few professionals took the top floor offices.

Moreover, the remainder of the Activity Corridor remained largely inactive, with the exception of a trend for churches to buy property there. High land prices in the other areas of Irvine that kept churches out of the city were beginning to become a public relations problem for the Irvine Company, which responded by discounting parcels in the Activity Corridor for religious groups, sweetening the deal with a generous loan package.

The headquarters of the Irvine Unified School District, Woodbridge High School, and medical office buildings followed churches and synagogues. In time, a variety of other uses, including two senior housing complexes, filled in the acreage. After the new owners of the company took charge in late 1977, the decision was made to never again locate retail in the center of a village; it was placed on one of the major north/south streets where traffic volumes are heavier and visual exposure is greater.

It took more than a decade to complete Woodbridge, with the final quadrant in the southeast sector slightly denser than the remainder of the village. Overall, 24 percent of the village is in parks and open space, with a gross density of 5.4 homes to the acre. The 1,250 apartments in the village account for 13 percent of the total units.

Figure 17.4 Woodbridge brochure site map

Source: Orange County Archives

Homes on South Lake in Woodbridge are selling for nearly $4 million, while those less than a quarter mile away, which were once moderate-income units, are now worth over $600,000.

Westpark

Donald Bren placed his aesthetic and business imprint on the first village built after he assumed control of the Irvine Company.

Like other villages, Westpark would display a strong architectural and design identity. As in Woodbridge, open space and parks were an integral element of the village, narrowly situated between Culver Drive and San Diego Creek.

But Westpark would set a new urban pattern for the villages that followed: much higher density and many, many more apartments.

For Bren, it was the second opportunity in his business career to shape an entire community. The first was his bold agreement with the O'Neill family to create Mission Viejo on lands that, like those of the Irvines, traced their origins to the Spanish and Mexican ranchos of the 1700s.

At just 31, in 1963, Bren was president of the Mission Viejo Company, carving new neighborhoods into the rolling foothills along Interstate 5 about ten miles south of the Irvine Ranch boundary. Bren was able to give Mission Viejo a look that echoed its historic mission-era heritage, but turning a profit proved elusive; he sold his interest back to the O'Neills and returned to his homebuilding company.

But Bren Homes was largely a production builder, churning out mid-priced three-bedroom homes aimed at first-time or move-up buyers. And while Bren was building in the master-planned community of Valencia, he had no input on that New Town's design. Bren had even missed a chance to build in the first phase of Woodbridge (before the 1977 sale of the Irvine Company to the partnership he was part of) because he had been unwilling to agree to the sales terms the company insisted on.[2]

As planning for Westpark began, Bren sent marketing executive John Martin and his newly hired vice president for planning, Roger Seitz, to Europe on a mission to absorb the atmosphere of Mediterranean communities. It was that look that Bren had decided on for his first village, one distinctly different from the prevailing Irvine design style of Woodbridge.

Bren even selected the cities his executives would visit: St. Paul de Vence, Port Grimaud and Tropez in France; Portofino in Italy; and Porto Cervo in Sardinia.

Westpark's dimensions are much more modest than neighboring Woodbridge and came with two unusual elements on its borders. The first was the community of Culverdale, built before any Irvine Company village in the early 1960s on a small parcel of land hugging Culver Drive and the San Diego Freeway. The other element was Irvine City Hall.

Company lore held that the land for Culverdale was sold in the early 1970s when cash was scarce, resulting in a project that, for decades, was orphaned, surrounded by empty fields. A similar project (the Racquet Club), also on Culver Drive, five miles to the north, was likely built on land sold for the same reason.

The only real opportunity to blend Culverdale's traditional look into Westpark was to entice homeowners to accept the company's offer to rebuild their property walls along Culver in a whitewashed plaster with a new design fancier than their existing cinder block.

The anchor for the new village's western boundary was a 50-acre parcel of land designated for the city hall and an adjacent community park. It was the fourth and final city hall location the city had considered since its incorporation in late 1971.

Company planners saw the site selection as a complementary design opportunity if the city would select an architectural look for the new civic complex in line with the Mediterranean vision they were busily sketching out for the new village.

The city, however, was not persuaded and took one of the few opportunities it had to demonstrate a degree of independence from the giant landowner by choosing a design that was modern and stark; three stories tall, with a huge glass facade and a concrete-block exterior. Some observers likened the building's appearance to that of a prison.

Stuck with a non-conforming institutional structure at the border, Bren focused on the interior of the new village. Like Woodbridge, streets would meander and curve, and in almost every section of Westpark, have a generous setback. At the village entrances, the open space at either side of the entry street was ever grander than in Woodbridge, featuring tall, mature date palms standing guard at each corner. Indeed, throughout the village, palm trees would dominate the landscape palette, with bougainvillea and bird of paradise interspersed.

The walls were plastered in a variety of pastel colors that matched the design of the homes, all of which featured red tile roofs. Reflecting the increased density of the new village, a few homes were front-loaded, with their garages barely six feet from the street.

No single development exemplified the new look of Westpark more than the Corte Bella condominiums. Hal Lynch, the president of RecreActions Group, had seen one of the presentations Martin had given to builders describing what his team found on their European trip. Lynch was so inspired, Martin said, he arranged his own tour and invited Martin along. The result was Corte Bella. Martin described why he believed it was a success:

> Corte Bella was innovative because attached homes, clustered around motor courts, had not been built (in Orange County). . . . Lynch attributes the images from the Mediterranean trip as inspiration for the design breakthrough at Corte Bella, which incorporated subtle Italian flavors into the site plan, architecture, and landscape design . . . Innovation in planning had taken linear buildings plotting on an uninteresting alley and sited them on romantic and functional motor courts.[3]

Architect Ernie Vasquez had teamed up with RecreActions to build Corte Bella and said,

> Instead of the ubiquitous and often ugly carports that are common to high-density projects, the units at Corte Bella will have tuck-under garages, with

> direct access to the townhouse . . . Californians love their cars, so we started our design concept with a two-car garage footprint.[4]

According to John Martin, the Corte Bella concept was the perfect combination of creative design and business success, as the 23-unit to the acre project increased the land residual profit to levels substantially higher than the company had ever experienced.[5]

While Corte Bella challenged existing design traditions, it was the multiplicity of apartments in Westpark that broke the mold from previous company villages and would set the standard for villages that followed.

There were seven apartment complexes in the first phase of Westpark, totaling 2,650 units, just over 51 percent of the entire number of homes in the village. They were a mix of two- and three-story projects but with a stunning array of architectural design elements and upgrades that still stand out more than three decades after they were built. Irvine Company planners and architects also used the large stucco walls of the apartments to experiment with a variety of pastel colors, splashing ochre, marigold, rose, and pink hues on large expanses as well as to highlight smaller nooks.

While Westpark was winning awards in the homebuilding industry for design, architecture, and planning execution, community acceptance came slowly. The initial rainbow of bright and not-so-bright colors of the Old World was tough for many residents to accept after a decade of beige. The reaction may have played a role in the resistance to the second phase of Westpark, which stimulated a referendum on the project in 1991, the first time an Irvine Company–owned project in Irvine had gone to the public for a vote.

In addition to a general objection to growth and traffic, the second phase of the village bordered about a mile of right-of-way owned by Southern California Edison. While all the electric utilities within the village had been undergrounded, along the right-of-way stood 60- and 100-kilovolt power lines atop towers that loomed over the landscape. Before long, Irvine Company and Edison officials were faced with assertions that radiation from the lines might be harmful to future residents.

In a November election, the utility and landowner bested the dissident residents but not by much. The contest all but ended organized citizen opposition to village development, with future concerns about growth turning to the El Toro Marine Corps Air Station.

Westpark grew into a village where, despite its density and accumulation of apartments, it presents a surprisingly open appearance, with grassy parks and mature trees softening the concentration of homes and the commercial centers at its borders. As was the case in previous villages, the Irvine Company made a concerted effort to build houses for sale at a variety of price levels.

Woodbury

This village is an almost perfect square, 700 acres in size, and it is the angles of a square, sharp and well-defined, that give Woodbury its identity.

Figure 17.5 Mediterranean architecture, bold pastels, and lush landscaping define Westpark
Source: Author's image

The team of architects that created the village

> envisioned Woodbury to be a series of traditional, smaller-scale neighborhoods within the context of several large, multi-use Districts. This neighborhood structure of higher-density attached housing defines critical edges, transitioning to lower-density single-family detached housing at the two core neighborhood parks. A central village park serves the entire community, and includes an athletic club and a variety of play fields that are within walking distance of every home.[6]

Looking at an aerial view of Woodbury, three key elements stand out. First, at the dead center of the village, the large park, including a pool and the playing fields previously noted. To the west of the central park, another unique element of open space, the Jeffrey Spine (see Chapter 7), runs the length of the entire property, forming a linear park along Jeffrey Road. At the western corner of the village,

Figure 17.6 Entry tower at Woodbury

Source: Author's image

Figure 17.7 The City Stroll in Woodbury

Source: Author's image

another open space, but not usable by residents, is the catch basin for urban runoff, part of a system created by the Irvine Ranch Water District. The last noticeable element of open space is a series of small parks scattered throughout the interior of the village, each one with a theme tied to landscaping.

At the eastern corner of the village is a 50-acre shopping center. It is highly unusual in the planning history of the Irvine Company, as it is the only village retail center with a direct connection to the village it is part of.

Woodbury Center has the traditional frontage at the intersection of two large streets, Irvine Boulevard and Sand Canyon Avenue. There is a main, signaled entrance off Irvine Boulevard, as well as Sand Canyon. But the other entrance is located at the western side, at the confluence of three streets that meet each other not in the preferred alignment of traffic engineers, but at 45-degree angles. The longest of the three streets features a promenade of towering palm trees, paired together, stretching from the central park to the retail center. Former Irvine Company vice president of planning Roger McErlane described an epic internal battle between planners and traffic engineers to achieve the unconventional design as well as with executives in the retail division, who advocated for a design totally focused on the automobile.

Woodbury's entrances feature 60-foot-high brick towers, built in a style described as neo-Romanesque or Renaissance, depending on the planner or architect writing about them. Anyone driving by might think they had taken an unexpected time-warp turn to a church or villa in Florence or Turin in the time of Leonardo da Vinci.

Interior streets in Woodbury conform to a series of classic grids, straight and true to the lines the planners drew. But entry streets have generous setbacks, and the landscaping along them and throughout the entire village is lush even by previous Irvine Company standards. Double rows of trees are everywhere, and along the windrows, thick shrubbery hugs walls separating homes from the street. When the village was first built, trees came in 60- and 80-gallon sizes, which meant faster growth into fuller, nearly mature specimens.

Woodbury came to life just after the 2008 economic catastrophe that staggered the national economy and sent homebuilding into deep doldrums in California.

Rather than wait for a recovery, the Irvine Company initiated what was labeled the Executive Builder system. Traditionally, a builder would purchase lots on an incremental basis, or if they had faith in the market, an entire section of land at a time. Homes would be built and sold as fast as teams of contractors could be employed.

But the aftermath of 2008 meant no bank would lend money for land purchase or construction to even the most experienced or successful builders.

However, Donald Bren would.

Under the Executive Builder program, the Irvine Company selected a handful of builders and paid them a fixed fee to construct homes. The company assumed the market risk and reaped the reward of the sales price; the builders had virtually no risk, just do what they always did, build a solid structure on an agreed-upon schedule. It was a win-win and led the Orange County homebuilding community out of the economic downturn.

Of the approximately 4,000 units in Woodbury, all of which are medium or medium-high density, 1,581 are apartments, resulting in an overall density of 8.2 units per acre. (Irvine Company planning documents combine Woodbury and the adjoining Stonegate village as a single unit for numerical purposes.)

Tustin Ranch

At just over 1,900 acres, Tustin Ranch is the largest village ever built by the Irvine Company; it is also the longest, the narrowest, and perhaps the least representative of the planner's dream of a cohesive development with a distinct identity.

About a mile wide and just over six miles long, the village is an amalgamation of homes and apartments touching a golf course, existing older neighborhoods,

Figure 17.8 Row houses in Woodbury

Source: Author's image

and a ridgeline of rugged open space between Jamboree Boulevard and the historic western boundary of the Irvine Ranch.

The most distinctive element of the village with the most interesting planning history is the Marketplace, the first "big box" development undertaken by the Irvine Company.

David Mudgett was in charge of the retail division when plans for the Marketplace evolved. As negotiations with large retailers like IKEA and Home Depot started, the question of the appearance of a collection of huge buildings was under discussion as well. "So the challenge is, what does it look like? (Local officials) could have nixed the project because they didn't like the looks of it," he recalled in a 2025 interview.[7]

Mudgett said that in determining the overall appearance of the center, "a big box building has a lot of shade and sunlight on it. So, how do you get these planes that have no design features to them? How do you create something that . . . is aesthetically pleasing?"[8]

Roger Seitz, the company's vice president of planning, made the bold decision to hire Mexican designer Ricardo Legorreta to give the hundreds of square feet of massive walls an identity.

Figure 17.9 Intense colors at Tustin Marketplace in Tustin Ranch

Source: Author's image

"Legorreta, who is known for dealing with light and shade . . . came up with a palette of colors that were very vibrant, and in fact was the best palette for the sun and shade," Mudgett said.[9]

The colors selected were unlike any ever seen in Orange County, stunning, vibrant shades of yellow, purple, red, and a deep ochre. As it turned out, purple was the shade that put the Tustin City Council over the edge, led by Councilwoman Ursula Kennedy, who railed against the project until the Irvine Company decided to alter the look slightly.

> While the terra-cotta on the buildings remained intact, the Irvine Company agreed to repaint the freeway-fronting wall from deep purple to the basic reddish-brown. As a show of civic harmony, the repainting ceremony was led by Kennedy and company planning chief Seitz.[10]

Another unusual element planned but never fulfilled was a laser light show designed to emanate from a series of tall pillars that dotted the 4,000-space parking lots in the center. Budget considerations were cited most often as the reason.

Today, the brilliant hues of the Marketplace draw little to no comment from shoppers or merchants; the biggest change is taking place as elements of the outlying ring of buildings are being replaced by apartments (see Chapter 26).

Shady Canyon

The Irvine village that blends most effortlessly into its natural surroundings is almost invisible to nearly every resident of Irvine.

Shady Canyon has two entrances, one completely hidden from view, the other sheltered a few yards from the intersection of two roads on the periphery of the city's urbanization border. At the base of the low-slung San Joaquin Hills paralleling the San Diego Freeway, the Quail Hill Shopping Center would appear to be at the end of Sand Canyon Avenue (and the only traffic circle in Irvine). But a motorist continuing southwest on Sand Canyon to the crest of the hill finds an entry gate to a landscape dotted with rocks, sagebrush, and a scattering of homes.

Shady Canyon spans just over 1,000 acres, featuring 400 homes and homesites scattered throughout its ridges and valleys. And while the 300-acre golf course is larger than most similar designs, not a single home is built on it (a handful overlook the course).

SWA, responsible for dozens of Irvine Company projects in past decades, said in considering its plan for Shady Canyon,

> SWA's concept was to minimize grading, work with the site's natural topography, and preserve the rustic, unspoiled character of beautiful coastal canyons, creeks and ridgelines.[11]
>
> The [Irvine Company] worked closely with environmental organizations and [government] agencies to . . . protect and perpetuate biological diversity. The landscape architect [plan] ensured that the goal of preservation was carried into all areas of the project by using native plants, hydroseeding extensive areas of native grasses and coastal sage scrub, preserving significant rock outcroppings, and respecting unique landforms.[12]

Figure 17.10 Homes and golf course in Shady Canyon

Source: Author's image

Built out in the early 2000s, Shady Canyon is a stunning achievement unlike any other area in Irvine or Orange County, for that matter. Tom Fazio, who designed the golf course, said it was among his best career offerings; *Golf Digest* in 2021 named it one of the 40 best golf courses in California, the only one in Orange County, besting the Fazio course on the Newport Coast.

But it is less the challenge of playing the golf course than how it integrates into the natural environment and how a relatively small community of homes blends into a landscape that, to the naked eye, remains largely unchanged since the days of the Early California ranchos. Residents remark on the solitude of a place within a few short miles of an interstate freeway and one of California's only toll roads.

Irvine Business Complex

The plans for the Irvine Industrial Complex (IIC) were clear: a location for manufacturing and businesses to support it ONLY. No exceptions.

The initial acreage of the IIC grew around what was in the 1960s Orange County Airport at the western edge of the Irvine Ranch. Access was problematic until 1966 when the San Diego Freeway punched past Red Hill Avenue and connected with the Santa Ana Freeway.

Before long, large, low-slung buildings with names like Rockwell, Bertea, and Edwards began to pop up. Allergan, a locally-grown pharmaceutical start-up, birthed a headquarters building and a series of lab structures. The IIC was on its way to fulfilling its planned role as one of two huge job centers at either edge of the city of Irvine.

As it did with residential villages, the Irvine Company imposed a set of strict rules defining uses in the IIC as well as the appearance of what was built there.

Many lots were large and well-landscaped; there was no barbed wire fencing and the color palette was restricted as was the style and size of signage.

In early 1985, a group of Texas investors purchased a parcel on Von Karman Avenue at Kelvin Street. It proposed building the first apartments in the IIC, hiring Geoff LePlastrier to make it happen. Interviewed in 2024, Le Plastrier said that by the early 1980s, pressure was building on cities like Irvine to provide additional housing. Council members, unconcerned about the strict IIC limits, signaled a willingness to allow housing in the industrial area. (By that time, office projects had begun to creep into the IIC as well).

Le Plastrier's biggest hurdle was to obtain approval of 65 percent of the IIC landowners to allow the residential incursion. Considering that virtually every decision-maker he needed to convince was a man, he recalled that "I'm embarrassed to say that I hired the most beautiful, voluptuous woman, which I couldn't get away with today, to knock on doors" and obtain the signatures.

She got the approvals. The result was the construction and renting of the Charter Apartments, a "four over two" design of underground parking below 4 stories of 430 units (the complex is now named Sofi Irvine).

The race to transform the Irvine Industrial Complex—just renamed the Irvine Business Complex (IBC)—was on.

By 1992, the *Los Angeles Times* was reporting on an effort by Grimmway Development to build 87 apartments "surrounded by toxic chemical users," including a carpet mill and laboratory. Despite the developer's promise to install alarm warnings in the event of a toxic spill, it was a bridge too far for the city planning commission, which nixed the plan. William Rice, the project proponent, was apparently mystified by the decision, saying, "Other residential projects in the Irvine Business Complex are doing quite well."[13]

He wasn't wrong; by 2006, a city staff report identified 3,628 apartment or condominium units in the IBC and another 3,763 under construction, and yet *another* 6,600 units being processed by the city planning department.[14] Of the projects under construction, the largest was a massive 1,380-unit complex at the corner of Jamboree Road and Michelson, aptly named Central Park. It was across the street from the former headquarters of the Fluor Corporation, which had sold off its giant property to save itself from bankruptcy in 1999. While much of the headquarters building was saved and reimagined for office space, elements were still available for new construction, leading Bosa Development to build two 18-story residential towers there.

The city was in catch-up mode in the IBC; the July 2006 staff report cited previously explained that "a number of factors have prompted the IBC transition from a predominantly low-density office and industrial park to an urban, mixed-use center, including:"[15]

- A strong residential market demand combined with a lack of available land for new housing.
- An existing stock of aging, low-density tilt-up industrial buildings.
- A key location at the confluence of several freeways and regional circulation systems.

Figure 17.11 Apartments in the Irvine Business Complex

To provide some semblance of order to the runaway development, the city adopted a Vision Plan for the IBC that sought to bring some basic amenities to the area, principal among them, sidewalks. The staff noted that when the IBC was initially planned, large streets to move trucks and employees meant no consideration for pedestrians. The streets were long, the blocks were large. Now, with residential projects being "inserted" willy-nilly throughout the area, the potential to achieve a walkable community was exceedingly low.[16]

As the former IIC, now the IBC, continues to transform in 2025, upwards of 16,000 to 18,000 residential units continue to spread throughout not only the Irvine portion of the IBC but the area that is within the City of Newport Beach (which annexed a hunk of the property well before the incorporation of the City of Irvine).

The low-slung two-story look of the pioneering Charter Apartments is long gone; every new complex is four stories at a minimum, sometimes as high as six stories. Some new sidewalks are being added but not in a fashion that ties every residential building to another. Additionally, a dream from the 2006 Vision Plan for "a system of pedestrian linkages, parks and urban open spaces (to) link to the San Diego Creek and San Joaquin Freshwater Reserve, (plus) a pedestrian Creekwalk system envisioned adjacent to the San Diego Creek is nowhere to be seen.[17]

Until early 2025, citizen activism in the residential neighborhoods of the IBC did not exist, but when plans were unveiled to tear down thousands of square feet of existing office buildings for a massive set of warehouses, the previously silent population descended on City Hall with the same anti-growth objections familiar in previous citizen uprisings, largely centered on potential traffic impacts.

Notes

1 "Eighty Percent of Orange County's Jews Within Fifteen to Twenty Minutes of Campus," *Los Angeles Times*, January 26, 2003.
2 Doug Gfeller, phone interview by H. Pike Oliver and C. Michael Stockstill, August 19, 2020.
3 John Martin, "Inspiration from the Mediterranean—Doubling Land Values with an Attached Motor Court Project: The Corte Bella Story," Martin and Associates LLC, 2020.
4 "Architect Ernie Vasquez Builds Corte Bella," *Los Angeles Times*, February 5, 1989.
5 Martin, Inspiration from the Mediterranean, 11.
6 JZMK Partners, "Woodbury Masterplan," accessed September 9, 2025, https://jzmkpartners.com/project/woodbury/.
7 David Mudgett, interview by H.P. Oliver and C.M. Stockstill, February 3, 2025.
8 Mudgett interview.
9 Mudgett interview.
10 "Marketplace Repainting Ceremony Led by Councilwoman," *Los Angeles Times*, April 7, 1991.
11 SWA Group. "Shady Canyon," accessed September 9, 2025, https://www.swagroup.com/projects/shady-canyon/.
12 SWA Group. "Shady Canyon."
13 "Toxic Spill Concerns Halt Apartment Project," *Los Angeles Times*, January 11, 1992.
14 City of Irvine, Community Development Department, Staff Report CFM149.1 (2006), residential unit counts for the Irvine Business Complex (existing, under construction, and in process). Unpublished City of Irvine staff report, on file with the City Clerk.
15 City of Irvine, Staff Report CFM149.1 (2006).
16 City of Irvine, "Irvine Business Complex Vision Plan," 2006, accessed September 9, 2025, https://www.cityofirvine.org/irvine-business-complex/ibc-vision-plan/.
17 City of Irvine, "Irvine Business Complex Vision Plan."

18 Swords to plowshares

Transforming military bases

The A4 Skyhawk fighter jet eased into position, its engine emitting a droning whine. Heat shimmered on the runway. Just over two miles away, the Lomas de Santiago foothills formed a dusky wall against the horizon.

At just 2,200 feet, the modest mountains were a potential impediment to the massive C-17 cargo carriers that sometimes lumbered into the skies above Marine Corps Air Station El Toro, but this jet jockey would rocket over the hills, the base, and the pull of gravity—moments after takeoff.

The pilot responded to the military air traffic controller's permission for takeoff with a crisp, "affirmative, sir," then glanced for the final time at the instrument panel. Within seconds, fully fueled and armored, the A4 screamed into the sky.

The Marine Corps pilot was on a training mission, learning how to land one of the nation's most sophisticated and deadly fighter jets on a small patch of runway marked to mirror the deck of a US Navy aircraft carrier.

The carrier deck outline was a minor spot on the 10,000-foot-long runway, a distance roughly equivalent to half of Central Park in Manhattan.

Pilots repeatedly flew what became known as the racetrack pattern to hone their landing skills on the asphalt and concrete, a much more forgiving training ground than a tiny island of steel pitching in the rolling waves of an ocean.

The racetrack commenced as the A4 hurtled into the sky. Climbing swiftly, the pilot's view revealed the Santa Ana River canyon winding back through Riverside and San Bernardino Counties and the sprawl of suburbia that covered them. Just before banking his plane west, the pilot glanced over his right shoulder at the new communities that had grown up to El Toro's borders: Lake Forest, El Toro, Leisure World, and Mission Viejo.

Swinging into a long roll, the A4 glided over a landscape dominated by row upon row of mature Eucalyptus trees, arrayed like a medieval army of pike-bearing soldiers. These were windrows, planted at the turn of the century to diminish the fierce Santa Ana winds and protect the blossoms of thousands of citrus trees. There were still orange and lemon orchards growing between the windrows, but they were methodically being cleared to make way for homes and businesses in the new city of Irvine.

Now the pilot was trying to skirt the edge of development along Jeffrey Road, heading south by southwest—the Pacific Ocean was visible just 10 miles distant,

DOI: 10.1201/9781003683087-18

Figure 18.1 An A-4 Skyhawk fighter jet at Marine Corps Air Station El Toro (1982)
Source: Courtesy U.S. Marine Corps via Wikipedia

sparkling above the skyline of Newport Center. Depending on the time of day, he might catch sight of an Air California Boeing 727 landing at Orange County Airport a few miles north.

Now came the critical approach to the runway. Dipping the A4, the pilot made a tight arc, passing over Lion Country Safari—an open-air wild animal park—and the confluence of Interstates 5 and 405, which formed one of the region's most valuable properties, the so-called Golden Triangle, owned by the Irvine Company.

The plane's engines thundered as it flew lower and lower. On occasion, motorists uninitiated to the racetrack pattern would bolt upright, first in fear, then in amazement, as one of America's mightiest warplanes roared past their windshields, occasionally close enough for pilot and driver to recognize each other's human form.

The pilot's target was now in sight. Focusing intently, he guided the jet over the green fields at the end of the runway, an area shown on planning maps as the Crash Zone. Flaps down, slowing to landing speed, the A4 dropped onto the white ribbon of aggregate with a thud, leaving behind another patch of burned rubber atop hundreds of others, hopefully within the dimensions of the landing zone.

There was no time to gloat over success or curse a miss. As quickly as the plane had landed, the pilot revved the engines and soared back into the sky. Another "touch and go" was notched into the records. On the nearby freeway, an auto sped by with a bumper sticker reading: "The Sound of Jet Noise Is the Sound of Freedom."

***

The nearly 5,000 acres of land that comprised El Toro MCAS was at the far eastern border of the Irvine Ranch. It was appropriated by the federal government in 1943, along with 1,600 acres of land on the western edge, for the Santa Ana Naval Air Station Lighter Than Air (LTA) project. The military bases would form the bookends to the giant ranch's midsection, taking some of James Irvine II's most productive lima bean fields.

Irvine died in 1947, still hoping he might reclaim the properties the government had paid him for, but by then, both bases had rooted into the ground, militarily and physically, adding infrastructure, buildings, and personnel that portended an extended stay. (Both bases were also leaking toxic wastes into their respective footprints, a problem that would eventually cost the federal government millions of dollars.)

As William Pereira surveyed the Irvine Ranch with his planning eye in the early 1960s, the twin bases at the ranch boundaries were far removed from any potential urbanization. Most of Pereira's focus was on the 10,000 acres immediately adjacent to the new University of California, Irvine, he was laying out, the area he hoped would become the first element of the New Town of Irvine.

A decade later, it was the newly incorporated City of Irvine that was charting a long-term plan, using the one devised by the Irvine Company as a starting point.

Both the company and city plans acknowledged the presence of El Toro and created growth scenarios anticipating that the bases would either remain in perpetuity or someday be closed down. But the distant and unknowable future of the base mattered less than the daily reality of noise emitted from the aircraft taking off and landing there.

Irvine was the community nearest the base, most impacted. But other areas had also been forced to deal with the consequences of noise.

In late 1976, there was a contentious hearing at the Orange County Board of Supervisors where home development in the foothills east of the base was imperiled by a county study that had identified restrictions calculated as a result of jet noise.[1]

The board was decidedly pro-growth, but faced with limits imposed by the county's own noise study, the supervisors created a potential loophole for developers, allowing them to conduct their own noise study in hopes of bending the limit line.

Twelve years later, development pressures had accelerated, and the Marines were now deeply engaged in land-use issues at the edges of their property—and beyond.

Early in 1988, a dispute that had been enmeshed in litigation for six years came to a head in Laguna Hills, where the developer of Leisure World, the large retirement community, reclaimed 180 acres of land, eliminating a Marine Corps deed restriction attached to it.

The property was a corridor designed to accommodate an aircraft on approach to El Toro, where a pilot might experience "some sort of in-flight trouble (needing) an option to try to set the plane down," said Marine colonel Jack Wagner.[2]

The Marine Corps agreed to waive the deed restriction in return for a payment of $8 million from Leisure World owner Ross Cortese.[3] The Marine Corps planned to use the money to build 100 units of housing at the base, which at the time had a waiting list of 1,500 families needing a place to live (ironically, the units were later built by Irvine Pacific, the Irvine Company's in-house home and apartment building arm).

Two years later, it was the County of Orange agreeing to a complicated land swap and purchase with the Marine Corps. Land was purchased adjacent to El Toro, as well as at the Tustin LTA base, to accommodate critical road construction in Lake Forest and along Interstate 5 near the junction with Interstate 405 (San Diego Freeway). The sale generated more than $17 million, which the Marine Corps again said would be used to build housing at the Tustin LTA base.[4]

***

Monitoring and battling boundary incursions and struggling with a persistent lack of housing for Marines and their families did not stop El Toro from operating at full strength, carrying out its "national defense mission on an around-the-clock basis."[5] It had been the case following the end of World War II, then intensified as the Korean War and later Vietnam ramped up the need for Navy and Marine Corps air actions. And as new versions of fighter aircraft were designed and built by companies like Douglas Aircraft and Lockheed, squadrons of the jets found their way to El Toro for pilot training. At its peak, there were 6,500 personnel at the base.

Recalling their time at El Toro, reminiscing veterans described it as a posting where military discipline was a bit more lax than at bases with longer histories. "Fewer salutes" was the way one former officer remembered it. He also recalled enjoyable encounters with young women from Laguna Beach and Newport Beach who were welcomed at the Officers' Club for drinks and dancing (the club interior was used for the film *Top Gun*).[6]

Few Irvine residents visited the base unless they were veterans with privileges at the Post Exchange. However, once a year, for two or three days, El Toro welcomed hundreds of thousands of visitors to its annual air show. Local residents didn't have to be reminded of the date of the aerial extravaganza, as the fabled precision flying team—the Navy Blue Angels—began practice runs over northern Irvine a few days before the big event.

And it was big. The attendees watched the nation's most sophisticated aircraft as they screamed over the massive landscape of the base. Some years, there were even re-enactments of Marine ground assaults, featuring helicopter drops and explosions.

***

Initial development in Irvine centered around UC Irvine and expanded towards the San Diego Freeway, where the Irvine Company developed the villages of University Park and Turtle Rock. North of the freeway, there were small pockets of homes that had been built before the city incorporated, followed by the Irvine Company's massive village of Woodbridge, which extended more than a mile in width from Culver Drive to Jeffrey Road.

It was the alignment of Jeffrey Road that marked what would become a critical and contentious boundary: the Community Noise Equivalent Line (CNEL) and its accompanying noise thermometer.

Jet noise from the base varied with the type of aircraft and the takeoff and landing pattern. When jets departed from the longest runways, which were generally aligned north-south, noise was relatively minimal in Irvine. When jets landed on the runways, residents of the retirement community in Leisure World bore the brunt of the impact.

If weather or maintenance issues were at play, the Marines switched takeoffs and landings to a shorter runway that ran east/west, sending departing planes directly over the northern section of Irvine. And when the racetrack pattern was used for aircraft carrier landing practice, noise would generally follow the entire circular path of the aircraft. And when that happened, complaints about the noise followed.

Recognizing that noise and residential development were incompatible, military officials and federal environmental planners had developed a system that identified noise levels on a geographic basis and plotted lines on maps of air bases to locate them. These became the CNEL.

The CNEL formula considered a variety of factors related to noise: how often flights took place, their duration, whether they were during day or night, and how the weather would impact the resulting jet engine blast. Using a numerical range, 65 decibels became the identification of a community equivalent noise level (CNEL) under which residential development would not be permitted. Thus, the "65 CNEL line" entered the planning and community vernacular in Irvine and surrounding communities. In Irvine, it looked like a circle that had been squeezed into a rectangle, with curving corners at the top and bottom coinciding with its overlay of the El Toro runways.

The CNEL only documented noise, not safety. El Toro had rarely experienced aircraft accidents; the worst took place when a giant cargo plane, struggling to clear the Lomas de Santiago east of the station, didn't make it, killing all aboard. Fighter jets had gone down only a couple of times during El Toro's lifetime, first in 1974 when a jet lost control shortly after takeoff, forcing the pilot and his passenger to eject before the plane crashed into the horse stables on the base. The second crash came in 1988 during the annual airshow when a jet plunged into the base chapel.

Eight years after the city had incorporated, jet noise had become an increasingly pressing issue for Irvine's new leaders to wrestle with. In June 1979, Councilwoman Mary Ann Gaido was calling for the city to sue the Marine Corps. "What I'm concerned about are the operations that have far exceeded routine operations and constitute an environmental insult to our city," she told the *Los Angeles Times*.[7]

The Marines acknowledged that at the time "Under orders from the Navy command . . . two A-4 jet fighter squadrons at El Toro were ordered to take intensive carrier landing practice runs."[8]

Those runs prompted more than 1,007 complaints from residents during a two-month period.

A year later, after a series of negotiations, the Marine Corps and the city adopted a memorandum of understanding that limited the circumstances under which the runways nearest Irvine could be used for takeoffs and landings. City leaders called it a win, while acknowledging that had a lawsuit been filed against the Marines, the potential for prevailing over national defense needs was remote.

It wasn't just the city experiencing conflict with the base and its operations. In late 1981, the Irvine Unified School District (IUSD) identified five possible sites for new schools in the fast-growing district and sent them to the California Division of Aeronautics for review.

Building schools to keep up with home development was critical to the Irvine Company and to elected leaders in the city. The district's reputation as a place with excellent teachers and facilities was a key selling point for homebuyers with young families.

So school leaders were surprised when the state officials rejected all the sites, pointing to the flights of KC-130 air tankers at El Toro, which, according to department spokesman William Bronte, "need every inch of runway to take off."[9]

***

As might be expected, operations at El Toro were closely followed by the Irvine Company.

On a personal level, senior executives in the government relations department made it a point to develop and nurture relationships with the base commander and senior military staff. As command of the base routinely turned over every two years, meeting, socializing, and occasionally negotiating with a new brigadier general became routine.

But however positive and cordial the relationship became, it rarely resulted in any concrete changes to Marine Corps policy. One of the Irvine Company's Washington, D.C., contacts familiar with Navy land use policy and personnel noted that for most base commanders, "if it wasn't going to break during their time in command, they weren't interested".[10]

The company also made sure its local member of Congress was aware of how base operations were impacting its plans and outlook for the future. Republicans John Schmitz (1970–72), Andrew Hinshaw (1973–1977), Robert Badham (1977–1989), and Chris Cox (1989–2005) all acknowledged the base's mission and importance to national defense and the local economy. However, none went out of their way to help the company with the larger issue of how the base affected land development.

Internally, as development plans on the periphery of El Toro proceeded, the question of the ultimate fate of the base was always on the minds of company planners and executives. The Irvine Company usually fielded an opinion survey of Irvine voters every other year and often included questions probing the attitudes of residents about the base and its operations. Consistently, a majority of Irvine voters viewed El Toro favorably. The only time the needle moved to the negative was when it was pointed out that jets flying overhead sometimes carried live bombs.

The company also conducted its own noise studies around the base, using technical firms with the reputations of providing highly accurate findings that could be used in negotiations about the precise location of the CNEL line or that could be the basis for seeking operational changes at the base that might benefit development.

The Irvine Company's Washington, D.C. advocate, William Ferguson, had been learning more about El Toro issues as he began advancing the company's interests through legislation and by building relationships with key members of Congress.

Ferguson focused much of his efforts on cultivating members of subcommittees that authorized appropriations—the hundreds of millions of dollars of federal funds allocated every year to cities, counties, special districts (water and sewer agencies were a Ferguson specialty) for construction or operational improvements.

As El Toro loomed larger as an issue, Ferguson saw an opportunity in late 1986 to utilize the military construction bill to secure an appropriation for extending the runway at El Toro. Interviewed in early 2024, Ferguson recalled that the runway extension would provide a measure of noise relief beneficial to potential Irvine Company development near the base but was a low priority on the Marine Corps' list of desired El Toro funding.

Ferguson convinced a friendly member of Congress on the subcommittee to take $8 million that had been programmed in the bill for Marine Corps housing and switch it to the runway extension. "That's when the Marines recognized what the Irvine Company could do in Washington," Ferguson said. "But at the same time, we made it clear to the Marines that we desired to find ways we could work cooperatively on issues that affected both entities."[11]

Working behind the scenes with Ferguson was Rich Anderson, a planner in the Department of the Navy, whose responsibilities included analyzing the interplay between Navy bases and the surrounding civilian communities. Anderson was especially aware of the Navy's three large air bases in Southern California at El Toro, Miramar (in San Diego) Twentynine Palms (in the high desert), and the LTA base in Tustin.

Interviewed with Ferguson in 2024, Anderson—who retired from federal service in 2003—said that of the many commercial and local governmental entities he worked with on air base issues, the Irvine Company was one of the most professional in its approach. "The company was very transparent about its goals, and they didn't play games."[12]

Anderson described the relationship between the company and the Marines as one of "equilibrium," a contrast to the difficulties he dealt with at Miramar (at the time the home of the Navy's "Top Gun" fighter pilot training center). "The company recognized the CNEL lines," he said. The feeling of equilibrium, Ferguson added, helped to develop more mutual trust between the company and the Marine Corps.

Anderson's analysis of current and future operations at the bases in California and elsewhere in the nation meant that he had to understand how changes in aircraft operations, weaponry, noise impact and dozens of other elements of modern aerial warfare would impact the day-to-day base decision-making.

But Anderson was also increasingly aware that at places like El Toro and Miramar, where urban growth would continue to pressure aircraft operations, change was inevitable.

***

Until 1977, the decision to open and close military bases was largely left to the president of the United States, acting as commander-in-chief. In that year, Congress asserted authority for closures, citing its constitutional mandate to raise and maintain armies.

For many members of Congress, a military base in their district was the ultimate serving of political pork, bringing jobs, prestige, and the welcome opportunity to appear at parades and other military ceremonies. Any effort to diminish a facility so carefully located and nourished was sure to evoke fierce resistance.

In the 1980s, following the fall of the Soviet empire, a new system for base evaluations was created that severely limited the ability of legislators to impact the process: the Base Closure and Relocation Act (BRAC). An independent committee was established to assess the status of bases nationwide and recommend changes as necessary. Military preparedness was a key factor, balanced against potential cost savings and the status of the base in its environment. That meant facilities like El Toro, once rural but now experiencing urban growth on virtually all sides, would likely be identified for closure.

The first round of closures occurred in 1991, with El Toro being spared from the list. But in 1993, explaining its decision to finally shutter El Toro, the BRAC commission reported that the base had "the lowest military values, no expansion possibilities, is the subject of severe encroachment and land use problems, and has many of its training evolutions over private property."

BRAC's reasoning notwithstanding, state and local officials put up a fight to keep El Toro open. Governor Pete Wilson was joined by the Orange County Board of Supervisors—led by retired Marine Corps general Thomas Riley—in a stirring but ultimately failed effort to keep the jets flying. El Toro's closure was set for 1999; long before it arrived, the fate of the huge property would spark a political and planning firestorm in the county.

***

On the western boundary of the Irvine Ranch, bracketed by the city limits of Santa Ana and Tustin, James Irvine II planted lima beans in soil so fertile and well-suited for the crop that bountiful harvests resulted, year after year.

Like all of the land at the center of his vast holdings, the property was very flat, but what made it especially favorable for lima beans was a low water table near the Peters Canyon Wash, one of the two natural waterways that crisscrossed the ranch. Locals called the area "La Ciénega de Las Ranas," the Frog Swamp.[13]

So Irvine was first surprised, then worried when in 1928, he learned that a Navy officer was surveying the Tustin property, seeking a potential site to build a base for dirigibles. It turned out that when Irvine met with Rear Admiral W. A. Moffett, the chief of naval aviation, not only was a dirigible base on his mind but an airfield as well.

Dirigibles—or blimps in common terminology—had been used in warfare beginning in World War I. While their utility for combat had been replaced by piston-driven aircraft in the 1920s, their ability to fly silently above coastal waters was valuable enough for the U.S. Navy to maintain a small fleet. Moffett chose a site in the San Francisco Bay area for a blimp base that would ultimately bear his name, giving James Irvine II a reprieve in 1928. But when Pearl Harbor was attacked, the Navy was back in Orange County.

Captain C. E. Rosendahl arrived with records Moffett had maintained identifying more than 1,000 acres of land the Navy thought suitable for a new dirigible base.

Irvine was dismayed and, in a desperate attempt to retain ownership of the lima bean fields, offered to lease the property to the Navy for $1 per year. Irvine also wanted assurances that water wells on the property would be maintained and a suitable amount of water would be made available to him for farmland outside of the new base boundaries.

During March and April 1942, both the *Santa Ana Register* and the *Anaheim Bulletin* reported on negotiations between the Navy and the man the *Bulletin* referred to as "Rancher Irvine."[14] Irvine apparently recruited the Santa Ana Chamber of Commerce into the negotiations with the Navy, as the March 20, 1942 edition of the *Register* reported—erroneously it would turn out—that chamber officials announced a lease agreement of $1 per year Irvine had sought

> was revealed following a conference last night of chamber directors, Irvine, and Lt. Commander W. S. Lee, . . . it is understood that Irvine will meet for negotiations with the commander of the 11th Naval District at San Diego after the necessary papers for the giant project have been signed by Secretary of the Navy (Frank) Knox.[15]

The negotiations, if they happened at all, did not bring Irvine the relief he sought. The Navy insisted on purchasing the land and gave assurances that it would guarantee Irvine water from wells on the land where the base was to be built. Just before construction began on what was estimated as a $7 million project, Irvine managed to harvest one last crop of lima beans.

Irvine must have known he was holding a losing hand in the negotiations, as on April 1, 1942, even while he and Navy officials were still conferring, ground was broken on the site and construction began. In a few weeks, two giant concrete slabs had been poured that would become the floors of a pair of what was then—and still is—the most significant wooden structures ever built.

Soaring 17 stories into the air, more than 1,000 feet long and 300 feet wide, the two hangars created a surreal presence amidst the acres of empty land that surrounded them during World War II and well into the 1970s when development began to creep near the base boundaries. (The hangars were placed on the National Register of Historic Places in 1978).

By 1953, the dirigibles had been decommissioned, and helicopters were stationed at the Tustin base, a result of their increased and effective use during the Korean War. Sixteen years later, it was another war in the Far East, this time in Vietnam, that turned the base into Marine Corps Air Station (Helicopter) Santa Ana. Three Marine operating and training groups for helicopters were located at the base, one of only three helicopter training bases in the nation.

Foreshadowing what would become the base's ultimate fate, in 1976, the City of Tustin annexed the base property. Located at the western boundaries of the Irvine Ranch, much of Tustin had been covered in orange groves until the late 1960s; its leaders realized that urbanization of ranch properties in Newport Beach and Irvine meant similar development was likely in the small city. And as was the case with El Toro, those same leaders understood that a reckoning with the Marines was inevitable.

Helicopters, large and small, flew daily missions out of the Tustin base—east to the high deserts, south to Camp Pendleton, and to other points of the compass. And like the jets at El Toro, a fleet of choppers departing at a relatively low altitude could create a substantial impact. In Woodbridge, on nights when seasonal low clouds in the spring blanketed the huge community, vibrations from the whirling helicopter blades could literally shake a two-story home like a mild earthquake. Similar impacts were reported in communities under other departure patterns.

Particularly noisy were the Sea Stallions, weighing in at 50,000 pounds and capable of carrying an additional 8,000 pounds of troops or supplies. Tustin officials specifically asked that Sea Stallion flights be banned over the city—the Marines refused to do so. City of Irvine leaders were also complaining, going so far as to suggest the base ought to be closed.

However, while the complaining continued, it became increasingly apparent that, regardless of how long the base remained in use, its presence was hindering the extension of a roadway critical to development in northern Irvine and eastern Tustin, all of which was on land owned by the Irvine Company.

***

Looking at a 1980 map of Orange County tells the story.

Jamboree Road, named in remembrance of the 1953 Boy Scout Jamboree held at the site of what had become Newport Center and Fashion Island, was a robust, four-lane highway that began at Pacific Coast Highway, then angled past the Upper Newport Bay and through the growing Irvine Business Complex. It connected to the San Diego Freeway adjacent to the massive Fluor headquarters, then continued north until just above Barranca Parkway, where it stopped.

Engineers and planners drew alignments extending the long path of Jamboree from the dead end at Barranca through the base and past Edinger Avenue, which ran parallel to the AT&SF railroad tracks. Further north, an extension of Jamboree would need to connect to the other major freeway that crossed the Irvine Ranch, the Santa Ana Freeway, Interstate 5. From there, to accommodate what would become a large new village in Tustin a few years later, Jamboree was to be extended far into the Lomas de Santiago to Santiago Canyon Road.

But the lines on the maps meant little to the Marines. Base commanders at every military installation in the United States hewed to the concept of defending the perimeter from incursion of any kind. This rule had been followed at nearby El Toro and at Tustin. And beyond the boundary issue, Tustin had a critical training facility that stood directly in the path of the logical extension of the Jamboree: the heavy-lift pad. In this huge, circular concrete mass, pilots in training lifted large concrete balls to simulate a wartime reality of carting a tank or truck into the air and onto a battlefield.

Probes from the two cities and the Irvine Company on a potential deal for the Jamboree extension had been made for years, to little effect. However, in 1981, a new commander was appointed to oversee both Tustin and El Toro: General Richard M. Cooke. "General Cooke was the key" to breaking the gridlock, according to William Huston, city manager of Tustin at the time. "He has a very realistic perspective on the situation, and that is that development pressures are there."[16]

Cooke took the initiative, according to the *Times* story, "consolidating the Marines' needs into a small shopping list, which he sent to the two cities and the Irvine Company.".[17]

Within days, a letter of reply had been received, and negotiations among the Marines, the Irvine Company, and Irvine and Tustin began in earnest. They concluded in the spring of 1985 with the Marines giving up the right of way through the base for the Jamboree extension. They also agreed to shift the takeoff pattern for helicopters traveling over Tustin to alleviate noise impacts. In return, the Irvine Company gave up 39 acres of land for future development into housing for Marine personnel and families. The cost of the road extension, estimated by Tustin officials as close to $90 million, would be shared by Tustin, Irvine, and the Irvine Company.

Work began in 1987 on the new road and was completed in the summer of 1990. An editorial in the *Los Angeles Times* noted that at the breakfast ceremony, "officials likened the opening of the new stretch of Jamboree Road to the completion of the transcontinental railroad."[18]

The Jamboree Road agreement would prove to be anticlimactic in the history of the Tustin LTA base. Less than a year later, the federal base closure process would conclude that both El Toro and Tustin were no longer viable military installations and would be shuttered in 1999.

A story in the April 13, 1991 issue of the *Los Angeles Times* summarized the analysis that led to Tustin's fate:

> The construction of hundreds of homes, offices, and industrial facilities surrounding the 1,500-acre helicopter base has so compromised its military mission that Tustin is now ranked as the least valuable of the nation's eight domestic Marine Corps Air Stations, the Defense Department said.[19]

Like El Toro, Tustin LTA was destined to undergo a transformation. And while the process would be different than El Toro's, it would include its share of political drama and ultimately, tragedy.

According to OC Historyland, the authoritative website researched and written by the late county history expert Phil Brigandi, three other sites on the Irvine Ranch were used during World War II. The first was in Irvine Park, where the U.S. Army held training operations from 1942 to April 1943. The site was named Camp Rathke. In nearby Peters Canyon, another Army camp was established for both training and rehabilitation purposes, serving pilots in training at the Santa Ana Army Air Base (now the site of Orange Coast College in Costa Mesa). Finally, at Abalone Point on the Irvine Coast, a concrete emplacement was built for coastal guns that were never installed. All three properties were returned to the Irvine Company after the war.

Notes

1 "Jet Noise and Development Restrictions," *Los Angeles Times*, December 8, 1976.
2 "Leisure World Developer Pays Marines $8 Million," *Los Angeles Times*, January 27, 1988.

3 "Leisure Would Developer Pays Marines $8 Million."
4 "County Buys Land from Marines for Road Projects," *Los Angeles Times*, May 8, 1990.
5 "El Toro Base Operations Continue," *Los Angeles Times*, February 12, 1981.
6 "Top Gun Filming Locations," OCThen.com, accessed September 13, 2025, http://www.octhen.com/2006/11/top-gun-filming-locations.htm.
7 "Irvine Councilwoman Calls for Lawsuit over Jet Noise," *Los Angeles Times*, June 9, 1979.
8 "Irvine Councilwoman Calls for Lawsuit over Jet Noise."
9 "School Sites Rejected Due to KC-130 Flight Paths," *Los Angeles Times*, October 5, 1981.
10 Rich Anderson, interview with H.P. Oliver and C.M Stockstill, February 23, 2024.
11 William Ferguson, interview with H.P. Oliver and C.M. Stockstill, February 14, 2024.
12 Rich Anderson, interview with H.P. Oliver and C.M Stockstill, February 23, 2024.
13 RBF Engineering, The Tustin Hangars: Titan of History, n.d.
14 "Rancher Irvine Negotiates with Navy," *Anaheim Bulletin*, April 18, 1942.
15 "Santa Ana Chamber Confirms Navy Lease," *Santa Ana Register*, March 20, 1942.
16 "Jamboree Road Extension Agreement Finalized," *Los Angeles Times*, July 31, 1983.
17 "Jamboree Road Extension Agreement Finalized," 1983.
18 "Jamboree Road Opens," *Los Angeles Times*, August 12, 1990.
19 "Tustin Base Ranked Least Valuable," *Los Angeles Times*, April 13, 1991.

19 The path to the Great Park

Nine years passed between the announcement of the closure of El Toro Marine Corps Air Station (MCAS) and a final, definitive public vote that ended the bitter warfare it engendered.[1]

A detailed retelling of events that divided the county's economic and political classes during that time is worthy of an entire book. This history presents an analysis of the prolonged conflict to illuminate how the outcome led to the creation of the second major community on the Irvine Ranch that was not planned and developed by the Irvine Company and how a controversial regional park in the center of it dramatically changed the political landscape of the City of Irvine.

***

Within months after news broke that El Toro would be closed, public and private forces began to organize around two fundamental positions. The first was to convert the base from a military to a civilian airport; the second was to make it anything BUT an airport.

The Orange County Board of Supervisors had a solid, three-vote majority in favor of an airport. Accordingly, a special county task force was organized and staffed to begin sketching out plans for an airport and surrounding support district.

Led by real estate mogul George Argyros, political and business interests in the county collaborated on a public relations campaign to sell the benefits of a new airport to the voters. Polling showed that in general, residents in the northern part of the county (west and north of the 55 Freeway as a general dividing point) favored the airport concept. In the southern section of the county, a strong majority opposed it.

Opponents of the airport quickly formed three organizations: Taxpayers for Responsible Planning (TRP), the El Toro Reuse Planning Authority (ETRPA), and Project 99.

TRP was primarily a loose-knit collection of individuals and groups centered in the Laguna Hills area, headed by a former Marine Corps officer. ETRPA was a coalition of cities in the south county, well-funded and with professional staff to provide research and action, the big dog in the fight. Project 99 was created by Larry Agran, who had lost his seat on the Irvine City Council in 1990. Agran had

DOI: 10.1201/9781003683087-19

tried out several strategies to remain politically active, including a quixotic run for the Democratic nomination for president in 1992. Opposing the airport offered him a ready-made opportunity to revive his grassroots organizing skills, which he did on a full-time basis.

As might be expected, environmental groups were also united in opposition to the airport. It was from somewhere in their midst that, in early 1994, the idea of swapping Irvine Company agricultural and open space to some or all of the shuttered base was first broached. And it first came from an unlikely source.

Agran had jousted with the Irvine Company for nearly every day of his 12 years as an elected official. So no one expected his opinion piece in the *Los Angeles Times* in February 1994 to contain such a juicy offer for the giant landowner.

Agran proposed exchanging Irvine Company farm and grazing lands for all or part of the El Toro base. The company could develop anything except an airport, while the agricultural lands would be permanently preserved, a long-held Agran goal.

A couple of months after Agran's trial balloon was launched, reports surfaced of serious discussions between the company and the federal Department of the Interior on the concept. Many environmental advocates gave tentative endorsement to the idea, but a few others sharpened their pencils and—comparing the value of hundreds of acres of open land that could be quickly developed to existing strawberry fields—came to the conclusion it was a sweetheart deal.

Discussions continued, with press reports describing conflicting versions of what was taking place, attributed both to Irvine Company executives and officials from the Department of the Interior. In March 1995, the company said the discussions had terminated and there would be no deal.

The land swap proposal was an example of the variety of ideas and opinions proffered about what to do with the closed base. But while talk would continue for all nine years that transpired between the closure announcement and an outcome, the real action was at the ballot box.

County supervisors struck first, just 16 months after the closure decision. Thus, El Toro was still up and flying in November 1994 when Measure A passed, providing that the base would be designated for aviation use in the county general plan.[2] Two years later, anti-airport opponents gathered enough signatures to place Measure S on the ballot to repeal Measure A, but it was defeated.[3] Emboldened by the vote, the county released an environmental impact report defining an airport that would serve 38 million annual passengers and the closure of John Wayne Airport to commercial aircraft.

Embracing the political maxim that you can't beat something with nothing, ETRPA continued to rail against the airport before developing what was labeled "The Millennium Plan" for El Toro.[4] It envisioned a mix of commercial, residential, and open space uses. In 2000, the Millennium Plan had been translated into yet another countywide ballot measure, Measure F, which passed with a healthy 67 percent majority.[5]

But it proved to be a short-lived victory. Challenged in court, a judge found Measure F unconstitutional.[6]

While the political ping-pong of successive ballot measures played out, millions of dollars in public and private funds were being expended on the issue.

***

George Argyros was a poster boy for the story of Orange County's post-WWII business environment—a place where someone with drive and energy could thrive.

Arriving in the mid-1950s, the son of Greek immigrants, popular accounts of Argyros' early years bring to mind the Horatio Alger stories of the mid-18th century: a young man grows up poor, works hard, and becomes wealthy.[7]

He started work in a grocery store as a box boy. What would become a lifelong affiliation with tiny Chapman University in Orange began with his graduation there in 1959. Searching for a way to gain financial success, Argyros determined that real estate was the most profitable path to his goal. His first significant investment was a strip mall in Tustin.[8]

For the next two decades, Argyros worked largely under the county's business community radar to amass a portfolio of apartments and commercial properties, "nose to the grindstone" in the words of one former advisor.

Argyros was not shy about his politics. In 1972, he served as co-chairman of the campaign to re-elect President Richard Nixon, beginning a long set of affiliations with Republican candidates that would include Ronald Reagan, Pete Wilson, and George H. W. Bush. Argyros also found a political and personal soulmate in a young member of the state assembly, Bruce Nestande, who would play a key role in county politics as a county supervisor and, later, as an aide to Argyros.

In addition to his portfolio of apartments, office buildings, and retail centers, Argyros had teamed up with former U.S. Air Force major general William Lyon (owner of a homebuilding company) on the acquisition of Air California in 1981. Purchased out of bankruptcy for $61 million, the pair renamed it Air Cal. Stoked by marketing that promoted it as Orange County's hometown airline, Air Cal was soon successfully competing for customers flying to San Francisco, San Diego, and Sacramento. Just five years later, Air Cal was sold to American Airlines for $225 million, earning Lyon and Argyros a reported $30 million each in profits.[9]

Argyros was busy in 1981. He had also purchased the Seattle Mariners baseball team for $13.1 million. His nine-year ownership was marked by fan disdain and regular criticism of the team's management by Seattle sportswriters, some of whom complained regularly that Argyros had doomed the team to mediocrity by refusing to pay big dollars to star players. Argyros let the harsh words bounce off him; he sold the team in 1989 for $76.1 million.[10]

Argyros took command of the movement to convert El Toro to commercial use with gusto. He funded both campaign and public relations efforts and appeared at events to promote it. Never reluctant to advocate forcefully for something he believed in, Argyros could be pugnacious and combative in person. A city councilwoman called him a bully after a spirited debate.

Airport opponents saw self-interest in the battle, but political consultant David Ellis, who worked with Argyros from the initiation of the effort, disagreed. "George saw the airport as the right thing to do for the county."[11] (Left unsaid was the fact that Argyros lived on Harbor Island in Newport Beach, directly under the John Wayne Airport flight path).

Ellis was just one of many political operatives who toiled in the battle. But at the top of the scrum was one Orange County man who just a few years earlier had perfected an approach to campaigning that made him notorious and rich. He would go on to play a controversial role in the aftermath of the final El Toro campaign and the equally controversial project it would spawn.

***

Larry Agran had ridden his leadership of Project 99 back onto the Irvine City Council, first in 1998 as a council member, then in 2000 as mayor. Just a year later, the *Los Angeles Times* reported that ETRPA "had quietly stepped aside to let Irvine create the new initiative, a significant shift in the often fractious anti-airport coalition."[12]

"Many of those involved said Irvine was given the green light because they didn't feel like battling over El Toro, as long as it doesn't have runways," the *Times* story continued. "Irvine's plan 'is not going to permit an airport and that's all anybody really cares about,' said Tristan Krogius, an anti-airport activist from Dana Point."[13] Len Kranser, who operates an anti-airport webpage, added, "Irvine officials argued convincingly that the city's proximity to the base earned it the right to control what might be built there. That extended to drafting the new initiative."[14]

It was also hard for ETRPA leaders to ignore the fact that the consultants who had previously worked for George Argyros—and had been successful on his behalf—were now on the payroll of the City of Irvine, developing the strategy to once and for all drive a stake in the heart of the pro-airport position.

The dream team was Forde and Mollrich, composed of Arnold Forde and Stu Mollrich. Mollrich was the junior partner, a diminutive, brilliant strategist and wordsmith who was a veteran of Republican campaigns all over California.

But it was Forde who brought a checkered history racked up from the mid-1970s when he had worked with his first partner, William Butcher, in the partnership known to California politicos as Butcher Forde. The pair had begun scoring wins in individual races by their innovative approach of targeting direct mail to voters that identified issues that mattered most to them. Butcher and Forde understood computers and how to utilize them, while many other campaign managers were simply buying newspaper ads and sending generic mail to the same voters, over and over. The Butcher Forde approach might feature dozens of different messages on different issues targeted to voters by age, geography, sex, or a host of other variables.

Butcher and Forde entered the World Series of California politics when they applied everything they had learned to the 1978 Proposition 13 campaign headed by Howard Jarvis; it rewrote the property tax laws and created a political earthquake in the state.

For the next two decades, Butcher Forde had the hot hand in California campaigns, managing statewide initiatives and candidates for governor and other offices as well as providing strategic advice to corporations and businesses. But, in doing so, the pair developed a reputation of using anything-goes tactics. And while most of their peers in the campaign business would stay on one side of the political party spectrum or the other, Butcher and Forde were happy to work for the highest bidder no matter the philosophical outlook. For example, while both men had no personal feelings one way or the other regarding gays, lesbians, or same-sex issues, they didn't hesitate to take on John Briggs and Anita Bryant's sometimes vicious statewide initiative to ban homosexual teachers. Before long, they were labeled the "Darth Vaders" of direct mail politics.

When the pair sold their firm in 1995, Butcher departed for the Caribbean, leaving politics behind to try his hand at raising scallops. Forde became the face of the new partnership with Mollrich.

Forde struck some as the classic absent-minded professor. He dressed primarily in casual pants and polo shirts, and often eschewed socks. He wore his hair in a mop, short bangs hanging over his forehead and ears; for most of his life, he looked like he needed a haircut. In the last years of his career, he adopted the short shaved head look favored by Jerry Brown, a one-time client.

Forde had an open face and what some who worked with him described as a goofy smile. His eyes were intensely blue and radiated from his perpetually tanned face. He spoke with an odd cadence, sometimes stuttering slightly as he strung together sentences, his voice rising octave by octave.

But Forde could be cold and calculating, dismissive of people he considered peripheral to the task at hand. And while the blue eyes and goofy smile could charm, they could also turn intimidating.

Little of this mattered to the Irvine City Council. In Forde and Mollrich, they saw a pair of hired guns who could match the resources and sophistication of some of the county's most powerful political and business leaders.

Forde and Mollrich had spent $4 million of city funds, much of it to create and mail colorful surveys to local residents asking what they wanted at the base instead of an airport. Agran had developed his own set of alternatives while running Project 99, including a huge open space area that had been described as some sort of massive park.

It fell to Forde to write the persuasive language in what would become Measure W, slated to go before voters in March 2000. Testifying years later in a post-mortem on the issue, Forde said he had been reviewing The Millennium Plan, but after evaluating the public opinion polls the city had sponsored, he found it wanting. People said,

> Hell, with all the development and traffic and everything down here, they'd rather have an airport.
>
> So I looked at the whole 4,500 acres and I said, my God, how big is UCLA? USC? So I put everything that anyone could have ever wanted in open space, and had a park bigger than Central Park, and you had all these public-type facilities. So we had the city say, Look, we have a better use for this space than the Millennium Plan or an airport.[15]

Forde's brainstorm was translated into Measure W, with the official title of "The Orange County Central Park and Nature Preserve Initiative." The text of the proposal ran for two full pages, promising

> this initiative will allow for the creation of one of America's greatest parks, with open space, sports and recreation facilities, museums, libraries, arts and cultural attractions, and a home for major universities and research centers. It will also not generate traffic, congestion, noise and air pollution associated with the development of a commercial airport.[16]
>
> The initiative went on to promise that
>
> the central location and accessibility of El Toro will provide Orange County's three million residents with an opportunity to enjoy a park on a par with Golden Gate Park and the Presidio in San Francisco, Griffith Park in Los Angeles, and Balboa Park in San Diego.[17]

The initiative also contained a large map showing almost 90 percent of the El Toro footprint in open space or nature reserve.

Nowhere was there any mention of residential or commercial development. Voters would find out about that later, courtesy of the federal government's land management arm, the General Services Administration (GSA), which had a different vision for El Toro than Measure W.[18] (Arnold Forde died in 2018.)

Notes

1 "El Toro Closure Timeline," *Los Angeles Times*, July 3, 1999.
2 "Measure A Passes," Orange County Register, November 9, 1994.
3 "Measure S Defeated," *Los Angeles Times*, November 6, 1996.
4 "Millennium Plan Proposed," ETRPA Archives, 1999.
5 "Measure F Approved," Orange County Register, March 7, 2000.
6 "Measure F Ruled Unconstitutional," *Los Angeles Times*, August 12, 2001.
7 "Argyros Profile," Chapman Magazine, Spring 2003.
8 "Argyros Profile."
9 "AirCal Sold to American Airlines," Wall Street Journal, July 15, 1986.
10 "Seattle Mariners Sold," New York Times, August 1, 1989.
11 David Ellis, interview with C.M Stockstill, February 2024.
12 "ETRPA Shifts Strategy," *Los Angeles Times*, March 18, 2001.
13 Tristan Krogius, quoted in *Los Angeles Times*, March 18, 2001.
14 Len Kranser, quoted in *Los Angeles Times*, March 18, 2001.
15 Arnold Forde, testimony, Irvine City Council archives, 2003.
16 "Measure W Full Text," City of Irvine, March 2000.
17 "Measure W Full Text."
18 "GSA Land Disposal Guidelines," *U.S. General Services Administration*, accessed September 13, 2025, https://www.gsa.gov/real-estate/real-property-disposal.

20 A park, a plan, and new faces on the Irvine Ranch

The euphoria in Irvine and points south in Orange County came to an unexpected halt two days after the March 5 election. Measure W had passed, all but guaranteeing that El Toro Marine Corps Air Station (MCAS) would never see another airplane fly into or out of it.[1]

Then, the congressman for the area, Chris Cox, announced that the government would sell off the 4,700-acre base; the estimated value was in the millions of dollars. The sale would be consistent with the Measure W promises of a major park for the area, Cox said. As time passed, however, it became clear that the government contemplated significantly more commercial uses for much of the former airbase.

Irvine city officials quickly sprang into action. Visits to the Department of Defense were undertaken, followed by intense discussions with federal authorities delving into just what a sale might entail. As the talks progressed, the government released an environmental impact report showing that hundreds of sites on the base were polluted by chemicals that had been used for decades to clean airplanes and other military equipment. The list of specialized mixtures was staggering; it even included potential radiation exposure to people who had painted instrument panels for jets so they would glow in the dark.[2]

One especially troubling pollution discovery was a groundwater plume containing volatile organic compounds (VOC) that was almost three miles long.[3] The Navy joined forces with the Environmental Protection Agency and an alphabet soup of state and local health agencies, promising that all the toxins would be removed. The Navy also forged a partnership with the Irvine Ranch Water District to implement a plan to sink hundreds of wells along the plume that would extract the polluted water and spray it into the air, rendering the pollution inert.

The General Services Administration (GSA) was tasked with selling the property. The GSA had disposed of dozens of former military bases since the Base Reuse and Closure (BRAC) process had begun a decade earlier. But there were several methods to do so, the most common being a reconveyance of the property to another state or federal agency (essentially a land transfer), followed by a Public Benefit Conveyance or an Economic Development Conveyance. As the names implied, placing the property in the hands of a city or public agency would be done to promote economic development, benefiting the local area.

DOI: 10.1201/9781003683087-20

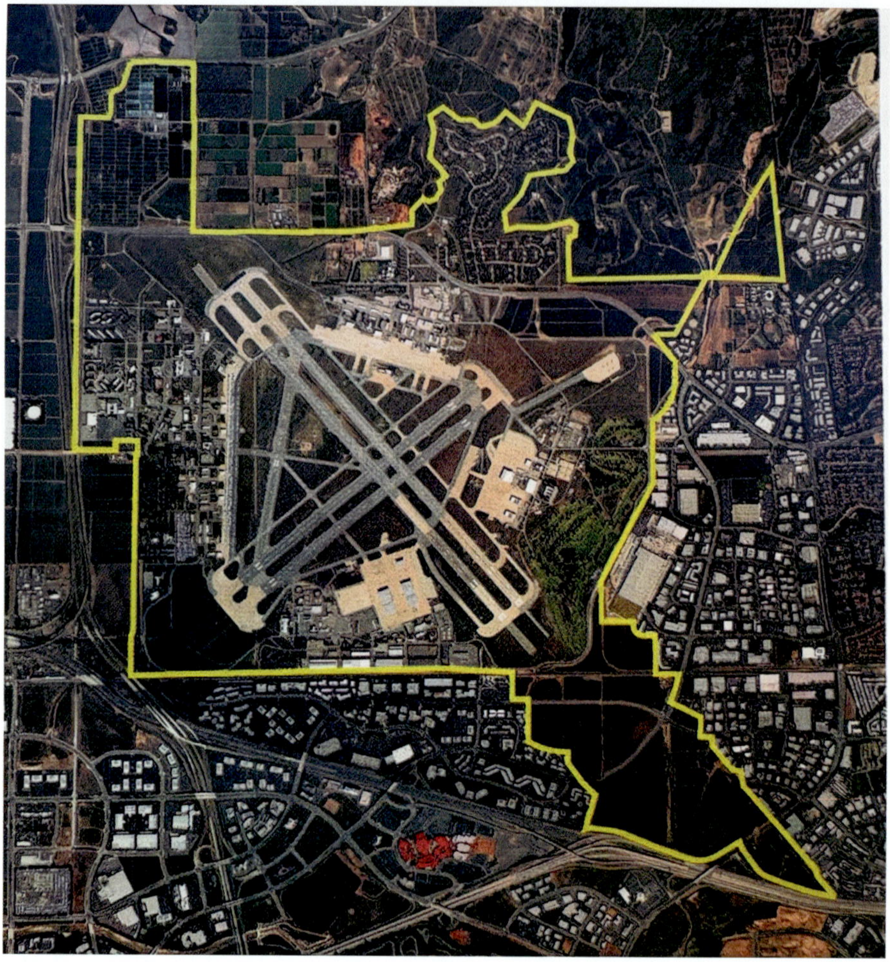

Figure 20.1 Map of Marine Corps Air Station property sold at auction by the U.S. Government

El Toro, however, offered the GSA the chance to sell the property directly. It was an unusual strategy; a 2021 report by the GSA looking back ten years at all the base transfers showed that of Navy properties totaling 177,121 acres, El Toro was the only property that was sold directly.[4]

Irvine managed to strike a deal with the GSA to preserve its hard-fought park option. The GSA agreed that 1,370 acres in the center of the closed base would be dedicated to the city as a condition of the overall sale, and that the successful buyer or buyers would cough up $200 million for the new park in addition to the final sales price.[5]

However, there was a catch.

The GSA knew the price of the property would be diminished if it contained pockets of pollutants. So when it came time to define the sale parcels, GSA planners drew around as many of the toxic sites as possible, leaving a vast stretch of area for the park that was about 85 percent devoid of pollution. Several sites within the newly defined park were so polluted it would take years—perhaps decades—to clean up. Playgrounds, baseball fields, or other recreational uses in those areas would just have to wait. But there were also a handful of sites within the area contemplated for homes and businesses that needed to be cleaned up. GSA decided to lease those properties to the ultimate buyer and transfer ownership upon removal of the toxins.

GSA hired a real estate sales expert in late 2004 and began its analysis of how to sell off El Toro. The decision was made to divide the property into four parcels and set a minimum price for each. The parcels would be auctioned, one by one, in what might have been the nation's largest silent auction: bidders sent in an email with a bid and waited.

It was soon apparent that while dozens of real estate developers had obtained and reviewed the GSA sales package, only one was likely to clear the field: Lennar. (The obvious missing bidder was the Irvine Company. Word filtered down that the company's owner, Donald Bren, dismissed any interest in El Toro, preferring to concentrate on the remaining acres he already owned and was committed to completing in his lifetime).

A publicly traded company, Lennar was one of the nation's largest homebuilders. In 2006, the company reported revenues of $16.3 billion and $148 million in profits. It built 49,568 homes and carried more than $4 billion in debt. It also had a sister company, LNR, that developed commercial and industrial properties. Lennar's regional president, Emile Haddad, told the *Los Angeles Times* the company spent $2 million to analyze the El Toro property before deciding to bid for it.[6]

The bidding began in February 2005; Lennar quickly demonstrated its intent to capture every acre of land for sale. For Parcel 1, the minimum bid was $60 million, which Lennar took without opposition. Parcel 4 also went for the minimum of $60 million to Lennar. The final two parcels were claimed by Lennar at prices higher than the minimum: $310 million for Parcel 2 and $154 for Parcel 3. The total cost was $649.5 million.[7]

In return, Lennar said it expected to build 3,500 homes and 3 million square feet of commercial and industrial space. One of the first deals it reported for the property was a tentative agreement to build a cemetery—something that was noticeably absent from the planned community of Irvine.

Lennar was just ramping up, but in the two years that had passed between the Measure W election and the bidding for the base, political leaders in Irvine had been busy preparing for the moment the sale took place. The city had created a nonprofit corporation to govern what had now become known as the Orange County Great Park.[8] City staff and a handful of other employees with government management experience had been hired. In addition to the five members of the city council, four "civilians" were named to the new board. The future looked bright.

***

The story of the Great Park spans two decades, with two major players: the City of Irvine (and the entity and governing board its council members and mayor created and controlled, the Great Park Corporation and the Great Park Board of Directors), and the consecutive iterations of Lennar's ownership of adjacent lands for development: Lennar/LNR, Heritage Fields, and Five Point Communities.

There was continuous interaction between both players; to tell the story clearly, this history tracks each one separately, then brings their stories together at critical points along the way.

Irvine/Corporation for the Great Park

Irvine had annexed the base shortly after the election. County supervisors, finally conceding defeat in the decades-long airport contest, were happy to be rid of it (officially, annexation came a few months later after state approval).

The city placed the base property into a redevelopment zone. Under California law, this created a special tax benefit, allowing an increment of property taxes collected there to be separated into a special fund for improvements to what by law was deemed to be a "blighted" area. The definition of blight had long been controversial in the state, with other jurisdictions stretching it to levels of imagination worthy of a Disney movie.

Looking forward to the planned Lennar development, it was estimated that in 45 years close to $900 million dollars in tax increment funds would accumulate to build the Great Park. The future money, plus the $200 million that Lennar was handing over, would prime the pump for immediate needs; once the future increment funds could be determined in detail, bonds could be issued by the city, providing even more money up front.

Thus, Irvine elected leaders began referring to the Great Park in what became their mantra: "the first great metropolitan park of the 21st century." Agran, largely responsible for the funding and strategy that had produced Measure W, embraced the Great Park in a political bear hug. In subsequent local elections, he would field a slate of candidates that he headed with the label of "The Great Park Team."

Initial plans for the governance of the Great Park—described in ballot language as a benefit for all of Orange County—called for a board of "no more than seven, and up to 30 directors, two of whom shall be officers or employees of the City of Irvine."[9]

Organizational efforts were underway when the Irvine City Council made a governance U-turn and rewrote the relationship between the city and the Great Park board of directors. When the bureaucratic dust settled in April 2006 (a year after the nine-member board began meeting), all policy and fiscal authority was vested in the city council; the Great Park board of directors was left with only an advisory role.

It might have been expected that after being publicly emasculated by the city, board members would rebel. But three of the four "public" members were close allies of Agran. While a couple complained about the coup in private, they remained silent in public. The fourth member was Richard Sim, a former Irvine Company executive with extensive experience in the requirements of large-scale land development. He hung on for six more months, then resigned.

One reason for Sim's departure was what he considered excessive spending on public relations.

Arnold Forde and Stu Mollrich, who had crafted the winning campaign for Measure W, showed up on a Great Park board agenda one day as the proposed recipients of a no-bid, $100,000 per month contract to provide public relations and strategic advice to the new park development organization. To those unschooled in the realities of Irvine politics, it appeared an odd choice. Forde and Mollrich had no expertise in urban planning, land development, or recreation. They had never worked on ballot measures related to parks or supported a candidate who ran on a platform of open space and athletic fields.

Forde and Mollrich were hired because Larry Agran wanted them hired. He had the votes to make it happen and, brushing aside some muted criticism, brought them on board.[10] Forde, in addition to bringing genuine expertise in shaping public opinion, also became a regular wingman for Agran at the many meetings he attended in the initial months and years of the park's development.

While bringing two of the county's most successful and notorious political hired guns into the Great Park mix was soon forgotten, the way Irvine had wrested control of the park into the city council was not. The Orange County grand jury decided to examine just how the park was coming along and found that among other things, the new park governance plan was "inconsistent with the spirit and intent of Measure W."[11]

While a grand jury criticism might seem ominous, in Orange County the grand jury was a paper tiger. District attorneys had rarely acted on grand jury actions; on occasion the panel's findings about some aspect of county government would spur a follow up, but in most cases government leaders found the grand jury little more than an annoyance. The report on Great Park governance was filed away and largely forgotten.

As the Great Park staff initiated the removal of dozens of buildings and military installations (including munitions bunkers made of tons of concrete), the board started a search for a firm to create a plan for the massive acreage.

Dozens of companies submitted proposals in what became a competition that was covered in depth by the local media. When a set of finalists was selected, board members and staff visited projects designed by them in New York City, France, and Spain. (Staffers privately called the tightly scheduled, multi-nation trek "the Magical Mystery Tour").

The design job was awarded to Ken Smith, a Harvard-trained landscape architect from New York City.[12] Smith brought a dash of Manhattan panache combined with personal quirkiness to Orange County, where the planning community tended to the informal, bordering on casual. Smith dressed entirely in black. He wore glasses with round lenses surrounded by thick black frames popularized by famous architect Le Corbusier (Charles-Édouard Jeanneret 1887–1965).

Park staffers who visited his home in Manhattan's Soho district reported a flat largely devoid of furniture with a single mattress resting on the floor of the bedroom. Smith's wife, who accompanied him to park social events, had hair dyed bright orange.

Smith's catalog of work was almost entirely urban. His best-known design at the time was a sculpture memorial done for American Express in memory of the 9/11

Figure 20.2 Ken Smith

Source: Courtesy American Society of Landscape Architects

victims. He was largely a one-man show, so to demonstrate the ability to ramp up a plan quickly, he recruited a team including well-known Los Angeles landscape architect Mia Lehrer, artist Mary Miss, and architect Enrique Norton from Mexico and brought in Gafcon, a project management firm from San Diego,

(The fact that Smith had never designed a project on the scale of the Great Park was pointedly illuminated by Dr. Diane Ghirado, a professor of architecture at the University of Southern California. She was a member of a design jury the city had hired to evaluate the three finalists for the job. In her summary report, she asked if "the Orange County Great Park is prepared to let Smith learn about designing large parks here?" Her report continued:

> to choose this team entails some huge unknowns, leaving the day to day design decisions in the hands of persons unknown at this time. A project of this scale . . . merits the nearly full-time attention of the chief designer, and that will not happen with Smith.[13]

Her warnings were ignored.

Figure 20.3 The orange balloon has become the symbol of the Great Park

Source: Author's image

About the time of Smith's selection, Lennar announced the gift of a giant helium balloon to be located at the grounds of the initial park development. The orange balloon quickly became the park's symbol, and riding it up 400 feet into the air became the first way most residents got a look at the enormity of the former airbase.

Park staff were less than enthusiastic when presented with the budget for operating the balloon. It was built and operated by the French firm Aerophile, whose staff had to be trained and licensed abroad.[14]

Within a few months, Smith unveiled a master plan for the park with a striking element at its heart: a two-mile-long canyon, sixty feet deep, that would be carved into the middle of the 1,300-acre park, culminating in a large lake at one end. Smith envisioned the canyon to be landscaped up its sides, with walking trails at the bottom.

The canyon was just the start. Also included in the plan was a forested bosque, a botanical garden, parking lots planted with orange trees—all supported by a system of sustainable water reclamation and energy generation. There would also be a grand entrance, with a giant sculpture arching over fountains. To accommodate museums and art galleries, the plan envisioned a cultural and historic district overlooking the lake. In recognition of El Toro's World War II heritage, a veteran's memorial was planned. Another key element in the plan was a sports park.

The plan drew wide praise and was honored by the American Planning Association, the American Institute of Architects, and the American Society of Landscape Architects.[15]

Translating the vision onto the ground, however, was not without its challenges.

After months of lofty rhetoric about what the park would eventually become, the reality of making it all happen proved bumpy. Focusing on the area around the giant balloon, Smith insisted that the plastic furniture in the balloon waiting area be ordered from Spain instead of purchased locally. He created a "bio swale" to drain the area that was ineffective. A shade structure he designed with canvas strips shredded in the strong Santa Ana winds that periodically roared onto the base (the balloon was tethered down by heavy chains affixed to anchors buried deep in its concrete base).

Other problems began to crop up. The popular golf course on the property was eliminated from future plans. The cemetery that Lennar had proposed was quietly erased, which prompted a lawsuit from its sponsor, the fabled California cemetery owner Forest Lawn. After Forest Lawn hired a high-powered Washington, D.C. law firm, the suit was settled out of court.[16] A woman died after falling at the park's seasonal ice skating rink. Her family sued the city and received an undisclosed payment.

The multiple firms that Smith had assembled were dubbed as the Great Park Design Studio. However, none had ever worked with one another before. Smith would typically spend a few days on site at the Great Park and then return to New York. The Design Studio soon clashed with Great Park Corporation employees over issues like contract terms, billing support documents, quality of work product, and a common approach to moving the park from paper to reality.

As would become more clear years later, one sticking point that caused continuous friction was the fact that initially there was no overall budget for the planning of the park. The city employees who were administering the process were used to creating a 12-month budget every year and measuring progress accordingly. At the Great Park, while there were some contracts based on time and materials, in the case of the overarching goal of designing and building the park, there was little to no budgetary detail.

After two years on the job, the park's CEO was abruptly transferred to a city hall job and was replaced by the city's public works director—he lasted just four months before retiring from the city. The assistant city manager was assigned to run the park.

None of the upheaval escaped notice of the local media. As time passed, mainstream newspapers began to focus on the Forde and Mollrich contract, plus an alleged sweetheart deal to supply trees to the park. The *OC Weekly*, the county's alternative paper, was especially aggressive in its coverage of the park. In late 2010, it ran a cover story illustrated with a caricature of Agran ascending on the balloon, in a wicked homage to *The Wizard of Oz*.[17]

In 2008, a semblance of stability in the organization was restored with the hiring of Michael Ellzey as the fourth CEO in five years. Ellzey, a former Marine, had worked in Northern California on large-scale construction projects.

Ellzey's timing was unfortunate, as 2008 ushered in one of the most catastrophic economic downturns in the nation's history, pummeling Wall Street and taking down venerable investment banks Bear Stearns and Lehman Brothers.[18]

It was the shockwaves from Lehman's destruction that hit hardest at the Great Park: Lennar had borrowed $775 million from the bank to finance its purchase of the property. Emile Haddad described the bizarre task of sifting through Lehman's financial rubble to determine who had inherited his debt. Lennar cut staff and hunkered down to preserve cash.

From 2007 until late 2009, when little was accomplished on the large-scale elements of the park, programming expanded, with concerts, funding for an artist-in-residence, gardening classes, and the introduction of a farmer's market. A used carousel was discovered by a board member and rehabilitated for use at the park. Board members defended the expenditures, explaining that they wanted to "energize" the park and encourage people to visit it.

Plans to demolish the huge runways that crisscrossed the base seemed to be in constant flux. In line with the park's sustainability goals, a contractor that had demolished and recycled Denver's Stapleton Airport runways was hired and produced plans to recycle virtually all the El Toro runway concrete and asphalt. Discussions ensued between the park staff, Lennar, and the contractors that dragged on for months.

In early 2008, the contractor quit, trucking its expensive equipment back to Colorado.[19] Yet, five months later, in one of the annual "State of the Park" addresses, the promise was made that "the actual breaking up on runways would begin next month."[20] Indeed, in the first few months the city controlled the park, a few hundred yards of concrete had been removed, broken up and used in various places around the park. The runways remained largely intact until 2010.

With public pressure mounting, Ellzey convened meetings with park staff and consultants, which led to the unveiling in October, 2009, of a $60 million construction program for about 200 acres of the park next to the 27-acre "Preview Park" that contained the balloon, a refurbished hangar, visitor center and a small playground.

The Western Sector Plan was approved in February 2010.[21] A few months later work began on four soccer fields, a 105-acre farm, and a new "Palm Court" building and plaza designed to showcase cultural programs.

In January 2011, the park got a boost when Haddad announced a revised funding agreement with investors and a promise to deliver lots to homebuilders by early 2013. Moreover, some of the $40 million in planned infrastructure improvements would start to be built within the park.[22]

Then, just weeks later, California governor Jerry Brown dealt Irvine and dozens of other cities a blow by shutting down redevelopment agencies and seizing unencumbered funds.[23] The nearly $1 billion in future revenue that Irvine needed to fund the Great Park development had vanished.

Heading into the election year of 2012, the Great Park had become the city's defining political issue. In November, Agran lost the mayor's race to Stephen Choi, and Christina Shea was returned to the council after a six year absence. While Agran would remain on the council for the remaining two years of his term as a councilman, he was now at the short end of a 3–2 council majority. And that majority had the Great Park in its sights.

***

Lennar/Heritage Fields/Five Point Communities

The Lennar team now working under the banner of Heritage Fields was headed up by Emile Haddad.

A native of Lebanon, Haddad was tall, slim, and intense, but he also had a winning smile. He liked to tell people the name "Haddad" in Lebanon was equivalent to "Smith" in the United States. Haddad was an engineer by training and had worked in war-torn Beirut before immigrating to the United States. Arriving with his family to live with his brother, Haddad struggled to put his degree to work, instead finding jobs in construction. A few years later, he caught on with a homebuilder that was acquired by Lennar.

Haddad staffed up with project managers familiar with both homebuilding and commercial development. He hired well-known local public affairs and community relations professionals and began making the rounds of local events to talk up the vision he and Lennar had for the new community that would be built around the Great Park.

One of Haddad's best-known project managers was Robert (Bob) Santos. Santos was soft-spoken and methodical. He had substantial experience in large-scale development, thanks to more than a decade of similar work with the Irvine Company on signature projects like Woodbridge. Santos was often the lead in presentations at the city council and other public bodies. Tragically, he died of cancer in 2012.

Like the Great Park owners, Lennar was faced with the seemingly endless chore of making the former base ready for development, which meant clearing hundreds of acres of land and demolishing buildings and other facilities while developing plans for sewers, water lines, roads, and other infrastructure.

Heritage Fields also had to negotiate both community-wide and neighborhood level zoning agreements with the city. Three major "development agreements" were hammered out, each one giving Haddad a few hundred more units in exchange for more promised funding for park development and operations.

As Heritage Fields focused on the Great Park, its parent company, Lennar, was wheeling and dealing in real estate all over California. Just after the BRAC process that resulted in El Toro coming on the market, the Naval Shipyard in San Francisco was transferred to the City and County of San Francisco for redevelopment; Lennar got the nod from the city to make it happen. Parcels were initially due to be transferred in 2006 and 2007.

Also in 2007, Lennar sold off a 68 percent interest in Newhall, a giant landholding at the far northern end of Los Angeles County adjacent to the new community of Valencia. The price was $970 million, the buyer was CalPERS, the giant pension system for California state employees. CalPERS was diversifying its investment portfolio into real estate, and what could be a better, safer, and more lucrative holding than land in the Golden State?

The decision backfired two years later in spectacular fashion, with the CalPERS real estate holdings plunging in value. Lennar stepped back in and happily paid $139 million to purchase 15 percent of Newhall, giving it a 47 percent holding, with financial partners owning the remainder (and with the understanding Lennar would direct the development).

With three large and diverse real estate development assets added to its corporate holdings, and with Heritage Fields recovering from the 2008 economic meltdown, Lennar made the decision a year later to create a new entity called Five Point Communities that would manage Heritage Fields, Newhall, and San Francisco Naval Shipyard (by then, land at the closed former San Francisco Giants Candlestick Park had also been obtained.) Haddad was appointed president of Five Point, whose owners included Lennar, LNR, and a host of private investors—including Haddad, who invested $1 million of his own money.

Of the four projects, Heritage Fields should have been the easiest to manage. Newhall was under the jurisdiction of Los Angeles County and in the district of long-time supervisor Michael Antonovich. An hour's drive from the county headquarters (and staff) in downtown Los Angeles, Newhall issues were often placed on the back burner of the district management apparatus. Additionally, the Los Angeles County planning bureaucracy was notoriously complex and ponderous.

San Francisco was even worse—city politics were defined by warring tribes of special interests reflecting a Rainbow Coalition of progressive, liberal and radical leftists. Planning decisions moved along at a sloth-like pace. To top it off, Lennar (to benefit Five Point) was suing Tetra Tech, the firm responsible for identifying toxic areas of the naval shipyard.

When Five Point went public in 2017, much had changed in the intervening eight years since it was created.

Notes

1 Orange County Register, "Measure W Passes, El Toro Airport Defeated," March 6, 2002.
2 U.S. Navy, *Environmental Baseline Survey for Former MCAS El Toro*, Prepared by Southwest Division, Naval Facilities Engineering Command, 2003. Available at City of Irvine archive.
3 U.S. Navy, *Environmental Baseline Survey*, 2003.
4 General Services Administration, *Post-Implementation Report on Base Realignment and Closure (BRAC)*, 2021, https://www.governmentattic.org/52docs/GSAbracPostImplementRpt2021.pdf
5 . General Services Administration, Post-Implementation Report, 2021.
6 *Los Angeles Times*, "Lennar Analyzes El Toro Site," February 17, 2005.
7 *Los Angeles Times*, February 17, 2005.
8 City of Irvine, "Formation of the Great Park Corporation," Council Minutes, 2003.
9 Ballot language for Measure W.
10 Irvine City Council Agenda, April 2006, Contract Approval for Forde & Mollrich.
11 Orange County Grand Jury Report, "Governance of the Great Park," 2007.
12 *City of Irvine Press Release*, "Ken Smith Selected to Design Great Park," July 2006.
13 Diane Ghirado, "Design Jury Summary Report," City of Irvine Archives, 2006.
14 *Los Angeles Times*, "Balloon Rises Over Great Park," October 2007.
15 American Planning Association, "Award for Great Park Master Plan," 2008.
16 Jones Day, "Lennar Settles 10-Week Jury Trial After $200 Million Damage Claim Excluded," *Experience Highlights*, December 2010. *Forest Lawn Memorial Park Association v. LNR California Investments, Inc.; Heritage Fields El Toro, LLC; Heritage Fields LLC; LNR Heritage Fields, Inc.; LNR Partners, Inc.; (formerly known as Lennar Partners, Inc.); LNR Property Corporation; Lennar Corporation; and Does 1–100*, Case No. 30-2008-00111856 (Superior Ct. of Ca., County of Orange, Central Justice Center), https://www.jonesday.com/en/practices/experience/2010/12/lennar-settles-10-week-jury-trial-after-$200-million-damage-claim-excluded.
17 *OC Weekly*, "The Wizard of Agran," November 2010.
18 Jeff Collins, "Mothballing of Lennar Home Projects to Continue," Orange County Register, January 29, 2009, https://www.ocregister.com/2009/01/29/mothballing-of-lennar-home-projects-to-continue/.
19 Orange County Register, "Runway Recycling Contractor Quits," February 2008.
20 C. Michael Stockstill, "Too Big to Fail," *Planning* (American Planning Association), April 2014.
21 City of Irvine, "Western Sector Plan Approval," Council Minutes, February 2010.
22 Orange County Register, "Haddad Announces Revised Funding," January 2011.
23 California Department of Finance, "Redevelopment Agency Dissolution," 2011, https://dof.ca.gov/Programs/Redevelopment/.

21 Building Heritage Fields

In the ten years that had passed since Measure W was adopted at the 2002 county elections, the Great Park had increasingly dominated Irvine's political atmosphere. As candidates were elected every two years, a pair of them would join Larry Agran in a ruling majority on the five-member council.

All that changed in November 2012. Agran, in the middle of a term, ran for mayor and was defeated by Stephen Choi. Former mayor and council member Christina Shea won a council seat and, with Jeffrey Lalloway, assembled a new majority that vowed to make big changes at the Great Park. They started in early January 2013 by dismissing the four individuals on the nine-member Great Park Board of Directors who were not council members.

In one swift action, Michael Pinto, James "Walkie" Ray, William Kogerman, and Santa Ana mayor Miguel Pulido were gone.[1] Of the four, Ray and Pulido were closest to Agran and the previous council majority; Pinto was a member of the Laguna Beach Conservancy, and Kogerman was a retired Marine who had been active in the anti-airport campaigns. Almost without exception—and even though officially their votes were meaningless—the four had rarely voted against the Agran majority as Great Park board members.

Next on the chopping block were Arnold Forde and Stu Mollrich, who were swiftly removed from the project by the new majority. Beth Krom, now on the losing end of the 3–2 council, termed the actions "a witch hunt," a term Agran would echo in the months ahead.[2]

The action that would ultimately prove the most consequential was the new council majority's decision to audit the management and construction of the Great Park by the Design Studio. A budget of $225,000 was approved to hire the firm of HSNO to conduct a comprehensive review of the Great Park's ten years of spending records.[3] (Actual work on the audit started on June 17, 2013.)

While the new council was reordering its park priorities by setting admission prices for the balloon and carousel ($10 for adults on the balloon, $5 for children, $2 for anyone on the carousel), it also approved an agreement with the U.S. Department of Energy to hold the Solar Decathlon at the Great Park in October. The contest brought together college and university students to design and build houses with maximum energy efficiency. More than a dozen schools would participate in

DOI: 10.1201/9781003683087-21

the event, which had previously been held annually on the National Mall in Washington, D.C.[4]

While the city's political drama played out in the initial months of 2013, Emile Haddad was preoccupied with two big issues: the opening of the first new homes in Great Park Neighborhoods and the hundreds of acres of the Great Park that remained empty.

Haddad's team had selected a parcel of land on the north side of Irvine Boulevard for its initial development. It made sense: the area was adjacent to existing infrastructure and free of debris. Heritage Fields marketed the 760-home community as Parasol Park, reinforcing its creation of a large park and community center at the base of the neighborhood. As was the case decades earlier when the Irvine Company opened the first phase of Woodbridge—with extra landscaping and enhanced amenities—little expense was spared to make the community center look like it had been built for a five-star hotel.

The neighborhood appeared to have existed for years, as many of the mature trees from the El Toro base had been carefully removed, crated up, and stored until they were replanted in Parasol Park.

Eight homebuilders vied for the privilege of selling their products in Parasol Park, using a variety of traditional styles to meet Haddad's desire that the new community resemble a small town anywhere in America. More than 20,000 people visited Parasol Park during its "open house" weekend; sales followed.

Across the street from Parasol Park—now connected by a striking wood and metal pedestrian bridge spanning Irvine Boulevard—the first element of new neighborhoods on the main property of the former base was taking shape. While it would have signature parks, trails and other community improvements, hundreds of the planned new homes would look out on a vast wasteland of weeds, deteriorating buildings, and military detritus.

Haddad conceived a bold plan to jumpstart the Great Park: he would take charge of constructing 688 acres of the park in return for a variety of financial arrangements related to previous infrastructure spending and future studies, plus the most lucrative incentive possible for any developer: permission to build 4,600 *additional homes* on top of what the city had previously authorized.

The cost to Heritage Fields would be $172 million (less some reimbursements for future infrastructure spending). Haddad specified that if accepted, he would be in charge of the development, with only regular building code oversight from the city. To assure his authority, he proposed to lease the land from the city, then return it when improvements were completed.

For the $172 million, the Great Park would receive[5]:

- Sports fields totaling 176 acres.
- Forty acres of the planned bosque, plus playgrounds and a farmer's market.
- Trails and open space on 36 acres of Upper Bee Canyon.
- An 18-hole golf course.
- A wildlife corridor of 178 acres.

In addition, $10 million would go to the city to operate and maintain the newly improved acreage for ten years and close to $20 million for studies and a promise to dress up areas of the park not planned for development.

Haddad floated the proposal in late spring, 2013, but it was not until November that the issue came to the city council for consideration. After a marathon council meeting highlighted by testimony from dozens of sports enthusiasts (some in soccer and baseball uniforms), the council agreed to the principal elements of the deal.[6]

Within a few months, the promised improvements were underway, followed by the first homes in the main section of Great Park Neighborhoods. While many of the new homes there displayed design elements familiar to Irvine—sloping tile roofs, stone facade accents, and an overall sense of a Spanish/mission look, with a scattering of ranch-style structures—the designs that drew the most comment were unlike any seen before in the city.

Some of the homes were three stories tall and narrow, creating a neighborhood that was dense and looked it. Roofs were flat, the sides of the homes sharply defined. Where in other parts of Irvine homes designed by the Irvine Company had a softer profile, in the Great Park Communities, the look was all edges. The accents

Figure 21.1 Baseball fields at the Great Park Sports Complex

Source: Author's image

Figure 21.2 Some homes in Great Park Neighborhoods were designed to differ from those found elsewhere in Irvine

Source: Author's image

on these homes accentuated their height. Some observers saw an element of 1930s Art Deco being recreated or sleek Streamline Moderne as an inspiration.

The color palette of the new homes trended towards shades of gray, with green, maroon, and blue highlights, a marked deviation from the variety of beiges common to Irvine Company villages.

Like other neighborhoods in Irvine, the entry streets were curved, meandering into the middle of the vast property and generously landscaped on the medians and parkways. But the plants and trees were drought-tolerant, giving the scene a somewhat desert feel.

Visitors to open houses for the new community were impressed, and sales across all product lines took off. Haddad and his team had also achieved success in the commercial sector, building and leasing a large office complex in the acreage on the opposite side of the giant landholding. And there was activity in some of the old military warehouses and adjacent properties, where a regional food bank established a distribution center and a recycling company was turning garden clippings into tons of compost.

***

The preliminary results of the audit were released in early January 2014.

The *Los Angeles Times* reported, "The audit faulted Great Park leadership for allowing contractors to use excessive change orders, not fully vetting major vendors and paying a communications and strategy team $6.3 million under contracts for the park's design." In addition, "Auditors said they were unable to answer some questions because several key players, including park designer Ken Smith and a former Great Park employee, refused to speak with auditors."[7]

The audit report also cited a simple statistic: of the 1,347 acres in the Great Park, "only 88.05 have been improved through construction activities conducted by the

Great Park. Another 26.59 acres have been activated for public use but are largely unimproved. These numbers exclude any leased agriculture space." The audit also spotlighted four examples of cost overruns:[8]

- Installation of the Great Park balloon. $351,833 over budget.
- Construction of Preview Park: $7.670 million over budget.
- Tree relocation, storage and maintenance: $518,428 over budget.
- Electrical power improvement project: $1.458 million over budget.

HNSO requested that the council authorize subpoenas to compel witnesses to testify as well as an increased budget to investigate past Great Park operations more thoroughly.

The *Times* went on:

> Councilmembers Larry Agran and Beth Krom, who helped steward the project from its beginnings, denounced the decision. "What I fear we are witnessing here tonight is the apex of a campaign of lies, distortions, and misrepresentations . . . Krom said." Agran dismissed the plan for additional work as a waste of money.[9]

With authorization to proceed with subpoenas if needed, HSNO partnered up with a law firm and brought on a retired judge to assist with depositions.

More than a year later, in late March 2015, HSNO released its "Orange County Great Park Forensic Contract Performance Review," noting that the findings were not an audit in the strict sense as defined by institutional accounting standards but as a consultant, where "the practitioner develops the findings, conclusions and recommendations presented."[10]

The report had pages of details reflecting the initial findings, but it was the depositions of 20 individuals associated with the project that drew the most interest from Irvine's political community and the press.

A remarkable 29-page statement from park CEO Michael Ellzey revealed just how the planning and development of the park really functioned. Ellzey began with a description of his first day on the job:[11]

> Chairman Agran met with me privately to welcome me aboard . . . I was told there were two things that would be critical to my success . . . First, that I would have two deadlines for completing construction projects—July 2008 (Great Park Anniversary) and November 2008 (Election Day). Second, I was advised Yehudi Gaffen was the most important player on the project; that he would be my counterpart; and that Mr. Gaffen "speaks for me" [meaning Chairman Agran] on project matters . . .
>
> The project team was also under extreme pressure to deliver projects in advance of Irvine's elections . . . Arnold Forde was the enforcer of these project dynamics, and Yehudi Gaffen was the implementer of the program—each pursuant to the stated authority of the Chairman.

> Over the next several months, as I became familiar with the project structure—Design Studio personnel (Gafcon, Ken Smith, Fuscoe staff), Great Park Corporation project staff and the City's program manager, Bovis, and of course a very involved Chairman Agran, Arnold Forde and Yehudi Gaffen (the "power structure")—it became increasingly clear that Mr. Gaffen was running the project and that City staff had been relegated to virtual irrelevance. The Design Studio was informing staff which projects would be undertaken.[12]
>
> The consultant was telling the client what to do. The city was not being treated like a client or owner, but more as a facilitator and expeditor of projects determined by others . . . Based on my experience managing large public projects, this practice was clearly unacceptable for a public agency. However, I was the chief projects officer and my responsibility was to build the park, so I found a way to work within the system, as dysfunctional as it was . . . through the efforts of good staff and consultants, projects were completed on schedule and within budgets approved by the City Council.[13]

Ellzey's statement continued with a description of how he fought to gain control of the project from "the power structure" and others and struggled to implement plans he considered incomplete or shoddy. He brought in two men he had worked with in the past to help oversee the Design Studio, which he described as empowered to the degree that they simply ignored or overrode the city and Great Park Corporation staff.[14]

Gaffen and Forde were singled out repeatedly in Ellzey's statement as his chief adversaries in the Great Park drama (observers of the process pointed out that Ellzey's description of events, where in the end he prevails over the pair, could be construed as self-serving).

Having been cast in such a negative light, it was not surprising that Forde and Mollrich retained an attorney to advise them on responding to depositions. Gaffen went further, hiring a well-known San Diego public affairs firm to create a sophisticated media campaign to counter claims that the firm had abused its role on the project.

Agran complained early in the process that the entire audit effort was a taxpayer-paid hit piece aimed at him. During the November 2014 city campaign, bright orange signs with his image and the words "Agran Audit" began popping up along major streets in the city. The sign had a link to a website where the entire audit could be read.[15] Agran lost his council seat (Krom was termed out of council races).

It was the beginning of a new act in the drama of the Great Park, but it would be years before the final curtain. HNSO surrendered its accounting license to the California Board of Accountancy and paid a $150,000 penalty. The board found the firm used "deficient findings" to justify performing additional audit work. The firm's attorney said "HNSO one hundred percent stands by its reports (on) the Great Park."[16]

Notes

1 "Irvine Council Removes Great Park Board Members," *Los Angeles Times*, January 9, 2013.
2 "Forde and Mollrich Fired from Great Park Project," Orange County Register, January 15, 2013.
3 "Audit of Great Park Spending Approved," *Los Angeles Times*, January 28, 2013.
4 "Solar Decathlon Comes to Irvine," *U.S. Department of Energy*, accessed September 13, 2025, https://www.energy.gov/eere/solar/solar-decathlon.
5 Emile Haddad, proposal summary to Irvine City Council, November 2013.
6 "Council Approves $172M Great Park Deal," Orange County Register, November 27, 2013.
7 "Great Park Audit Faults Oversight," *Los Angeles Times*, January 28, 2014.
8 "Great Park Audit Faults Oversight."
9 "Great Park Audit Faults Oversight."
10 HSNO Forensic Contract Performance Review, March 2015.
11 Michael Ellzey, deposition, HSNO Review, March 2015.
12 Michael Ellzey, deposition.
13 Michael Ellzey, deposition.
14 Michael Ellzey, deposition.
15 "Agran Audit Campaign Targets Council Race," Orange County Register, October 15, 2014.
16 Ben Brazil, "Orange County Great Park Auditor Surrenders Accounting License After Legal Battle with State Oversight Agency," Daily Pilot, January 29, 2020.

22 Joan at Peace

The trial that would determine the value of Joan Irvine Smith's Irvine Company stock was nowhere near its conclusion on the day she took a drive in San Juan Capistrano.

The small city famous for its beloved and historic mission had retained much of its charm in its downtown, with its collection of adobe and modern buildings. Tourists could wander along a cool alleyway and see the primitive water collection system that had been conceived by Franciscan padres and built with the labor of Indigenous men and women 250 years earlier. On Los Rios Street, across the railroad tracks, buildings dating back to the 1800s had been converted to tea houses and restaurants. Mexican food was available at venerable El Adobe and a cold beer at the colorful Swallows Inn.

In the hills that surrounded the city, huge mansions had been built, many not in Mission style but often in the faux French chateau look favored by the recently rich who desired an outsized home on an even larger lot.

On this day, Joan was not looking for a mansion but stumbled on something that would, with another passion, largely define the final years of her life in California.

It was fall 1984, and she was on her way to a water board meeting, she told the *Los Angeles Times.* Cruising along Ortega Highway, which ran east out of town, "Smith said she spotted a dilapidated stable on the roadside. 'I said, I could really shape that up,' she recalled. So last spring she bought the 12-acre stable and arranged to lease a few additional acres for her new enterprise."[1]

The article continues:

> Under Smith's close direction, workmen renovated the barns, installed new irrigation systems, planted bright coral-hued flowers and artfully trimmed the spreading oak trees . . . and they spruced up the show rings where horses can compete . . . The Oaks, with stable accommodations for 108 horses, takes in boarders but also has about 48 jumpers and hunters in training under horseman Jimmy Kohn. About a dozen of the horses in training are owned by Smith. "I am buying and selling show horses, same as I do back East,'" Smith said. "It sure as hell is no hobby," she added.[2]

Joan and her former husband, Morton "Cappy" Smith, rode, trained, and bought and sold horses at their home in Middleburg, Virginia, where Joan had moved when

DOI: 10.1201/9781003683087-22

they married in 1963. But, before long, she had also purchased her own home on a 215-acres estate down the road from the one they shared; she named it The Oaks.

While the Smiths' equestrian business involved them both, in Virginia, Joan lived in the shadow of her husband, whose career as a top competitor made him famous in the world of professional horse lovers. Also, Cappy had lived in Middleburg for decades, served as master of the prestigious Orange County fox hunt, and was regularly described as "dashing" and "movie star handsome."

So for Joan, establishing her own California riding center and creating and directing every aspect of its planning and operations was a watershed moment in her tumultuous life. It offered a welcome distraction from the litigation taking place in Michigan, where her attorneys were jousting with Irvine Company litigators in a case that seemed never-ending. And The Oaks was also an opportunity to leave her dream of owning and controlling the Irvine Ranch behind her—she had burned that bridge by agreeing to sell her stock years earlier.

The Oaks became a new dream, and Joan's wealth, which was estimated in her obituary to be $350 million, allowed her to create both a recognized center of equestrian excellence in that tony world as well as one of Orange County's most glamorous social hot spots.[3]

The debut of what would become the signature event for Joan, her mother, and her family came in June 1986, with the Oaks Classic, a Grand Prix jumping competition with a purse of $35,000. More than 800 friends were invited to the festivities. The *Los Angeles Times*' society columnist Ann Conway gushed about the Classic, complete with this description of Joan's outfit for the day: "Smith may have single-handedly invented designer equestrian events. She wore a Hermes dress matching the orange geraniums on the course."[4]

A year later, the purse for the Oaks Classic had doubled to $70,000; attendance was estimated at 3,000, with Joan proudly noting that 1,500 had paid $100 a seat, another 1,500 just $5. The *Times'* Mary Lou Loper chronicled the food: "Rococo catering provided shrimp and crab and piscine wonders for the first course, Mexican tostadas for the second, hot fudge sundaes for the third," as well as the scene: "The Oaks beauty is unchallenged, California green turf, California oaks, colorful, pretty jumps of brick and white latticing." And for the first time, the *Times* included a long list of socially and politically prominent visitors from California and around the country, putting San Juan Capistrano firmly on Southern California's map of where the rich and famous gathered for sport.[5]

In the lead-up to the 1988 Oaks Classic, the *Times* commissioned a profile of Joan featuring her business acumen as a breeder and trainer:[6]

> In a world that has not always seemed fair to her, Joan Irvine Smith, horsewoman, has learned a few things from Joan Irvine Smith, businesswoman. A gruff-talking, shoot-from-the-hip Katharine Hepburn type, she knows exactly what she likes, and what she doesn't like, whether in the boardroom or the tack room.[7] The *Times* story continued: "Smith . . . insists this is a commercial facility, and every horse is for sale, but the horsewoman admits that she recently turned down an $800,000 offer for Hermes, one of her favorite mounts."[8]

Figure 22.1 Joan Irvine Smith at her equestrian center—The Oaks
Source: Courtesy of Calisphere

The 1988 Oaks Classic did not disappoint, according to the lengthy *Times* account of the weekend event,

> a curious mixture of leather and lace, of Tanqueray and tallyho. Despite the Classic's mistaken reputation as an invitation-only soiree, patrons' tickets [sold] for $200 for the weekend. And for the brown-bag set, bleacher seating costs only $8 per day . . . but who's to complain? For a little more than the cost of a movie ticket, spectators get a full day's entertainment at a world-class sporting event—and a rare peek in the generally closed world of horses and high society.[9]

Photos of Joan at this and other events show that she was still vibrant, and though not the slender ingenue, married three times in her 20s, she had a figure she wasn't afraid to accentuate in jeans and riding boots. She was a natural blond, now showing streaks of gray and silver; when she wore her hair long, she sported braids or a ponytail, and as she aged, she favored bangs at the neck, her high forehead bare. Her eyes were piercing, with wrinkles beginning to crease the corners. Joan spent hours in the sun, which left her with a perpetual tan.

So, year after year, Joan enjoyed the role of hostess to the coastal elites, as well as former husbands, her children, their spouses, and a growing host of grandchildren, some of whom were competing in the junior riding events. But Joan's fascination with glamor took a new turn in 1992, leading her on a chase not on the dusty turf but in the refined world of the gallery.

***

Susan Trager had finally started to make enough money as an attorney that she could afford to indulge in her passion for art.

Living in Laguna Beach, she had become familiar with the works of *plein air* artists, many of whom had migrated to the seaside enclave in the 1920s and 1930s and used its craggy shoreline and dramatic coastal canyons as inspirations for their landscapes.

Trager had started to acquire a few small paintings to display in her office, but in early 1992, when she returned to a favorite gallery, the paintings that had been on sale were all gone. Odd, she thought, and drove to another gallery in Laguna to face another empty wall where a dozen *plein air* landscapes had hung just weeks before.

Trager found the store manager and demanded to know where the artwork had gone. Well, funny thing, he replied, a woman who looks a bit like you was in here last week and bought every piece of *plein air* art we owned. Trager, who wore her strawberry blond hair in short bangs, was aghast when she figured out the identity of the buyer and realized who she was up against in the burgeoning competition for *plein air* domination—the wealthiest woman in Orange County—Joan Irvine Smith.

***

Joan Irvine Smith's passion for art was not new. As a child, she was a talented painter. Early marriages and motherhood no doubt left little time for creative outlets, but now, nearly 60 years old and wealthy enough to indulge almost any passion, her desire to gather *plein air* art blossomed.

In April 1992, the *Los Angeles Times* reported that

> over the last six months, Joan Irvine Smith has been scouring art dealerships, auctions, and the storehouses of private collections in search of Southern California's scenic past . . . [she] is snatching up hundreds, perhaps thousands of paintings that capture the beauty of Southern California before developers, including the Irvine Company, rolled over its canyons, hills and coves.[10]
>
> Exactly how much Smith has bought and at what price is uncertain. Peter Fairbanks, director of the Montgomery Gallery in San Francisco, which sold her dozens of paintings, estimates she has spent $6 million on one thousand pieces of various sizes and quality over the last six months. Joan had an answer for the obvious question: what will she do with hundreds of paintings? She would create her own museum to display the paintings she said reminded her of ". . . the California as we will never see it again–California as it was when I was a child."[11]

The 4,000-square-foot Irvine Museum opened on January 15, 1993, on the 12th floor of an office tower near John Wayne Airport in Irvine. The *Times* reported that it would "allow for the exhibition of 60 paintings at a time, tops," serving as a temporary site until a permanent museum could be constructed.[12]

The story went on,

> Smith said she has amassed about 2,000 works but would not say how much she has spent. Various figures have been bandied about; one expert who wished to remain anonymous said her outlay was well over $15 million. In the spring, she also plans to open a commercial gallery of California Impressionism and contemporary art.[13]

Over time, the museum became almost as important to Joan as The Oaks. She hired a curator, published more than a dozen coffee table books of art with color reproductions of many of her paintings, as well as her autobiography, *A California Woman's Story*, filled with art and recollections of her eventful life. Her son James Swinden joined her as a director of the museum, helping to organize tours that showed off the California artwork in the United States and abroad.

Long term, Joan said she hoped to convince UC Irvine to lease her enough land on the campus to build a permanent museum of 40,000 to 60,000 square feet. Campus officials expressed positive thoughts about the prospect but cautioned that any decision would require review at the campus and, ultimately, by the UC Board of Regents.

***

In 1995, Joan was in an expansive mood when it came to her equestrian interests, which were increasingly both a personal passion as well as a business venture. She approached the owners of Rancho Mission Viejo with a plan to create a large riding center less than a mile from The Oaks—to be called the Oaks Rancho Mission Viejo Riding Park.

It would be a fitting combination of the heirs of two of the largest ranchos-turned developments in the history of Orange County. Rancho Mission Viejo was the lower half of a vast rancho that originally bordered the Irvine Ranch to the north and ended in Oceanside, more than 30 miles to the south.

Rancho Mission Viejo patriarch Richard J. O'Neill Jr. shared Joan's passion for horses although he preferred to race them instead of show them. O'Neill's nephew, Anthony "Tony" Moiso, who was the chief executive officer of the company, enjoyed riding, roping, and sustaining the cattle-ranching heritage that was rapidly being replaced with the homes in the new communities the company was building in the hills beyond San Juan Capistrano.

According to the *Los Angeles Times*,

> The park will be about three times the size of her central show ring at The Oaks and able to accommodate about 5,000 spectators. She also suggested the fields could be leased for polo and soccer games, car and dog shows, and concerts.[14]

Around the same time, Joan purchased a 43-acre horse farm in Valley Center, near Escondido, to expand her breeding operations. While the Oaks Indian Hill Ranch prospered, it was the site of a tragic and mysterious death that sent Joan reeling.

A prized stallion, South Pacific, had to be euthanized after he was infected with a rare disease. In an echo of her past bouts of paranoia, Joan seized on the notion that someone had deliberately killed the valuable horse by injecting it with the fatal parasite. She summoned law enforcement officials, hired a private investigator, and set up an 800 number to accept tips from anyone who might have a clue to solve the crime.

Joan theorized that a rival breeder was responsible, but no evidence ever surfaced to determine if foul play was the reason for the stallion's death. In the end, Joan was left with some frozen semen from South Pacific's stud career she vowed never to sell.

***

In the earlier years of her life, Joan was regularly entangled in legal actions, most relating to the Irvine Company and almost always at her instigation. With her sale of company stock, some observers believed she might curtail her litigious ways.

Such was not the case.

A few months into her *plein air* art buying spree, Joan sued a former art agent for $24 million, claiming she never got paintings he was supposed to have purchased for her collection. According to an Associated Press story in the September 14, 1993 edition of the *San Francisco Examiner*, "Agent Michael Johnson said he didn't keep the paintings, and his attorney claimed Smith hadn't paid Johnson's fees."[15]

Smith explained that $4 million was for missing paintings and $20 million was punitive damages. Johnson's attorney replied that he was due $750,000 in finder's fees still unpaid. (No record can be found in print about the outcome of the suit).

Four years after the art agent episode, newspapers were reporting on another legal action that featured the tried-and-true subject that sends reporters and editors into a journalistic frenzy: sex.

From the *Los Angeles Times,* February 2, 1997:

> It all began in 1985, when Smith . . . hired (James "Jimmy") Kohn, a nationally renowned handler and trainer, to work on her horse ranch, The Oaks . . . At The Oaks, Kohn and Smith developed an intimate as well as working relationship, according to court documents and to friends and co-workers who know them both (in 1985 Smith was 51, Kohn 38).[16]

The story continued:

> But in April, 1989, attorneys close to the case say, Smith became furious when Kohn brought (Meg) Howard to The Oaks. Smith learned that day that Howard was Kohn's "live-in girlfriend" and that "he planned to marry her," according to court documents. Smith responded by terminating her

> association with Kohn, which she described in court records as "more than a business relationship."[17]

Kohn then sued Smith and obtained a restraining order against her, the *Times* story continued.

> Smith responded in July, 1989, by filing a cross complaint against Kohn and Howard (the pair married in April 1990 and divorced two years later). Howard later sued Smith for malicious prosecution, a matter that received attention just last month from the 4th District Court of Appeal. That court reversed a decision . . . to dismiss Howard's suit (which) has been returned to Superior Court for further litigation.[18]

"In 1993, Kohn and Smith agreed to a confidential settlement."[19](Thirty-two years later, the authors were unable to determine the outcome of the Smith-Howard litigation).

The last major lawsuit initiated by Joan was against a fellow equestrian entrepreneur, Robert "RJ" Brandes, owner of Blenheim EquiSports in San Juan Capistrano. Smith and Brandes had teamed up in a partnership to build yet another horse show arena. But within months, Smith was accusing Brandes of bad faith dealings and sued him.[20] Interestingly, the lawsuits connected with Smith's art and equestrian conflicts were handled by new teams of lawyers—her old standby Howard Friedman of Loeb and Loeb, who had represented her and her mother for decades, had retired in 1997.

The lawsuit was heard by a former federal judge retained as an arbitrator, who ruled in 2008 that Brandes was in the right and the partnership owed him more than $2 million.[21]

***

In late December 2007, *Orange County Register* reporter Norberto Santana was hanging around the fifth floor of the county executive building, where offices of county supervisors were located. It was a slow season for news, prompting Santana to poke around for a possible story or feature.

Within a few minutes of his arrival, Supervisor John Moorlach came through the elevator doors, pushing a cart filled with what Santana quickly determined was artwork.

Interviewed in 2025, Santana recalled that Moorlach looked worried as soon as he spotted him; Santana had a reputation as a bulldog reporter, eager to pounce on any story he considered with a whiff of scandal or bad governmental behavior. Moorlach explained that the paintings were on loan from the Irvine Museum, the property of Joan Irvine Smith. He said they were worth about $5,000, had not cost the taxpayers anything, and would be returned a few months later. Hanging on them on the walls of the county office would be a positive for visiting taxpayers, he assured Santana.[22]

The reporter was not convinced. The next day his story about the paintings ran in the *Register,* where he argued that a loan of any sort to an elected official had value and needed to be reported (the county legal staff had signed off on the loan, declaring no conflict or need to report).[23]

At that point, Joan Irvine Smith entered Norberto Santana's life.

"My phone rings," he said, "and this woman is screaming at me, telling me my story is all wrong. Then she says I need to come to see her and she can explain why I am wrong."[24] Santana was born in Cuba and came to the United States at age 6 as a refugee from the Castro regime. The experience, plus his naturally combative personality (he was an amateur boxer), rankled him when challenged on a story.

"I drove down to San Juan and knocked on Joan's door. Her house looked like John Wayne lived there. All the furniture was leather, and there were colorful horse blankets all over. The walls were covered with artworks."[25] Joan proceeded to lecture Santana about the paintings, then went on to explain their history and value. She loaded him up with a number of the books about *plein air* art her museum had published. Then she sent Santana to a local deli to buy sandwiches so they could continue their discussion. Thus did a modestly paid reporter begin what would become a series of lunch meetings with the wealthiest woman in Orange County, with him paying for lunch.

As other people who came into Joan's orbit had experienced, Joan took the floor of her modest home (a very nice double-wide trailer set among the oaks and landscaping in a corner of The Oaks riding center) and regaled Santana with her views on the Irvine Company, her family, local events, and art. After a couple of meetings, he recalled, the two achieved a genial standoff of sorts, with Santana becoming increasingly fascinated by his one-on-one exposure to the commanding personality of the colorful, vibrant Smith. "She reminded me of my mother, who would not back down from anyone," he said.[26]

Apparently, there were few if any limits to Smith's view as to what was appropriate for she and Santana to discuss or experience, as one of the most unusual (and slightly macabre) incidents was when Smith demanded that Santana accompany her to visit Richard J. O'Neill, patriarch of the family that developed Mission Viejo and Rancho Santa Margarita.

It might have been a productive interview but for the fact that O'Neill was on his deathbed, Santana said. Joan was apparently unfazed.

The Smith-Santana relationship continued, with weeks, then months passing between visits. Eventually, he said, they just stopped meeting.

***

Horses and art were two foundations of Joan's life after parting ways with the Irvine Ranch; philanthropy became the third.

Joan had given to various charities throughout her lifetime, but after receiving $190 million from the sale of her Irvine Company stock and likely inheriting a similar amount after the death of her mother, her ability to expand her donations increased exponentially.

One of her first major charitable initiatives was related to a tragic accident in Culpepper, Virginia.

On equestrian grounds there in May 1995, actor Christopher Reeve, famous for his role in the *Superman* movies, was riding in a competition when his horse balked mid-jump at a fence, throwing Reeve headfirst to the ground. He broke the first and second cervical vertebrae on his spine, rendering him paralyzed below the neck and unable to breath without a ventilator. Joan learned of the accident and was horrified: "I've ridden for years, had horse accidents and four or five concussions . . . it could have happened to me."[27]

But she was also impressed that Reeve refused to blame or destroy his horse after it happened and maintained a positive attitude about his future. It inspired her, she told *ABILITY* magazine, to found a research center at UC Irvine devoted to the study of neurological disease and injuries.[28]

Joan and her mother, through their foundation, made the initial $1 million gift to UC Irvine to establish what would become the Reeve-Irvine Research Center. She donated $80,000 from her Oaks Classic the following year. Reeve's friends in entertainment—including Robin Williams, who roomed with Reeve early in their careers—appeared at benefit events.

The Reeve-Irvine Center is now home to dozens of researchers seeking ways to help individuals with spinal injuries.

Four years before her efforts on behalf of Reeve and neurology, Joan's interest—some would say obsession—with water supply in southern California led her to establish the National Water Resource Institute. The institute would conduct high level research and, using her initial gifts to prime the fundraising pump, grow into self-sufficiency as time passed. In 1993, Joan and her mother's foundation endowed a fund to create an annual $50,000 award in water research named the Athalie Richardson Irvine Clarke Prize.[29]

In the final two decades of her life, Joan Irvine Smith most often turned to UC Irvine as a recipient of her charitable giving, donating to establish a law school there, to the medical school, and to other academic areas. And in 2016, Joan donated her art collection, valued at $17 million, to UCI to be housed in a museum on university land.

In October 2025, UC Irvine announced it was acquiring the existing Orange County Museum of Art in Costa Mesa. The huge structure will likely replace plans for any similar museum on the campus and, at some point, will contain both Joan Irvine Smith's art collection and others.[30]

Notes

1 "Joan Irvine Smith Finds New Passion in Equestrian Center," *Los Angeles Times*, January 12, 1986.
2 "Joan Irvine Smith Finds New Passion in Equestrian Center."
3 "Joan Irvine Smith: Obituary," *Los Angeles Times*, December 20, 2019, https://www.latimes.com/local/obituaries/la-me-joan-irvine-smith-obituary-20191220-story.html.
4 Ann Conway, "Oaks Classic Draws Elite Crowd," *Los Angeles Times*, June 6, 1986.
5 Mary Lou Loper, "Oaks Classic Expands with $70,000 Purse," *Los Angeles Times*, June 7, 1987.
6 "Joan Irvine Smith: Horsewoman and Businesswoman," *Los Angeles Times*, February 13, 1988.
7 "Joan Irvine Smith: Horsewoman and Businesswoman," *Los Angeles Times*, 1988.
8 "Joan Irvine Smith: Horsewoman and Businesswoman," *Los Angeles Times*, 1988.
9 "Oaks Classic Weekend Blends Sport and Society," *Los Angeles Times*, June 3, 1988.
10 Leslie Berkman, "California Collector on an Art Spree," *Los Angeles Times*, April 15, 1992. https://www.latimes.com/archives/la-xpm-1992-04-15-ca-248-story.html

11 "California Collector on an Art Spree," *Los Angeles Times*, 1992.
12 Zan Durbin, "Irvine Museum Paints Early California's Portrait," *Los Angeles Times*, January 14, 1993, https://www.latimes.com/archives/la-xpm-1993-01-14-ca-1678-story.html
13 "Irvine Museum Paints Early California's Portrait," *Los Angeles Times*, 1993.
14 Leslie Berkman, "Joan Irvine Smith Plans New Riding Park," *Los Angeles Times*, March 6, 1995. https://www.latimes.com/archives/la-xpm-1995-03-06-me-39423-story.html
15 "Joan Irvine Smith Sues Art Agent for $24 Million," San Francisco Examiner, September 14, 1993.
16 "Trainer's Lawsuit Reveals Rift with Joan Irvine Smith," *Los Angeles Times*, February 2, 1997.
17 "Trainer's Lawsuit Reveals Rift with Joan Irvine Smith," 1997.
18 "Trainer's Lawsuit Reveals Rift with Joan Irvine Smith," 1997.
19 "Trainer's Lawsuit Reveals Rift with Joan Irvine Smith," 1997.
20 David Reyes, "Irvine heires says partner in firm diverted millions," *Los Angeles Times*, August 4, 2007.
21 "Arbitrator Rulse on Brandes, Irvine Smith Dispute," *Orange County Business Journal*, October 8, 2008.
22 Interview with Norberto Santana by C. Michael Stockstill, March 10, 2025.
23 Interview with Norberto Santana.
24 Interview with Norberto Santana.
25 Interview with Norberto Santana.
26 Interview with Norberto Santana.
27 "Reeve-Irvine Reseach Center," *ABILITY Magazine*, 1996, https://abilitymagazine.com/smith_reeveirvine.html
28 "Reeve Research Center," *ABILITY Magazine*, 1996.
29 UC Irvine Foundation, "Establishment of the National Water Research Institute and Clarke Prize," 1993.
30 UC Irvine Press Release, "UCI Acquires Orange County Museum of Art," October 2025.

23 Dirigibles no more

It had been unseasonably warm on November 7, 2023; temperatures into the upper 70s during the daytime, dropping to the low 60s in the early morning hours.

The sky was clear, so when the first flames at the base of the huge northern hangar on the former dirigible base appeared, they were easily seen from the homes just 300 yards away.

The fire began inside the towering cavern of the 17-story hangar, probably among materials stacked at the sides of the plywood covering that stretched across the giant girders arching from one side of the structure to the other.[1]

Firefighters arrived around 4 a.m. Mobilizing to attack the fire from the ground and with two water-dropping helicopters, they soon discovered fire hydrants near the hangar were not connected to a water source. After emptying the water in their engine tanks, the firemen decided to let the flames consume the hangar and burn out, a process that would take almost three weeks.

When it was over, the fire had completely demolished the wooden elements of the hangar, leaving four concrete towers sticking into the sky at the corners of what had once been one of the largest stressed wood structures in the world. The hangars were so unusual that they had been designated a historic landmark by the federal government.

And even though the former Lighter Than Air Station (LTAS-Tustin) had been gifted to the city of Tustin in 2002, the U.S. Navy had retained ownership and control of the hangar. The Navy would spend nearly $100 million to clean up toxic residue and other debris left behind after the fire.[2]

The tragic accident brought a new focus on the state of redevelopment of the former military base, by 2023, more than two decades in transition. While real strides had been made, there were still huge gaps of open space between islands of development—residential, commercial and institutional—that had sprung up in the interim at the new community named Tustin Legacy.

***

Unlike the process that resulted in the El Toro Marine Corps Air Station being sold at auction, the transfer of the Tustin base was accomplished under a federal law that allows surplus properties to be given over to stimulate economic development, officially known as an Economic Development Conveyance.

DOI: 10.1201/9781003683087-23

Figure 23.1 The remaining hangar at Tustin Legacy

Source: Author's image

After conducting a public opinion survey of Tustin residents, the city created a master plan for redevelopment of the 1,600-acre base in 2001, paving the way for the initial transfer of 997 acres in 2002 (the Navy sold 235 acres of the base directly to developers).[3] After creating what was essentially an outline plan, the city sought implementation partners through a request for proposal (RFP) process that attracted submissions from Lennar, Trammell Crow, and a partnership of Centex Homes and Shea Properties, with Shea/Centex the eventual winner.

The RFP process and land transfer had proceeded as a legal battle with the Santa Ana Unified School District moving towards a trial date. A relatively small portion of the base property was within the school district, prompting district officials to demand both land for a school and funding for it. The city wanted to give land to other districts, leading to charges that the city was making decisions based on race because 97 percent of Santa Ana students were Hispanic. After more than six years of conflict, the city and school district settled, with the district accepting $38 million and a 22-acre site for a school.[4] To date, the Santa Ana district has not constructed a school on the base property.

While the comprehensive planning effort was underway, the city council decided to fast-track a retail project at the southeastern corner of the base. Vestar Kimco Partners was selected to build what would become The District. Opened in 2007, it featured a 14-screen AMC theater, a variety of restaurants, both casual and upscale, and a large number of small shops.[5] Within a few months of its opening, a Costco, Whole Foods, Lowe's, and Target opened on the outer ring of the massive parking lots built to accommodate the expected crowds of shoppers.

However, the repercussions of the 2008 financial crisis on Wall Street dealt The District a nasty blow. Initial shopping numbers dropped as the shockwaves of the economic downturn spread across Orange County. At the time it opened, there were almost no homes built within half a mile of The District. The closest concentration of homes was in Westpark in Irvine, which had a robust set of its own retail centers already open and operating. Moreover, the odd configuration of traffic ingress and egress at The District made it challenging to access the center. As the years passed, stores and restaurants opened and then closed regularly.[6]

The economic calamity also led to the collapse of one of the master planning partnerships when Centex petitioned the city to withdraw from the plan it had crafted with Shea Properties.[7] Soon thereafter, the remaining partnership, which had been structured as Tustin Legacy Community Partners, pulled the plug on its promise to create a master plan, leaving roads and other infrastructure half-built. Unwilling to wait for another developer, the city took control of 335 acres in the middle of the base.

By 2010, Tustin Legacy was slowly coming back to life.

The first set of apartments and condominiums was nearing completion, bringing a genuine population base to the community. At the far edge of the former base, the Orange County Rescue Mission was moving into the Village of Hope, a facility designed to house and assist homeless individuals and families. Within a few months, its 192-bed facility was more than 90 percent occupied.[8]

The Tustin Unified School District completed a new elementary school, and another ambitious educational center sponsored by the South County Community College District was taking shape as the Advanced Technology and Education Park (ATEP). The facility had a specialized curriculum focused on engineering and science. Not far away, the Orange County Sheriff's Department built a state-of-the-art training center.

By 2014, the city had decided to assume the task of administering a master plan for the property. One of the first deals it made was with Standard Pacific Homes, which paid $73 million to build 375 single-family homes and a local park on 78 acres of land.[9] The city also created an assessment district on the acreage, with future revenues slated to reimburse the city for infrastructure and to fund future schools.

In the months that followed, additional residential projects—including apartments and condominiums—began to rise from vacant lots, most of them in sight of the huge south hangar that loomed over the area like a mammoth Sphinx.

The most creative office project at Tustin Legacy was completed in 2019, with construction by Lincoln Properties of FLIGHT, a campus designed to promote a feeling of interaction and informality for individuals and companies that leased offices there.

FLIGHT's twin four-story glass towers plus two smaller incubator buildings border more than 100,000 square feet of community space, including a large dining hall that overlooks the winding walking paths in Tustin Legacy Park.[10] FLIGHT's timing was unfortunate; just a year after opening, the COVID-19 epidemic brought leasing to a standstill. The few visitors to the dining area (named The Mess Hall), which contained a number of creative food offerings, had the place to themselves for months.

At the other side of Tustin Legacy, a small retail center and office complex came online, bringing residents the first stand-alone grocery store (Stater Brothers) and pharmacy (CVS), plus an assortment of traditional retail outlets. Just across the street, the Legacy Magnet Academy opened, focusing on a TIDE (Technology, Innovation, Design, Entrepreneurship) curriculum.[11] The school has a striking design that mimics the outline of a hangar.

A half mile away, Veterans Sports Park is open, a 30-acre facility with soccer and baseball fields, tennis and basketball courts, a playground, and a veteran's memorial.[12]

As the commercial developments began to fill in vacant spaces, residential construction continued to thrive, with deal after deal (including a second one from the Irvine Company to match an earlier project) adding to the total of residential units, now over 4,000.

But like El Toro and the Great Park, the sheer mass of Tustin Legacy continues to dominate the landscape. A view of the property on Google Maps highlights the huge open parcels at the eastern and western corners of the property, with a collection of smaller lots connecting between them. Along Edinger Avenue, which forms the northern boundary of the property, motorists gaze out at a view of hundreds of new vehicles stored on an otherwise empty lot. On the other side of Tustin Legacy, parcels awaiting development are given over to weeds.

And then there is the remaining hangar.

With its historic twin now reduced to four concrete pillars and a lingering pile of ashes, the south hangar remains a quandary for the city. Over time, various architects and developers have come up with possibilities for reimagining the colossal structure, including making it a movie studio, a concert venue, or an indoor recreational complex.

Because the hangar is a national historic landmark, the potential for any remodeling or reconstruction there is much more complex. Cost to maintain the hangar ranges into the half-million-dollar figure annually.[13] And now knowing that the other hangar might have been saved had sufficient water been available in fire hydrants, additional funding may be spent to install such a system.

For now, the hangar endures, its seven acres of interior space sometimes used as a setting for commercials or films. Occasionally, one of the few remaining lighter-than-air ships still flying has alighted on the tarmac for a sentimental reenactment of the wartime past.

But in recent years, those visits have ceased. The gargantuan hangar doors are closed. A visit to the inner space of the immense structure is a silent, ghostly experience, humans reduced to mere specks on the polished concrete floor. Amid the stillness, it is not difficult to imagine that the spirits of sailors and aviators from the Greatest Generation still inhabit the hangar, and that they might awaken to assemble should morning reveille ever sound again.

Notes

1 "Historic Tustin Hangar Destroyed by Fire," *Los Angeles Times*, November 8, 2023.
2 "Navy to Spend $100 Million on Hangar Cleanup," Orange County Register, November 15, 2023.
3 "Tustin Legacy Redevelopment Plan Approved," City of Tustin Archives, 2001.
4 "Santa Ana Unified Settles Base Dispute," *Los Angeles Times*, August 22, 2008.
5 "The District Opens in Tustin," Orange County Register, August 15, 2007.
6 "Retail Struggles After Financial Crisis," Orange County Business Journal, February 2, 2009.
7 "Centex Withdraws from Tustin Legacy Partnership," *Los Angeles Times*, October 12, 2009.
8 "Village of Hope Nears Capacity," OC Weekly, March 10, 2010.
9 "Standard Pacific Buys 78 Acres in Tustin," Orange County Register, May 14, 2014.
10 "FLIGHT Office Campus Opens," Orange County Business Journal, October 1, 2019.
11 "Legacy Magnet Academy Opens," Tustin Unified School District News, August 2021.
12 "Veterans Sports Park Dedicated," City of Tustin press release, October 2022.
13 "Preservation Challenges for Tustin Hangar," National Trust for Historic Preservation, accessed September 13, 2025, https://savingplaces.org/stories/tustin-hangar-preservation.

24 The other Donald

Don Koll had a vision.

It was a vision as clear as the endless blue skies he saw from the cockpit of the Air Force fighter jet he flew as a young pilot.

Orange County, Koll knew, was destined to grow, especially the area around the Orange County Airport, which in the early 1960s was just beginning to take shape as the Irvine Industrial Complex (IIC). Here, Koll believed, Irvine Company plans for the area focusing entirely on manufacturing were too limited. MacArthur Boulevard, in Koll's vision, would come to resemble Wilshire Boulevard in Los Angeles, lined with gleaming glass and steel towers.

In 1962, seven years after graduating from Stanford with a degree in economics and his service in the U.S. Air Force completed, Koll started a construction company. He had been working for his cousin building custom homes in and around Newport Beach but believed he would prosper on his own.

Koll's big break came in 1967 when he drove by the airport on Main Street. He was familiar with airport logistics and layout, so it was immediately apparent that the land at the end of the runway where planes landed could never be developed for anything other than a few scattered low-rise buildings; it was in and around the crash zone, a description that would hardly appeal to most real estate developers.

Koll saw something different. He envisioned an area where he could build single-story tilt-up concrete buildings—nothing fancy, low on amenities, but at a cost low enough that they could be leased to small businesses, start-ups, and other retail and commercial orphans that couldn't afford more spacious or elegant surroundings.

The Irvine Company held the property but, at the time, was more focused on the easy pickings of hundreds of acres it owned without the complexity of aircraft noise or potential hazards. When Koll offered to purchase 100 acres of the land, the company agreed, and the Koll Airport Business Complex was born.

Built in phases over a decade, lots were located around an internal street labeled Skypark Circle. At build out, there is more than one million square feet of multi-tenant, incubator, industrial, low-rise office and support retail buildings. Today, many people refer to the area not by its official name but as Skypark. No matter what it was or is called, the collection of print shops, one-person consulting firms,

DOI: 10.1201/9781003683087-24

Figure 24.1 Don Koll

dance studios, micro-manufacturing centers, and dozens of other enterprises is a thriving ecosystem of commercial activity. Koll once told a real estate consultant that Skypark was the most successful and profitable development he ever owned.[1]

The Koll Airport Business Complex marked the beginning of Koll's march through the IBC and much of its surrounding territory, making the Koll Company, after the Irvine Company, one of the most prolific commercial office builders, owners, and property managers on the Irvine Ranch.

***

Bill Dennis had been enjoying a successful and lucrative career at Coldwell Banker in 1985 when he was recruited by Richard Ortwein to join the Koll Company. The company, which up to that time had concentrated almost exclusively on office buildings, wanted to expand into southern California's Inland Empire—the counties of Riverside and San Bernardino. The target was industrial development.

In his self-published memoir, Dennis combined his personal recollections of the professional roller-coaster ride he took during his four years at Koll, along with a comprehensive history of how Koll had achieved such a prominent place in the history of Orange County commercial/industrial development on the Irvine Ranch.

Dennis explained how early decisions made by the Irvine Company—when it leased or sold land—planted the seeds for a harvest by Don Koll.

The Irvine Company's primary goal, he wrote, "was attracting major corporate users to the area with attractive land pricing and sophisticated master planning."[2] Several initial users, Dennis pointed out, were firms involved in defense or electronics. But the Irvine Company neglected to think decades into the future to consider what might happen to the land if the initial buyer's business circumstances changed. "Most of these early transactions contained no restrictions on resale or reuse of the parcels and inadvertently provided highly attractive development opportunities that were exploited as the area matured and companies' plans changed."[3]

The first example of this phenomenon took place on a 177-acre parcel of land at the intersection of Jamboree Road and MacArthur Boulevard, when, in 1959, Collins Radio, a subsidiary of North American Rockwell, leased the property:

> Koll's signature Orange County project, Koll Center Newport was started on the Collins sublease. Ten acres had been sold to the County of Orange for the Harbor Justice Center, 20 acres were deeded back to the Irvine Company in return for the right to acquire fee title to the rest of the land . . . at full build-out in the late 1980s, Koll Center Newport consisted of 1.5 million square feet in fifty mid-rise and high rise office buildings, restaurants, the Pacific Club and the Fairmont Hotel [later the Meridian Hotel], which added a touch of international flair to the area, as it was owned by Air France. Pilots and flight attendants were bussed from Los Angeles International Airport for overnight stays at the hotel, filling the lobby with lilting chatter *en Francais*.[4]
>
> Koll Center Newport was followed in 1981 by . . . Koll Center Irvine. The first phase was a five-story office building. In the mid-1980s, two additional high rise office buildings and a short-lived restaurant completed (the development). The second phase . . . was developed in the mid-1980s, 2.4 million square feet in four distinctive black granite-clad high rise office buildings, Prego Restaurant and a Marriott Hotel.[5]

Marriott purchased its site. The entire property had previously been owned by Bertea Corporation, which purchased it in 1966, and was subsequently taken over by Parker-Hannifen, an aerospace company.

> Also in the late 1980s, Koll Center Irvine North was created across the 405 Freeway at the corner of MacArthur and Main: 1.5 million square feet in four high-rise office buildings, a hotel, fitness center and restaurants. Columbia Savings was the financial partner, and the project was lost in foreclosure to the federal government during the savings and loan crisis of the early 1990s.

> Two of the project's buildings were completed by another developer who acquired the parcels from the Resolution Trust Corporation.[6]

The Koll Irvine North property is one of the area's most unusual commercial projects; it features a series of water elements that cascade through the center of the common areas, which are also the site of generous grassy parklets. And unlike almost any other commercial facility in the Irvine Business Complex, the development features a number of colorful structural artworks. The architectural variety of the buildings also gives the area a distinct flair.

As the boom times of the 1980s accelerated, the financial rewards for Koll and his associates swelled. Dennis recalled being motivated to perform

> by the company's manifestations of affluence . . . two private airplanes, meetings at Don Koll's Palmilla Resort in Cabo San Lucas, and company condos in Aspen and on Maui, weekly free car detailing at the office, elaborate black-tie Christmas parties for employees, partners and families, and extravagant, raucous open houses to introduce the next great project to the real estate community.[7]

Dennis continued:

> Early in 1985 I watched in awe as Don and some of his original partners sold a majority interest in several mature early projects to a major financial institution for $75 million, with several key Koll partners receiving checks in the range of $5 to $10 million, a transaction known thereafter as "Koll Big Deal."[8]

Dennis offered an insight into the inner workings of the Koll operation.

> The huge number and size of Koll projects was made possible by joint-venture financing . . . not only with Aetna, but also Northwestern Life, National Life of Tennessee, Connecticut General, and often with aggressive banks led by Wells Fargo . . . In exchange for this equity and debt financing, the financial partners would retain anywhere from 50 to 75% of project long term value.
>
> If a project fell short of the pro-forma, Koll's share would be reduced . . . the trade-off was that Koll partners usually incurred little to no personal financial risk in the event of poor project performance, and the company earned development fees (vital to operating cash flow) . . . mature projects would then generate property management fees.
>
> Koll also operated a major construction company . . . annual revenue at its peak was in the $400 million range, and fee income was generated both from Koll projects as well as those of third party clients. Combined construction and development fees for Koll developments ranged from seven to ten percent, well above industry standards . . . This was the basic financing structure commonly used throughout the industry by major developers.[9]

Flush with capital and energy, by the late 1980s Koll had established offices throughout the United States and was thriving in development as well as construction and property management. A signature Koll project that made even the most successful Orange County developers envious was the Palmilla Resort and Golf Course Koll undertook in Cabo San Lucas. Located on a beautiful bay, Palmilla became a favorite retreat for Koll and his family. Irvine Company vice president Thomas Wilck spent a weekend there and, upon his return, told associates that the people at the resort "treated Don Koll like a god."[10]

Koll had also teamed up with an unlikely set of partners to develop an equally unlikely project at the site of Lion Country Safari. Veteran music industry executive Irving Azoff and entertainment business executive Robert Geddes joined with Koll to build and manage Irvine Meadows Amphitheater on a hillside above the formerly thriving wild animal park.

At the time, there were just a handful of small venues in Orange County that hosted rock music. Irvine Meadows changed the game, with more than 5,000 seats (more on the grass area above the structure), it was large enough to justify a stop on a concert tour. Some of the biggest names in rock music (and the occasional classical performer) played at Irvine Meadows.

(How often Koll himself attended concerts there is unknown, but a brief conversation with an Irvine Company executive at the Hollywood Bowl provided some insight. Koll wandered into a reception before a concert by world famous tenor Luciano Pavarotti. The event was a fundraiser for then-U.S. senator John Seymour. Spotting Koll, at the time a member of the Irvine Company board of directors, the Irvine executive introduced himself. Koll didn't know who was performing, he said, he just wanted to help Seymour by making a contribution. Koll then pulled out a pair of earplugs. "I use these when I'm hunting, but they also come in handy at the Irvine Meadows concerts—some of the bands they have there are really loud." Whether Koll used the ear plugs to accommodate Pavarotti's powerful arias is not known).[11]

Koll's empire was shaken by the real estate recession of the early 1990s. Projects that had been too aggressively structured began to falter. Nor did it help that Koll's real estate arm had been bleeding money on plans for construction of a 3,500-home community on the Bolsa Chica area of Huntington Beach. Environmental groups had battled Koll and a subsequent developer to the point that of the more than 3,000 homes originally envisioned, less than 200 were ultimately constructed.

Faced with a rapidly eroding real estate market and previously generous financial partners fleeing the sector, Koll restructured his holdings into four business units, cut costs, and laid off staff.

In 1991, as the downturn was deepening, Koll found a new partner when Peter Bren, brother of Donald Bren,

> came to Koll with an opportunity to work with him in implementing an investment strategy for one of Peter's clients. The client's fund had committed $200 million to be invested in non-performing commercial loans . . . Don and Peter formed a partnership initially known as Koll Bren Realty Advisors to executive the plan.

> The new business was led by Peter Bren and Chuck Schreiber from Koll. The initial venture . . . became the foundation of a growing investment and advisory firm which became Koll Bren Schreiber Realty Advisors in 1995 and thereafter as KBS Realty Advisors.[12]

In early December, 2001, six years after suffering a stroke that left him almost entirely incapacitated, Don Koll died at age 77. Associates and others who had worked and played with Koll remembered him as a man who liked to hunt and fish, was easy-going, and loved to laugh. Bill Dennis offered the observation that Koll left "an indelible imprint on the real estate industry that includes many successful alumni, a source of pride, especially in Newport Beach and Orange County."[13]

Notes

1 Author's recollection Bill Dennis, *California Dreaming: Family Foundations and a Marriage Lived Well* (Self-Published, 2019).
2 Dennis, *California Dreaming*, 209.
3 Dennis, *California Dreaming,* 209.
4 Dennis, *California Dreaming*, 217.
5 Dennis, *California Dreaming*, 218
6 Dennis, *California Dreaming*, 218.
7 Dennis, *California Dreaming*, 219.
8 Dennis, *California Dreaming*, 219.
9 Dennis, *California Dreaming*, 221.
10 Author's recollection.
11 Author's recollection.
12 Dennis, *California Dreaming*, 231.
13 Dennis, *California Dreaming*, 233.

25 DBA the Irvine Company

Like nearly every business entity, the Irvine Company has experienced a series of cycles.

The first was almost entirely dominated by agriculture, lasting from the days of the California Gold Rush until just after World War II.

The death of James Irvine II in 1947 marked the second cycle, as majority ownership passed to a charitable foundation Irvine had established a decade earlier. The foundation ownership was a transition, during which time two landmark decisions would set the company on a transformative journey from a giant farm to a New Town with a university at its heart.

Forced by federal law to sell the company in 1977, the foundation turned over ownership once again to a partnership but one that would last for less than seven years. During that interregnum, the company's commercial soul was redirected from planning to a leaner and more profit-driven enterprise.[1]

The current cycle began in late 1983 when partnership members sold their positions to Donald L. Bren. For the third time in more than a century, a single individual controlled the Irvine Ranch.[2]

***

Bren exerted his vision for how the company would operate with an immediate order to identify and monetize company assets he considered underutilized, undervalued, or in some cases, ignored.

In Newport Beach, two small strips of land barely wide enough to accommodate construction equipment were zoned and sold off to a homebuilder, who jammed a few condominiums onto the site. A larger parcel in Harbor View Hills that had lain fallow enjoyed a similar fate.

In Woodbridge, two six-acre rectangles of land adjacent to the South Lake that had been considered for a variety of uses, including a boutique hotel, were sold to Standard Pacific, which built large homes there. A dozen of the homes backed onto the lake, commanding a premium sales price, a portion of which the builder shared with the Irvine Company.

Bren looked out of his corner office on the ninth floor of 550 Newport Center Drive to consider how the outdoor retail center, Fashion Island, might be reimagined. Just a year after he became majority owner, a $40 million facelift and redesign

DOI: 10.1201/9781003683087-25

Figure 25.1 500 and 550 Newport Center Drive in Newport Center. Irvine Company headquarters

Source: Author's image

were underway, with a variety of architectural and landscaping touches from Bren himself. And within 550, Bren ordered a remodel of every floor, tearing out teak-wood paneling adjacent to the elevators, replacing it with marble. On the ninth floor, the lobby outside his office was expanded and repainted, with flowing, billowy drapes covering the floor-to-ceiling windows. (Bren made similar changes to the executive floor of a building he owned in West Los Angeles. Company executive Jack Flanigan said the lobby there looked like "the waiting room for Heaven").

As activity ramped up on the business side, legal problems cropped up within a few months of Bren's purchase. As described in Chapter 11, Joan Irvine Smith and her mother contested the value of the shares they sold to Bren, setting off a legal action that would take years to finish.[3]

A similar lawsuit came as a result of the sale, this one from the former president of the company, Peter Kremer.[4]

Dismissed at the end of his five-year contract, Kremer received $450,000 in severance pay plus $141,000 for his interest in two limited partnerships. He also owned 50½ shares of the company he received when he joined as president in 1977. When he departed, he was paid $7.6 million, $150,000 per share.

Just four months later, Kremer learned that Bren had paid $200,000 per share to buy out his partners. Kremer told the *Los Angeles Times* he tried to contact Bren to discuss what he considered an unfair discrepancy in the price, to no avail. He sued Bren and the company in September 1983.[5] Just over two years later, Kremer and the company settled the lawsuit. The amount of the settlement was not revealed at the time, but as a result of a later, unrelated legal action, it was reported that Kremer received another $5.5 million.[6]

***

When Bren took control of the company, the transition from agriculture to urban development was well underway.

In 1977, the new owners of the Irvine Company found that just over 36,000 acres of land being farmed or grazed was protected by the Williamson Act, which lowered property taxes on parcels that were enrolled in ten-year preserves. The company began to non-renew 12,514 acres of the preserves from 1978 to 1982, meaning that ten years after the notice, the land would lose its tax reduction. Bren continued the process. From 1984 to 1989, the company filed notices of non-renewal of the Williamson Act contracts encompassing 23,524 acres.

In June 1985, the company announced it would no longer raise cattle on the ranch.

> Within 60 days, the mammoth land development firm has sold off the vast majority of its herd, including 1,200 breeding cows, 1,000 young calves, 1,000 yearlings and about 65 bulls. The Irvine Company plans to lease to three other livestock operators the 38,000 mostly rugged acres it has been using to graze its own cattle. The lessees, in turn, are expected to purchase about 1,000 head of the company's breeding cows that remain.[7]

About the same time as the cattle were heading for the last roundup, management of the company's remaining 16,000 acres of land, being farmed for row crops, including strawberries and tomatoes, among other vegetables, were turned over to Treasure Farms, owned by Howard Marguleas. He was a member of the partnership that purchased the company in 1977 and had retained his shares after the Bren buyout. In the summer of 1988, it was 5,000 acres of avocados, Valencia oranges, grapefruit, and lemons that were transferred to Treasure Farms as well, leaving just 14 Irvine Company employees tasked with agricultural management.

In 2022, strawberries would be the final field crop to be planted and harvested on land owned by the Irvine Company. (Some strawberries are still grown on land owned by Five Point Communities around the Irvine Great Park). Citrus and avocado trees remain in production on Irvine Company land in the Lomas de Santiago foothills as this is written.

***

With the wind-down of agriculture and animal husbandry in clear sight, Bren began to reshape the company's organization and personnel.

At the head of the company was Thomas H. Nielsen, whom Kremer had recruited as a vice president. While Nielsen's initial responsibility was for residential development—the key profit engine for the company at the time—his concurrent assignment was to repair relations with many of the city and community leaders who felt either ignored or unappreciated by the company. When Kremer departed, Nielsen moved up to the president's office.

Nielsen had worked with Kremer at Newhall Land and Farming and later with Bren's former partner, Lou Malone, in the San Diego area. Educated as a civil engineer, like many Newhall alumni, Nielsen had an MBA from Stanford. He had also spent a year as an assistant secretary of the Air Force, appointed by Lyndon Johnson, where he analyzed Air Force operations and management issues.

Nielsen arrived at the company with a storied local past. He grew up in Fullerton, where his father ran a clothing store. At Fullerton High School, despite his modest frame (five feet seven inches, around 150 pounds), he had starred as the football quarterback and placekicker, making what became a legendary field goal at the conclusion of a rainy game to clinch victory.

Nielsen's personality alternated between gentle and gruff. He had little patience for long meetings or presentations; his body language in such settings sending the message, "get to the point." If he thought someone was off base, he didn't hesitate to say so. He once dropped in unexpectedly at an internal company meeting with a dozen employees scattered around a table. Poking his head inside from the doorway, he glanced around the room. "Lots of money at work here," he said sardonically, then popped back into the hallway. At another meeting, a company vice president made a less-than-charitable remark about how lucky Dwight Eisenhower had been to become president. Nielsen looked up and said quietly, "Well, he had just marched through Europe at the head of an army of liberation that saved democracy."[8]

Nielsen's humanity equaled his no-nonsense business acumen and demeanor. He was patient and accommodating at community events, leaving elected officials and others believing he truly cared about what they were saying. Nielsen made it a point to attend public meetings and, if necessary, stay to the conclusion. It meant hours of what other executives may have considered wasted time, but it gained him respect.

Nielsen also wore his heart on his sleeve in private and could display deep empathy. One former employee was struck by his grief at the funeral of a company executive's wife, who had died of cancer. Nielsen began sobbing during the service and continued to do so for long minutes.

Nielsen's wife, Marilyn, often accompanied him at the many events he was invited to and felt compelled to attend to show the company flag. The pair, unknown to most employees, had a long history of community involvement in Fullerton before his Irvine Company assignment. They regularly served Thanksgiving dinner at a local soup kitchen and had made generous contributions to a variety of local charities.

With Nielsen doing double duty as a manager and community relations leader, it soon became apparent that a management restructuring was needed to accommodate Bren's mission to make things happen faster and with greater profit. (Richard Sim and Brad Olson both recalled in interviews being asked by Bren how long it would take to build out Irvine Company holdings—residential, retail, industrial, and commercial. Both made estimates of 60 to perhaps 100 years. Bren told them his timeline was 40 years, leading to an internal mantra many employees whispered to one another: "build it in Don's lifetime.")[9],[10]

Sim had hired Olson to oversee all residential development, whether planned or underway. Moving up from director to vice president in two years, Olson found the burden of managing housing entitlement and construction in six jurisdictions overwhelming. With Sim moving up to oversee both land development and commercial/industrial operations, the company split the housing element into three new companies: Irvine Community Builders, Coastal Community Builders, and Foothill Community Builders. At the same time, separate departments were merged into the Irvine Office and Industrial Company.

In October 1984, Bren transferred William McFarland from his job at Bren Homes to the Irvine Company to oversee operations.[11] While McFarland reported to Nielsen on the organization chart, employees soon understood that his long relationship with Bren made the lines on paper relatively meaningless.

McFarland was much like Bren. Just over six feet tall, slender, he dressed formally and had a reserved demeanor. Soft spoken to the point that employees sometimes had to lean forward to hear him, McFarland kept his own counsel in meetings and briefings, rarely asking detailed questions or making lengthy comments. Like others, McFarland was appalled at the company's addiction to meetings. "Pre-meetings, morning meetings, I was there from 7 a.m. to 7 p.m.," he recalled in a 2024 interview.[12]

In his low-key manner, McFarland was soon quietly taking control of operations and budgets. With a combination of logic and gentle nudges to personnel and day-to-day activities, the company began to function with greater efficiency and focus. McFarland said he tried to find employees who would accept a more entrepreneurial approach to business and, once identified, "give them responsibility and direction, and let them make decisions." McFarland could, like Bren, make tough decisions with cool efficiency, but he also displayed a warm smile and friendly manner in small meetings or gatherings. Those who got to know him well called him Mac.

In addition to encouraging McFarland and others to reformulate the company's business DNA, Bren began to inject more of his own views into the decision-making process. At the same time, his managers were on a hiring spree to implement the "build it in Don's lifetime" vision. Olson recalls being chided by McFarland when he resisted beefing up staff. "I told him I am not an empire builder," he said, only to have McFarland retort that the company had billions of dollars in potential profits awaiting the bodies necessary to make them appear. Olson increased hiring; the company would expand to more than 1,000 employees within a few months.[13]

In June 1986, the *Los Angeles Times* was reporting that Nielsen would lose his administrative duties as president to concentrate on community relations. "The giant firm . . . which has gradually decentralized operations during the past two years, will now be more closely run by an executive committee of five company directors."[14] If Nielsen saw the writing on the corporate wall, he didn't say anything to other employees; however, just three months later, the office of the president was eliminated by Bren, who then named himself chairman of the executive committee.

(A month after the executive ranks were reconfigured, the company laid off 100 people.[15] It would be the first in a series of other layoffs that would follow in the years ahead. The cuts were sometimes explained as necessary due to market conditions or management restructuring, other times seemingly at the whim of Bren, his management team or for other, unknown factors. Similarly, to the frustration and puzzlement of the rank and file, on occasion, a new senior executive would unexpectedly parachute into the management structure, their tenure lasting a few weeks or months before disappearing in a similar mysterious fashion).

One beneficiary of the reorganization was Bren's closest advisor, Gary Hunt, who would now head up urban planning and design, policy management and entitlement, and corporate affairs. In July, Hunt had replaced Robert Shelton as the company's chief political strategist and operative, directing what, by then, had become one of the largest political contribution budgets in California. (Shelton retired).

Whatever title Hunt now held, or had held in years past, it was clear to everyone, from vice presidents to mail boys, that Bren and Hunt had cemented a relationship over the years that transcended business.

Bren and Hunt had initially clicked in the first months after meeting one another. Bren was fascinated by politics and its kaleidoscope of personalities and posers, all straining to achieve office, power, and influence. But to succeed in that arena, a person needed to submerge themselves into the swirling waters of handshaking, schmoozing, and relationship-building—politics was a personal, intimate business.

Bren wasn't comfortable in crowds, didn't do well at small talk, and was certainly never going to expose his inner feelings or principles to build trust with someone who probably just wanted his money.

Enter Hunt.

Hunt, too, was intrigued by the dynamics of politics and government; a career there offered him a path out of a lower-middle-class childhood in the high-desert town of Lancaster, where he grew up with two brothers and a sister in a three-bedroom home. Hunt took to the machinations of political life easily. With a quick mind and instincts for what motivated people, he found he could influence events almost without thinking.

But Hunt could think as well, and though his formal education record was spotty (he had bounced from one college to another, finishing law school while working but never taking the bar exam), while still in his 20s, he displayed a creative, sophisticated strategic intelligence. Hunt was also articulate: ideas, arguments, suppositions, and proposals flowed easily as he would analyze an issue or situation and verbally frame alternative approaches.

Hunt's outgoing and politically attuned persona made him an ideal counterpart to Donald Bren's reserved intensity. As Bren's ambitions expanded beyond land development into civic influence, Hunt became the translator—turning strategic vision into public engagement. Their relationship evolved from professional alignment to personal trust: Hunt joined Bren for tennis matches, ski trips, and increasingly intimate roles within the company's orbit. By 1984, Hunt was fielding unsolicited correspondence from admirers addressed to Bren, a reflection of his gatekeeping role. When journalists sought insight into Bren's business dealings or personal life, it was often Hunt who responded—managing not just public relations but the public narrative surrounding Orange County's most private power broker.[16]

Hunt was almost always at Bren's side. At the opening festivities for the remodeled Fashion Island, Hunt hovered closely to Bren for more than an hour, whispering the names of dozens of people as they came up to shake his hand. One former employee said that the two spoke almost every Sunday evening to plan the coming week's events.

Hunt was also the ultimate gatekeeper, granting access to Bren for appointments or outside events like the agent for a Hollywood superstar. Through it all, Hunt was mindful of Bren's insistence on privacy, never sharing any gossip or nuggets of—as he called him—"the chairman's" personal life.

Hunt would manage some of the most sensitive and critical issues the Irvine Company faced in the decades he worked there. In doing so, and with Bren's implicit approval, Hunt would build and maintain a stature that in some ways outshone his mentor. Throughout the 1980s and 1990s, as the company expanded into a commercial juggernaut, Hunt was recognized throughout Orange County and California as one of the key designers and implementers responsible for its success.

No more telling example of how the Orange County business and political community viewed Hunt's talents came in the aftermath of the bankruptcy of the County of Orange.

On December 6, 1994, the county declared bankruptcy after having "backed into a massive financial crisis," explained Dr. Mark Baldassare in his 1998 book *When Government Fails*, an analysis of the events and personalities that led up to the financial catastrophe.[17]

Orange County treasurer Robert Citron had a sterling track record of investment success, returning consistently high returns to the county, which had to invest its funds in the "pool" of assets. Dozens of other public agencies, school districts, and special districts, drawn by Citron's past performance, had also deposited funds in the Orange County Investment Pool.

Baldassare explained that Citron "borrowed money and invested it in derivatives, inverse floaters and long-term bonds that paid high yields."[18] The strategy worked until Citron bet the wrong way on the assumption the Federal Reserve would lower interest rates. When it didn't, the value of the pool—which had grown to more than $20 billion—began to decline. "By November 1994, county officials found out that (Citron) had lost about $1.64 billion in government funds . . . the county did not have the cash on hand to withstand a run on the money."[19] After the first bank that had lent money to Citron threatened to seize the assets held as collateral, the pool collapsed, and the supervisors declared bankruptcy.

With billions of dollars frozen by the bankruptcy, the affected agencies began legal actions and lobbying efforts to secure some of their funds. A three-man committee of public officials was convened to manage the process, but it proved difficult.

Watching from the sidelines, the Orange County Business Council (OCBC), which represented the county's largest employers and most substantial businesses, quickly concluded that to maintain order, a settlement with the dozens of depositors was needed:

> The (OCBC) appointed a three-person task force to assist (in) negotiations . . . The members included Gary Hunt, George Argyros of Arnel Development, and Tom Sutton (CEO) of Pacific Mutual. Hunt . . . had taken the lead role in engaging the business community in the financial crisis from the beginning. He was highly knowledgeable about city, county, and state government. He also had strong ties to Governor (Pete) Wilson.[20]

The OCBC team hammered out a deal with the pool participants in just three weeks. Hunt would segue from that success to organize a temporary sales tax campaign to fund county operations and other obligations designed to prevent cuts in county services. While it failed, Hunt's leadership on both the recovery strategy and the attempt to raise funds burnished his reputation as an effective leader who could get things done.

***

A year after the elimination of the role of president and the formation of the five-person executive committee to guide company operations, development had ramped up across the business lines of the company, including at the Irvine Spectrum and Irvine Center (see Chapter 15).

With a reported price tag of close to $200 million, the company embarked on an ambitious sprint to build seven retail centers, including three in the heart of Irvine.

The largest would be The Crossroads, 330,000 square feet of retail anchored by a Target store at the corner of Culver and Barranca, one of the busiest intersections in the city. Crossroads was a joint venture with Diversified Shopping Centers, one of three firms the company would partner with to both develop and later manage the center.

Partnering was an untried strategy for the company. David Mudgett had recently joined the company to head up retail development and operations. He was thrust into the internal debates about how best to make the new strategy a success. Mudgett remembers being pleased to work on retail projects "where we had the ability to implement and not be under short-term pressure, (which) is not often the case."[21] That situation, he said, reflected Bren's insistence on high-quality development and a long-term outlook.

The more formidable challenge was managing the transition to firms like Diversified, Donahue Schrieber, and SDC development, which meant seeing his staff shrink from 120 people to just 8. "The shame of it," he recalled, was how the reduction process took place. "The first reduction was just, 'we're trimming.' So you reduce, and you've got half your people left, and suddenly, (you are told) you ought to trim some more."[22] Just when a handful of employees were left after the layoffs and felt they were safe, Mudgett remembered, the announcement came that all the projects would be outsourced, leaving a retail department with fewer people than the fingers on two hands.

The firms selected had battled hard to win, as "everyone and their brother wanted to be the (partner of the Irvine Company), because they had a built-in base and no competition," Mudgett said. As it turned out, Mudgett noted, "The product was good, at least from a retail standpoint, and the quality was good."[23]

Bill Kenney was a partner at Donahue Schrieber at the time. He recalled in a 2025 interview that the firm had two contracts—one on a fee basis to oversee improvements at Fashion Island and the other to help build and manage the Marketplace "power center" on the border between Tustin and Irvine. In that partnership, he explained, Donahue Schrieber received equity in the project as well as being paid for operations. After two or three years, Kenney said, the Irvine Company bought out their equity share and took over operations.[24]

Building two other centers—one new on Jeffrey Road near the I-5 Freeway and a second that was an addition to the stepchild of the company's retail efforts, University Town Center—proved to be relatively easy, Mudgett recalled. But not Crossroads.

Councilwoman Barbara Weiner had won a seat on the council in large part because of her record of community service. She was also from Turtle Rock, which had the highest voter turnout in the city. Diminutive, with short dark hair and eyes, Wiener had strong opinions about many local issues, and one of them was that Crossroads needed a Saks Fifth Avenue store.

"She was adamant," Mudgett recalled:

> Her point was, you gave this stuff to Newport Beach. Why don't we in Irvine get the same respect and treatment? We spent many, many months . . . and it took all of the government relations skills and design skills to reach approval from the entire council for the project. The winning strategy was to refer to position Target as if it were located in Paris or the Côte d'Azur, and thus, Tar-jay.[25]

The ploy drew laughs at the council, Mudgett said, but underscored the company's commitment that the Irvine Target would create an upscale version of their otherwise mundane offerings.[26] The new Tar-jay won't be Saks, the company argued, but it will be the best Target store in Southern California. As it turned out, the store was a hit, making Crossroads a hit as well.

As 1987 ended, the company's pivot to what it described as the "executive developer" partnership was nearly complete. Even Fashion Island, the jewel in the company's retail crown, was turned over to a partner on a one-year trial basis. Employee head count was down to about 640 people from a high just 2 years earlier of 1,300.[27]

***

As a new decade approached, Donald Bren was seven years into ownership of the Irvine Company and had a clear vision for its future.

Management had been streamlined, with a small executive committee that Bren chaired, making the big decisions. A board of directors studded with some of the county's most prominent real estate, finance, and business executives, convened regularly to provide advice and counsel (one member was Don Koll, who competed successfully against the company in the commercial office market). Bren enjoyed a friendly relationship with Koll, but after Richard Sim complained that Koll was learning too much about company business strategy courtesy of his board position, less information was shared with the board.

The company workforce had been trimmed significantly, with much of the day-to-day operations and management being conducted by business partners in the "executive developer" mode. While some employees were bitter about the layoffs, others found similar work with the companies that were new partners. In some cases, the new jobs paid more than their previous positions at the Irvine Company (as the decades passed, it was not unusual that some employees laid off during slow times were recruited back to the company to tap into their institutional knowledge and relationships).

The company dodged a dangerous bullet in mid-1988 when county voters rejected an anti-growth measure; two years later, a sales tax increase for transportation improvements passed, meaning millions of dollars in future spending for roads and highways would be paid for by the county and cities instead of the company.

Also, the time-consuming lawsuit filed by Joan Irvine Smith and her mother had been settled. While financially painful, at least the uncertainty it cast on the company's future was over.[28]

The company's fiscal year ended on June 30, 1990. The next day, according to multiple sources, a year-long recession began. There are dozens of analyses attempting to answer the question of why; here is one that is authoritative and concise:

> the causes (include) pessimistic consumers, the debt accumulation of the 1980s, the jump in oil prices after Iraq invaded Kuwait, a credit crunch induced by overzealous banking regulators, and attempts by the Federal Reserve to lower the rate of inflation.[29]

Whatever the causes, the resulting recession hit the company hard. According to Richard Sim, it was only the second (and last) time in his 20-year career there that the company lost money. To shore up finances, Bren ordered a painful offloading of assets. First to go was the Tustin Golf Course, less than a year old. Apartments near the golf course with an underlying condominium map were put up for sale; tenants who had signed rental agreements one day and were moving in were told, in some cases, the next day to re-pack and find new lodging (the company gave them generous relocation assistance).[30]

At the time, the company held hundreds of ground leases for commercial and industrial properties. Bill Dennis, who was working at Coldwell Banker at the time, recalled in a 2025 interview walking into his office and being told by his supervisor that the firm had been selected to offer a variety of those leases for sale, something he thought would never happen.[31]

Sim estimated that sales of the commercial/industrial leases raised about $300 million, and a similar amount was obtained for shopping centers and leases on residential property. The money had to be raised, he said, because at the time, domestic banks pretty much stopped making loans for land development or construction. Sim is proud that during the cash crunch, he was able to convince the Industrial Bank of Japan to loan the company money "when nobody else would."[32]

Homebuilders suffered a similar fate, boxed out of funds to purchase raw land or finished lots. In 1992 or 1993, the company sold just a single parcel of land—not to a homebuilder, but to the Church of Latter-Day Saints for a new church building.

Layoffs continued, a handful year after year.

In 1993, the company made a move that shocked many Orange County real estate observers: Bren created a real estate investment trust (REIT) to hold the company's apartments. It was called Irvine Apartment Communities (IAC) and was listed in December of that year on the New York Stock Exchange, debuting at a price of $17.50 per share.[33]

Bren was careful to retain control of IAC, giving the Irvine Company about 60 percent of the stock, but to meet federal securities law, the new REIT had to have an independent board of directors and was required to file detailed public reports about its finances, something Bren and the Irvine Company had never done. Moreover, the executives operating the company were held to a high standard of responsibility to all shareholders, not just whomever had the majority ownership.

The REIT's first year started well, but the stock price took a hit in December when the Orange County bankruptcy shook the business and political communities to the core. It turned out IAC was holding $327 million in tax-exempt bonds, most of them issued by the county, which drove the stock price down to $14.75 a share. It turned out the bonds were guaranteed by the issuing banks, not the county, but the aftermath of the bankruptcy kept IAC from refinancing the bonds.[34]

Two Irvine Company executives had joined IAC: Patrick Smith as president, Richard Moran as chief financial officer. Smith pretty much changed his business card logo, as he had been in charge of the company's apartment portfolio; Moran had a larger role in company finance but was familiar with the apartment side.

Just 17 months later, Smith announced he was leaving as president; six months later, Moran exited. While their departures were accompanied by typical company press releases praising their performance and wishing them well, rumors spread that Bren was dissatisfied with how IAC was performing. In a 2024 interview, William McFarland confirmed that there were internal differences of opinion among executives about the value of land. "It was hard to make new land deals on (Irvine Company property) that would create accretive value for IAC shareholders," he said, one of the reasons for the turnover of the top executives.[35]

Next in the IAC president's office was Steven Albert. He lasted just about as long as Smith had, departing in early February 1997, with Bren stepping in the leadership role himself. In reporting his departure, the *Los Angeles Times* quoted "analyst Craig Leupold of Green Street Advisors (who) said Bren's appointment should heighten some investors' concern about conflicts of interest, given that IAC purchases land from the Irvine Company, which owns 54 per cent of IAC stock."[36]

Five months later, McFarland became the president of IAC. His appointment to the job, he believed, was intended to show the investment community that Bren was committed to making the REIT work. McFarland was aware of the conflict over land prices that had colored the first years of IAC's operations; he didn't want his responsibilities as CEO of a public company to make a dent in the positive relationship he had with Bren. "I told him, I hope we can remain friends," McFarland said, to which he said Bren replied, 'We'll work it out.'"[37]

However, the same issues recurred. In addition to purchasing property in other parts of California to develop apartments, Bren wanted IAC to buy land owned by the Irvine Company. Had IAC paid the price Bren was offering, McFarland said, the deals would not have been accretive for IAC shareholders. Moreover, the decision was not just McFarland's—two other IAC advisors had to agree to the sales price. They demurred, citing, among other factors, appraisals of the properties.[38]

Land prices were not the only point of contention. According to a Bloomberg News report reprinted in the *Los Angeles Times*, at the company's board meeting, when IAC managers "projected a rental growth rate of 5% for the year, Bren . . . thought the growth should be closer to 12%."[39]

In early 1999, Bren moved to take IAC private.[40] He initially offered shareholders $32.50 per share, but when the offer was criticized as too low, he bumped it to $34 a share in cash, totaling $569 million, and closed the deal. It was a nice payday for shareholders, as Bren's offer was nearly $7 above the stock's closing price.[41] At the time, IAC owned 18,758 apartments, most of them in Orange County. After the sale closed, McFarland resigned.[42]

In the aftermath of taking IAC private, another of the company's executive developer arrangements expanded. Western National Properties signed on to both a joint venture a new apartment complex with the company as well as to manage many of the existing apartments.

In March 1996, the *Los Angeles Times* ran a lengthy article combining a profile of Bren with the news that

> He is planning to sell major parts of his . . . private company to the public through a so-called real estate investment trust, or REIT. Access to investor cash would give the company the ability to expand outside the limits of the "ranch" and continue to grow locally. . . .
>
> Among the properties that may be offered are the Fashion Island shopping mall and surrounding office towers in Newport Beach and the entertainment center in Irvine.[43]

The story quoted Bren: "The Irvine Ranch is slowly being transformed. Our long-term goal is to transform what once was an agricultural company to a development company and to . . . the next, final step to create a large real estate investment company."[44]

Bren made similar comments to the *Orange County Business Journal* two years earlier. In that interview, Bren had speculated that there might ultimately be four separate REITs holding company properties in each. In addition to the value the

REITs would bring to the company—greater access to capital markets and less dependence on commercial banks and East Coast insurance companies—they would also provide "staying power, giving a permanence which is really important."[45]

Bren went on to argue that public ownership of at least some elements of the Irvine Ranch would promote stability. "The people who are shareholders of a company like Irvine Apartment Communities have ownership of the Irvine Ranch. As we broaden and create new REIT, there will be more public ownership of specific property on the Irvine Ranch."[46]

The REITs were never formed, nor was there ever again any public discussion about doing so. Neither McFarland or Sim recalls any internal analysis done in anticipation of REIT formation. Was Bren just floating a trial balloon to see if it would shake loose money from the banks or insurance companies? Indulging in public speculation for his own amusement? No one knows.

What Bren had done was articulate in no uncertain terms his ultimate goal. Gary Hunt once told an associate that Bren viewed the Grosvenor Trust as the model for a long-term real estate investment vehicle. The venerable organization headquartered in the United Kingdom certainly had demonstrated staying power, formed in 1677 based on properties owned in London. It is now valued at over $10 billion and manages assets worldwide.

***

The Irvine Company was, and is, very protective of its financial information. One of the few detailed examinations of its finances and business strategy can be found in a copy of the 2002 Business Plan, located in Ray Watson's papers in Special Collections at the UC Irvine Langson Library.[47]

Just 25 pages, the plan laid out objectives for the two principal company revenue generators, Irvine Community Development Co. and the Investment Properties Group, which included retail, office, and hotel properties. It also spelled out expectations for new entitlement, setting targets of 2,553 housing units and 10 million square feet of commercial space.[48]

A single theme occurs throughout the document: absolute specifics of what was to be achieved for residential lots to be sold, the amount of square feet to be leased, revenues for both, and anticipated percentage increases for land prices, apartment rents, and other revenue sources.

The numbers are precise. For example, room rates at the company's two hotels were set at $280 per night for the luxurious Four Seasons (no change from 2001) and $132 per night at the Hyatt, an increase of 3 percent. (The hotel occupancy rates at the time, 72 percent for the Four Seasons and 67.6 percent for the Hyatt, reflected an observation that Richard Sim made in his 2024 interview. Bren, he said, never really appreciated how much affinity points from the hotel management company impacted traveler decisions about where to stay. When the Irvine Company assumed management of its hotels from Four Seasons and Hyatt, occupancy reportedly dropped. Eventually, both hotels were sold.)

The business plan set an increase in apartment rents of 4.2 percent. It set a goal of selling 66 acres of commercial land at an average price of $26 per square foot and sales of 152 Shady Canyon Golf Course Equity memberships. Retail occupancy was to be at 95 percent by achieving a tenant retention rate of 75 percent while increasing average rents to $32 per square foot, an increase of 8.3 percent more than the previous year.[49] (According to the Bureau of Labor Statistics, the national rate of inflation at the end 2002 was just under 2.5 percent.)[50]

The revenue section was four pages long, summarized as "the primary objective in 2002 will be to effectively finance the Company's growth while increasing balance sheet strength and liquidity to prepare for potential changes in business cycle conditions."[51]

At the close of its fiscal year in 2001, the company had $4.1 billion in debt, $1.61 billion of it unsecured, $2.44 billion secured. As for liquidity, the business plan said "The Company maintains $850 million of unsecured corporate liquidity . . . the Company expects to use (the liquidity) throughout the year to fund development and acquisition activity, and projects a fiscal year-end balance of $52 million."[52]

The Financial Projections section of the business plan showed that at the time, as fast as the company was making money, it was also spending it: "Fiscal 2001 is expected to be a record year with revenue projected to increase 13% from $1.5 billion in 2000 to $1.7 billion."[53] For 2002, the report noted that "the Chairman's goal" was $1.8 billion. On the spending side, the report noted "relatively aggressive development in the office and apartment portfolios together with the $348 million acquisition of Fox Plaza and relatively moderate land development have resulted in capital expenditures of nearly $1.3 billion."[54]

The business plan text also reassured readers that the plan "has been tested for compliance with the financial covenants contained in the Bank and Private Placement documents and is comfortably within the required limits."[55] In other words, for all its awesome reputation as the owner of almost one-sixth of Orange County, the company had to convince its lenders that it would repay its loans.

Tom Rogers was treasurer of the company from 1990 to 2000. His job was to cultivate the banks and other financial institutions willing to lend billions of dollars to fuel the company's business operations and, increasingly, Bren's appetite for expansion beyond Orange County.

Rogers brought representatives from all the banks and other financing entities to Newport Beach for three days once a year to brief them on operations, spending, and profits. Bren wasn't eager to appear and speak at one of the meetings, Rogers said in a 2023 interview, but he always did so.[56]

In addition to long sessions replete with facts and figures, Rogers said he organized golf outings at Pelican Hill to show off the spectacular coastal views, followed by refreshments at the restaurant and bar at the clubhouse. But many of the bankers, Rogers recalled, wanted instead to take hikes in some of the thousands of acres of open space Bren had donated or set aside throughout the ranch property. The scale of the acreage, Rogers reported, impressed even the most jaded bankers.[57]

Rogers recalled his career at the Irvine Company as rewarding professionally and personally. With a twinkle in his eye, he said to underscore the unique nature of the company's value, he had gift T-shirts printed up with the company logo in red on the front, and on the back, the words "The Irvine Company, There are No Comps."[58]

The final four pages of the business plan looked five years ahead, predicting that from 2002 to 2006 the company would build or acquire an additional 10,417 units, build an additional 750,000 square feet of office space, and achieve cash flow sufficient to fund 90 percent of total capital.[59]

The plan concluded with an observation of what had become obvious: "The transformation of the Company into an investment property-dominated business is apparent in the composition of its assets. Investment properties and apartments are expected to comprise 83% of total assets in 2002, up from 52% eight years ago."[60]

***

As the company entered the new millennium, a number of veteran executives with expertise in land development and entitlement began to leave. Monica Florian and Richard Sim departed. Preceding them, Larry Thomas had left, as had one of his key public relations assistants plus public affairs manager Franz Wisner. (Wisner's departure to get married proved serendipitous; when his bride-to-be left him at the altar, he took a year off and went around the world with his brother on the travel plans for his honeymoon. The unlikely experience led to a successful book, *Honeymoon With My Brother*. Wisner never worked in corporate life again.)

It was the sudden departure of Gary Hunt that drew the most speculation about what internal machinations were underway at the company. Hunt ended his 25-year career at the company in early February 2001. The *Los Angeles Times* offered its analysis of Hunt's departure as well as others: "Hunt . . . had wrangled six months off to act as finance chairman for former state attorney general Dan Lungren's $30 million bid for governor. Insiders and local political leaders say Bren was initially reluctant to allow Hunt time off. Lungren was trounced."[61]

The *Times* then peered into its crystal ball to divine the aftermath of Hunt's departure, noting that "amid local battles and expansion beyond the ranch, Michael McKee emerged as vice-chairman—second in command to Bren—the corporate lawyer who joined the company in 1994 helped engineer a restructuring last year. McKee consolidated his power. In December, he became part of the three-member executive committee, with Bren and Richard Sim."[62]

Ray Watson was nonplussed about the changes. Bren, he told the *Times*, "had intended mainly to rearrange management, not trigger so many departures . . . [Watson] characterized Irvine Company's upheaval as no more extreme than that at many companies."[63] Joe Davis, who worked at the company for 16 years and eventually headed up community development, remembered the situation differently. Interviewed in 2025, he said that "one thing you could depend on was that every December and July there would be a reorganization."[64]

In the early years of the new decade, the company was still basking in the glow of the warm feelings Bren had engendered with his donations of thousands of acres of open space. But there were several other dynamics at work in the county that combined to slowly diminish interest in company developments and to reduce the company's profile in the media.

First was the continuing saga of El Toro Marine Corps Air Station and plans for its redevelopment. With the passage of Measure W in mid-2000, a decade of discord in the county was winding down; a conceptual plan for a huge park apparently drove a stake in the heart of plans to create an international airport there. But the federal government's decision to auction off the property around the park site for development soon garnered press attention, and the subsequent drama surrounding the City of Irvine's controversial efforts to create the Great Park dominated local media reporting.

The major environmental dramas of past years that had cast the Irvine Company as the villain were over. Laguna Canyon had been saved, thanks to Bren's decision to pull development plans there. Thousands of acres of open land had been set aside to protect the gnatcatcher as part of the innovative Natural Communities Conservation Program. The entire beach along the Irvine Coast was a state park. The toll roads were being built, a victory for development that had devastated opponents.

Thanks to the millions of dollars raised by the Measure M sales tax election in 1990, major freeway and arterial highway improvements throughout the county were underway or completed, including the massive rebuilding of the El Toro Y at the confluence of the 5 and 405 Freeways, adjacent to the Irvine Spectrum. Driving was relatively comfortable, especially in Irvine, thanks to the Y and the new toll roads, which opened in 1995 and 1999.

In Irvine, the 1988 Open Space Agreement permitted intensified development in exchange for donations of open space, with minimal city oversight. If anyone objected to seeing the bulldozers carving out new villages in places like Eastwood or Oak Creek, they were reminded that the ship of approval had already sailed. Furthermore, while new homes were being built in West Irvine and Northpark, the locations were far removed from the other villages; the majority of Irvine residents had no idea that development was even taking place.

Orange County's media landscape was diminishing. The Orange County edition of the *Los Angeles Times* closed, which dramatically reduced the paper's coverage of Irvine Company events. The *Orange County Register* went bankrupt in 2010, ending a long decline. New owners pumped up reporting at the paper in 2012, but four years later, they were forced to sell the *Register* to a partnership that became the Southern California News Group. The scale and sophistication of business and real estate reporting declined with the fate of the two papers, leaving the *Orange County Business Journal* to try and keep up.

Sim, Hunt, and Ray Watson had often been the face of the company and served as spokesmen for big-picture issues. Larry Thomas, with his experience working for a governor and—briefly—for Vice President George H. W. Bush—brought a level of sophistication to media strategy unmatched by other county business entities. All were now gone, and it would be another decade before the public affairs and public relations functions at the company again matched the reputation that Hunt, Thomas, and Jack Flanigan had built.

The internal atmosphere at the company was also changing. Employees who worked there reported that as the emphasis transitioned from development and entitlement to investment property acquisition and management, the mood shifted—people were more competitive. As one former employee noted, "the Wall Street people" began to fill new jobs. Another echoed that observation, using the term "bankers" instead of "Wall Street." More and more, what mattered most to the executives on the ninth floor were details on revenues and profits.

Michael McKee became the company's chief operating officer in 2007, retiring two years later at age 62.[65] In 2010, Ray Wirta, who had been part of the company's upper ranks of executives, assumed the title of president of the company, the first time the company had designated someone in that role since it was eliminated in 1986. Wirta stayed on until 2019, when Charles "Chip" Fedalen, the company's chief financial officer, became president. His tenure lasted until 2025, and today, two men are sharing the title of co-president of the Irvine Company: Frank Abeling and Jonathan Brinsden.[66,67,68]

***

In the run-up to the national financial meltdown of 2008, the Irvine Company initiated a dramatic off-ranch buying campaign. It began quietly in 2005 when Bren began purchasing an interest in one of the most iconic skyscrapers in New York City, the MetLife Tower at 200 Park Avenue. Rising to an imperial height of 987 feet over Grand Central Station, the building was built in 1963 as the headquarters of Pan American Airlines. When the airline went bankrupt, the building was purchased by the Metropolitan Life Insurance Company. Bren has since bought out minority owner Tishman and owns the building outright.

Two years later, the company bought 17 office buildings in San Diego County from an affiliate of the Blackstone Group. The haul included buildings in University Town Center near UC San Diego, as well as one in Del Mar and another in Mission Valley. In total, the buildings had 2.1 million square feet of space.[69] The buildings joined ten others owned by the company at the time, primarily in downtown San Diego.

Similar purchases were taking place in Silicon Valley, which Bren had identified as another logical target for expansion. Irvine Apartment Communities (IAC) had, by 2001, acquired two apartment complexes in Redwood City and Cupertino.[70] Thereafter, it was large retail and mixed-use properties being scooped up using what observers assumed was the growing pile of cash the company was accumulating. Of the properties it owns, Santa Clara Gateway is considered the premier holding.

In 2014, Bren paid $850 million for a 60-story office building located at 300 LaSalle Street in Chicago. The building has 1.3 million square feet of space.[71] In the following years, Bren purchased two other Windy City high rises at 1 North Wacker, 51 floors and 1.4 million square feet, and 71 South Wacker, 48 floors, 1.4 million square feet of space.[72],[73]

While looking across the country for opportunities, Bren also found a plum in his backyard, buying four office buildings from Transwestern Investment and Cigna in the heart of the Irvine Business Complex for $325 million. The buildings have 900,000 square feet of space. In 2025, the Irvine Company sold all of its downtown San Diego office buildings at a loss.

Notes

1 "TABI Partnership Wins Bid for Irvine Company," *Los Angeles Times*, May 21, 1977.
2 "Irvine Company Board Approves Bren Buyout," *Los Angeles Times*, October 15, 1983.
3 "Irvine Co. and Its Directors Have Been Sued for $5 Billion by Dissident Minority Shareholder Joan Irvine Smith," United Press International, October 19, 1983.
4 "Peter Kremer Sues Over Share Valuation," *Los Angeles Times*, September 22, 1983.
5 "Peter Kremer Sues Over Share Valuation."
6 "Kremer Settlement Revealed," *Los Angeles Times*, May 6, 1987.
7 "Irvine Company Ends Cattle Operations," *Los Angeles Times*, June 25, 1985.
8 Author's recollection.
9 Brad Olson interview with H.P. Oliver and C.M. Stockstill, March 25, 2024.
10 Richard Sim interview, H.P. Oliver and C.M. Stockstill, January 18, 2024.
11 William McFarland interview, H.P. Oliver and C.M. Stockstill, January 8, 2024.
12 William McFarland interview.
13 Brad Olson interview.
14 "Irvine Co. Restructures Management," *Los Angeles Times*, June 11, 1986.
15 "Irvine Co. Lays Off 100 in Restructuring Move," *Los Angeles Times*, July 9, 1986.
16 "Donald Bren Profile," Orange County Register, April 12, 1987.
17 Mark Baldassare, *When Government Fails: The Orange County Bankruptcy* (Berkeley: University of California Press, 1998).
18 Baldassare, *When Government Fails.*
19 Baldassare, *When Government Fails.*
20 Baldassare, *When Government Fails.*
21 David Mudgett, interview by H.P. Oliver and C.M. Stockstill, February 3, 2025.
22 Mudgett interview.
23 Mudgett interview.
24 Bill Kenney, interview by C.M Stockstill, 2025.
25 Mudgett interview.
26 Mudgett interview.
27 Author's recollection.
28 "Judge Awards $170 Million in Irvine Co. Stock Dispute." *Los Angeles Times*, May 6, 1987.
29 Carl E. Walsh, "What Caused the 1990–1991 Recession?," *Economic Review (Federal Reserve Bank of San Francisco)*, no. 2 (1993): 33–48.
30 Sim interview.
31 Bill Dennis interview with C.M. Stockstill, October 31, 2024.
32 Sim interview.
33 "Irvine Apartment Communities Sets IPO Terms," *Los Angeles Times*, October 3, 1993.
34 "County Bankruptcy Casts Shadow Over Irvine Apartment REIT," *Los Angeles Times*, December 12, 1994.
35 McFarland interview.
36 Partial Privatization of Irvine Aparment Communities (IAC) in 1997, "Bren Takes IAC Private," *Los Angeles Times*, February 6, 1997. This article covers Donald Bren's strategic move to increase his ownership stake in Irvine Apartment Communities, consolidating control but not yet removing it from public trading.
37 McFarland interview.
38 McFarland interview.
39 "Irvine Apartment Communities Reports Solid Growth," *Los Angeles Times*, February 6, 1999.
40 "Bren Offers to Take Irvine Apartment Communities Private," *Los Angeles Times*, February 6, 1999.
41 "Shareholders Approve Sale of Irvine Apartment Communities," *Los Angeles Times*, June 11, 1999.

42 McFarland interview.
43 "Irvine Company Plans REIT Expansion," *Los Angeles Times*, March 11, 1996.
44 "Irvine Company Plans REIT Expansion," *Los Angeles Times*, 1996.
45 "Bren Discusses REIT Strategy," Orange County Business Journal, May 23, 1994.
46 "Bren Discusses REIT Strategy."
47 "Irvine Company Business Plan, 2002," Ray Watson Papers, UC Irvine Special Collections.
48 "Irvine Company Business Plan, 2002."
49 "Irvine Company Business Plan, 2002."
50 U.S. Bureau of Labor Statistics, Consumer Price Index – All Urban Consumers (CPI-U), annual inflation rate for 2002, accessed January 14, 2026, https://www.usinflationcalculator.com/inflation/historical-inflation-rates/
51 "Irvine Company Business Plan, 2002."
52 "Irvine Company Business Plan, 2002."
53 "Irvine Company Business Plan, 2002."
54 "Irvine Company Business Plan, 2002."
55 "Irvine Company Business Plan, 2002."
56 Thomas Rogers, interview by H.P. Oliver and C.M. Stockstill, November 21, 2023.
57 Rogers interview.
58 Rogers interview.
59 "Irvine Company Business Plan, 2002."
60 "Irvine Company Business Plan, 2002."
61 "Irvine Company Executive Changes," *Los Angeles Times*, February 7, 2001.
62 "Irvine Company Executive Changes."
63 "Irvine Company Executive Changes."
64 Joe Davis, Interview by H.P. Oliver and C.M. Stockstill, April 10, 2025.
65 Michael McKee's Retirement (2008): "Irvine Co. CEO McKee Retiring," Orange County Business Journal, August 3, 2008.
66 Ray Wirta's Appointment and Tenure (2010–2019): "Ray Wirta Retiring from Irvine Company," *Connect CRE*, March 26, 2019.
67 Charles "Chip" Fedalen's Presidency (2019–2025): "CHARLES FEDALEN JR." Orange County Business Journal, August 17, 2020.
68 Frank Abeling and Jonathan Brinsden as Co-Presidents (2025): "OC Insider: Newport Center News," Orange County Business Journal, July 7, 2025.
69 "Irvine Company Acquiring Equity's San Diego Buildings," Orange County Business Journal, February 14, 2007.
70 Irvine Apartment Communities, Inc., Form 10-Q for the Quarter Ended September 30, 2001, U.S. Securities and Exchange Commission; see also *Irvine Apt Communities vs. Jamie Williamson, San Mateo County Superior Court,* filed March 22, 2001.
71 "Chicago Record: 300 N. LaSalle Sells for $850 Million," *Crain's Chicago Business*, May 19, 2014.
72 "PwC Renews Massive Lease in Irvine Company's Chicago Tower," *The Real Deal*, October 28, 2024.
73 "71 S Wacker—Chicago Commercial Real Estate," *Irvine Company Office Website*, https://www.irvinecompanyoffice.com/locations/chicago/west-loop/71-south-wacker.html.

26 The transformation continues

As the 41st year of Donald Bren's ownership of the Irvine Company concluded, a series of events took place that underscored the company's strength and resilience, overshadowing the fact that a change in ownership was increasingly inevitable.

Land for new residential development had diminished to two areas—an expansion of the Orchard Hills community in the foothills of the Lomas de Santiago in Irvine and a virgin tract of land in the city of Orange near Irvine Lake.

Preparations for the 500-lot Orchard Hills tract had been underway for several months but were threatened by resistance from existing residents.

Their objections centered on the presence of an asphalt plant. Hidden in a small cleft of land about a mile above Portola Parkway, All American Asphalt had operated for more than two decades, its presence known to just a handful of Irvine Company employees who worked in the nearby avocado groves. All American had purchased the land from the Irvine Company in anticipation of supplying asphalt for the Eastern Transportation Corridor.

The first phase of Orchard Hills had been built and largely occupied by 2020, when residents began to complain about noxious fumes wafting from the plant and into the backyards of their expansive homes, valued at $3 to $4 million. Frustrated by the South Coast Air Quality Management District's (AQMD) slow response to complaints, residents began a public display of their displeasure, carrying signs along Portola Hills Drive and Culver Drive, and making increasingly vocal appearances at city council meetings.

Lawsuits followed, a class action by a resident, as well as one by the city. As the clamor continued, a solution that seemed obvious, as well as expensive, began to circulate; the city would buy out All American and shut down the plant.

The Irvine Company was watching the trouble brewing and quietly calculating the potential impact if new homes were built even closer to the asphalt plant than the ones in existing neighborhoods, seething with frustration and anger.

Then, without any indication discussions had ever taken place between the city and the company, on February 28, 2023, came the stunning announcement that the city would buy the plant for $284 million and close it, using money from an Irvine Company gift of 80 acres of land at the corner of Jeffrey Road and Portola Parkway. Additionally, the shuttered asphalt site would be remediated and become part of a new element of city open space to be called the Gateway Preserve (the

DOI: 10.1201/9781003683087-26

80 acres was the most easily developable of a total of 475 acres gifted to the city by the Irvine Company as part of the deal; also, the acreage was identified on planning documents as a park).[1,2]

At the time, the value of the 80 acres was given as about $330 million.[3] As it turned out, that was a lowball figure of epic proportions, as Brookfield Homes purchased 70 of the 80 acres for $654 million, or just over $9 million an acre. Brookfield expects to build 1,200 homes on the site.[4]

The gift and sale arrangement reflected the awesome power and flexibility available to the Irvine Company not only in 2024 but what it had tucked into its corporate pockets since 1960, when the planning and development of the Irvine Ranch into a New Town began.

Unlike developers or builders who typically have much of their net worth riding on the outcome of a single project, the 93,000 acres the Irvine Company started with in 1960 was essentially free land without significant debt. It was located in five, and soon to be six, government jurisdictions that granted permission to develop, plus school districts, a water district, and dozens of other special districts and joint powers authorities. (In fairness, after Donald Bren paid approximately $864 million in 1983 to buy out his partners, the land basis changed.)

Time and again, the company used its land leverage to entice, reward, or ignore the cities and county on the ranch with a variety of development options, all of course within the legal confines of the general plans each jurisdiction had adopted (and in the case of Irvine, a plan that was close to a duplicate of the one created by the Irvine Company).

Moreover, with so much land, the company could, with minimal impact on its balance sheet, create parks, open space, lush landscaping, setbacks, decorative entries, and streetscapes that ate up acreage any other developer could never afford to give up. Add to the physical inventory of land Donald Bren's exacting and uncompromising eye for detail and quality, which he demanded on every company project, it created a formula unique in Southern California.

The Irvine Company's near-monopoly on retail properties in Irvine would also pay unexpected dividends in the same timeframe as economic and political forces combined to plunge California into a housing crisis.

In 2020 and the following months, state housing officials began to enforce a state law that required cities and counties to accept higher numbers of new housing units in their general plans. Failure to do so could subject the jurisdictions to penalties or, in some cases, trigger a law known as "builder's remedy," which stripped the jurisdiction of its zoning authority on a project-by-project basis. The draconian law would simply allow a builder to move forward on a project without any government oversight.[5]

Many cities were virtually built out and complained bitterly about the law. Irvine resisted as well, even though at the Great Park Neighborhoods and within the Irvine Spectrum and Irvine Business Complex, the potential for some new development existed, along with the potential for redevelopment.

For the Irvine Company, it was another opportunity, one that could benefit the city as well.

Figure 26.1 Apartments on a former retail site at Irvine Marketplace

Source: Author's image

The opportunity was officially memorialized in a Memorandum of Understanding adopted unanimously by the city council on March 14, 2023. It authorized construction of 4,500 new apartment units at six sites in the city, 1,025 of them affordable to low and moderate income individuals and families. Moreover, in lieu of building parks as required by state law, the council accepted the Irvine Company's offer to pay $14,500 per unit to the city, a windfall that would eventually total $65 million when all the units were built.

All of the sites for the new apartments were in the Spectrum area, with the exception of one at the edge of the 2,200-unit Promenade community directly across from the Spectrum Entertainment Center. But it was construction of a huge project in The Irvine Marketplace that drew the most attention.

The Marketplace is a retail center that straddles the cities of Irvine and Tustin, the square footage divided roughly half and half, with Jamboree Boulevard bisecting the property. Stores are built in a semi-circle facing massive parking lots, resulting in a retail footprint on an enormous open space. Additionally, and providentially for the Irvine Company, the COVID-19 economic hangover left several retail spaces empty or just barely hanging on in the Irvine quadrant of the Marketplace adjacent to the one the company was considering for new apartments.

With some changes to leases and relocation assistance, 79 acres were freed up in the easternmost quadrant to make it feasible for a teardown, which began in earnest in early 2024, followed by initial construction of two giant parking structures to accommodate the vehicles for the residents expected to move into the 1,261 new apartments in late 2025.[6,7]

As the transformation of retail to multi-family takes place in Irvine, plans for more traditional residential development were scheduled to begin in mid-September of 2024 in the hills east of the city of Orange. That 7,700-acre property had originally been zoned in the late 1980s, with more than 12,300 of housing units scheduled to be built.[8]

The company sat on the zoning until late 2023, then returned to the city with a plan that reduced the total number of units to 1,180, freeing up thousands of acres to be preserved in open space.[9] Despite the fact that all permits had been issued, a group of Orange residents persisted in opposing the project, lamenting the loss of open space and potential impact on wildlife; they filed a lawsuit in late 2024 in an attempt to stop it.[10] As this is written, it remains unresolved.

***

At the Irvine City Council meeting on Tuesday, May 25, 2024, council members found a surprising agenda item.[11]

The Irvine Company was proposing to build an entirely new village in the city, the first since Woodbury nearly 16 years earlier. It was called, with a sterile designation unlike the city's other colorfully named residential communities, "Spectrum District Village." The plan outlined a number of benefits, starting with the company explaining that it would construct 3,100 homes and apartments on the 200-acre parcel, fewer than the 5,000 allowed in the city's general plan. The company also proposed to grant the Irvine Unified School District a 10- to 13-acre site for an elementary school at no cost and the standard array of parks and trails common to other villages.[12]

The school site was first among a cascade of proffered benefits that would flow to the city if the village was eventually built.[13] Also listed were:

- The existing Oak Creek golf club, restaurant, and parking area would be gifted to the city for use as a community center.
- Zoning restrictions on a parcel of Irvine Company land four miles away would be lifted by the company so the city could build a public safety operations center.
- The Irvine Company would help establish a city Innovation Center to promote the creation of new small businesses.
- Parks would be built in the new village, as well as connections to the Jeffrey Open Space Trail on the western edge of the village.

Then the icing on the benefits cake: $90.4 million in community benefit fees to fund several existing City priorities, including a total of $26.2 million for "community projects," with $10.2 million for the Jeffrey Open Space Trail and the Walnut Trail. Additional funding would also pay for:[14]

- Expanding Irvine Connect, the city's free bus system.
- Enhancing urban forestry activities.

- Enhancing food security and non-profit support programs.
- Implementing bike center operations.
- Instituting trail and park improvements.
- Advancing vocational youth job training programs.
- Funding library system-related costs (the city had just broken away from the county library system to create and operate its own).

Normally, a village of the proposed size would have been subject to city requirements to build apartment units with rent rates calculated to meet the ability of low- and moderate-income families to afford them. In the case of the proposed village, that would have meant 465 units.

Mayor Larry Agran championed a different idea: issuance of 2,000 rental vouchers for low-income families that could be used to rent a market-rate apartment from the Irvine Company at 45 cents on the dollar, or a 65 percent reduction in rent. In return, the company would be relieved of the requirement to build the permanent units. Agran argued that the need for low-cost housing was so great that an immediate solution was needed.[15] The voucher plan was little more than a headline at the council meeting; closer examination showed the concept is part of a more comprehensive program designed to eventually move families into permanent low-income units the company is building around the city.

Preservation of open space was also promised if the new village was approved, in a series of parcels in the Lomas de Santiago hills that ring the northeastern boundary of the city, in total 315 acres. Much of it was in citrus production, and in an unprecedented arrangement, the city offered to assume operation and maintenance of the citrus from the Irvine Company.[16]

Some residents found this aspect of the proposal particularly galling; open space in their neighborhood was about to be razed for new homes while open space above Orchard Hills, Portola Springs, and near a new community known as The Summit—where houses were priced from $6 million and up—would get the benefit of additional open space. They also pointed out the city's spotty stewardship of past citrus management, noting that in 2009 some 5,000 orange trees were donated to the Great Park, but just two years later, 2,400 were given away because the city couldn't find a place to plant them.[17] Six hundred were planted initially, but the fate of the remainder, though planned for planting, remained hazy.[18]

The presentation map showed that the new village would be built at the corner of Irvine Center Drive and Jeffrey Road, extending north to the I-5 Freeway and east to Valley Oak Drive. Directly across the street on Irvine Center Drive was a small community of condominiums, Orangetree, and the campus of Irvine Valley College. Virtually all of the acreage was in current use as the Oak Creek Golf Course, an 18-hole facility with a driving range and golf club and restaurant. It was owned and operated by the Irvine Company.

The Oak Creek Golf Course was not shown on the city's original general plan. It had been placed there in late 1987 when the city council, under Agran's leadership, developed a plan to preserve open space in the city in return for allowing greater development on selected parcels of land owned by the Irvine Company but not yet developed. The plan was adopted by the city and submitted for a public vote in June 1988 under the title of 88–1 Irvine Open Space Initiative. It passed with a large majority (see Chapter 14 for more background).[19]

It didn't take long for residents who had lived or worked in the city in 1988 to recall that the ballot measure contained a map showing open space areas to be preserved—forever, the initiative promised. The golf course was clearly identified as open space, and as the initiative had been adopted by a vote of the Irvine electorate, changing it would require another public vote. So when will we get a chance to vote on this, they asked the council.

Well, maybe never, was the answer from city attorney Jeff Melching, who opined that because the 1988 initiative was actually placed on the ballot by the council, not as a result of a signature gathering process—that legal distinction allowed the council to change the zoning without a public vote.

Residents descended on council meetings loudly denouncing not only the proposed development but also the possibility that the public vote promised in the 1988 initiative would be ignored. Two former mayors, a former council member, and even a former Irvine Company executive who negotiated the agreement, Michael LeBlanc, reminded council members the promise made in 1988 had to be honored.[20]

The outpouring of opposition was, by Irvine standards, intense. Moreover, opponents formed an organization to focus their anger and opposition, raising money and retaining a law firm to prepare a challenge to the zoning if necessary.

None of the council members, except Agran, was familiar with the history of the open space initiative. Council member Michael Carroll prefaced his query to Melching on the subject by saying, "I was eleven years old when this happened." After sitting through hours of criticism from the audience, Agran allowed that Melching's legal views notwithstanding, a vote on the matter should probably take place.[21]

Melching drafted several ballot measure options and proposed a special election on November 18, 2025, to settle the matter. But then-governor Gavin Newsom created a redistricting plan for the state's Congressional seats to counter a similar effort in Texas. The legislature placed his item on the regular November 4 general election ballot, checkmating the city's plan for a special election. The matter was put over until early 2026.

The council action did not slow down Irvine Company efforts to build support for the new village. During the weeks before the May 25 unveiling at the council, the company had conducted public opinion polling and dozens of invitation-only briefings for residents who signed up via the company's website promoting the project. As this is written, those briefings continue, while opponents have created a non-profit entity, Save Irvine Open Space, to organize their opposition.

At the head of Save Irvine Open Space is former council member and mayor Christina Shea. With a total of 28 years of service on the city council, she is second only to Agran, who, at the end of his current term as mayor, will have served 40 years.

Facing Shea from the council dais must have been like a bad dream for Agran.

Shea had vaulted into Irvine politics decades earlier when she, her husband, and a group of conservative residents objected to the council's adoption of a human rights ordinance that among other elements, offered specific job protection to gays and lesbians. Shea led a referendum on the ordinance, which passed by a narrow margin. Agran was incensed with the decision. His disappointment was revisited in 1992 when Shea won a seat on the council.

On and off for 23 years, Shea and Agran clashed on virtually every issue before the city. And it was Shea who came out of political exile in 2012, her name identification and popularity still solid, to oust the council majority controlling the Great Park, a prelude to Agran being voted off the council two years later. (Shea recanted her position on protection for gays and lesbians in 2020 when the council symbolically removed the ordinance from city laws).

In mid-January of 2026, the Irvine Company made a major planning pivot, proposing, to replace the Oak Creek Golf Club with more than 50 acres of open, natural park space. The proposed Nature Park will be a "balance" to the lower density village to be built on the remainder of the course and adjacent properties.[22]

In previous months, the company had sought out community opinion through a variety of outreach efforts. Apparently, the feedback was substantial enough to provoke the new approach.

***

At Great Park Communities, Five Point continued to take down hundreds of acres of land even as high interest rates and potential tariffs on building materials slowed down similar housing development elsewhere. Making deals for land sales approaching $7 million an acre, Five Point was poised to take advantage of the city's requirement to build more housing with draft plans to intensify density as development pushed toward the last acreage owned at the site of the former El Toro Marine Corps Air Station.

Five Point also announced plans for development of a long-awaited retail center called The Canopy, designed by Almquist Development. To be anchored by a large grocery store, the center will have more than 80,000 square feet of tenant space. Renderings show buildings surrounding a large open space similar to a successful Almquist center in San Juan Capistrano.

Unlike the Irvine Company, Five Point is publicly traded. Great Park Neighborhoods is just one of three major landholdings the company owns. The others are in Valencia, a New Town in northern Los Angeles County previously developed by Newhall Land and Farming, and two properties in San Francisco, Candlestick Park and San Francisco Naval Shipyard. The San Francisco projects have been mired in bureaucracy and controversy ever since Five Point acquired them. Additionally, much of the shipyard property must be remediated to remove contamination leftover from past military activities. Valencia, with a successful critical mass of past development, is enjoying a steady stream of home sales.

Determining just how successful Five Point has been since being formed more than a decade ago is difficult. The financing arrangements for each of the communities are complex and opaque. The December 2024 10-K report filed by the company declared that there is a total of 10,500 homesites available for development at Great Park Neighborhoods and that 8,683 had been sold. Five Point also said that it has the potential to develop 4.9 million square feet of commercial space.

Ironically, the president and CEO of Five Point is Dan Hedigan, a former senior executive at the Irvine Company. He replaced Emile Haddad in February 2022.

Great Park Neighborhoods received a boost in community values when the city created two new seats on the city council, with the Great Park and surrounding neighborhoods entirely within a single council district.

***

Surrounding Heritage Fields, the Orange County Great Park had been renamed the Irvine Great Park as the early 2020s arrived. The Sports Park, funded by Lennar, opened in 2017, followed by Great Park Ice, a twin ice skating facility in partnership with the Anaheim Ducks professional ice hockey team. But the successes of the new park facilities were overshadowed in the final decade of 2020 by a lingering series of fights over creation of a long-promised veterans' cemetery and memorial.[23]

Between 2014 and 2018, an issue that would have seemingly garnered unanimous support from elected officials, developers, and community members alike instead became a political football. Encouraged by state legislation to fund an initial study of a 133-acre site with the acronym ARDA for a cemetery, the city council proposed to assist it with a $30 million allocation.[24]

The ARDA site was not included in the original Great Park acreage but had been transferred to Lennar as part of one of a series of development agreements between the city and the developer.[25] City planning for the area focused on recreational uses. Lennar, however, saw the potential for commercial development.[26] There were also reports, first whispered and then voiced in private conversations, that a cemetery anywhere near residential neighborhoods would discourage homebuyers, particularly some of Asian descent.[27]

In 2017, Lennar proposed to relocate the site of the cemetery from ARDA to an open area adjacent to the San Diego Freeway in what had been part of the El Toro base airplane crash zone. Lennar had leased the property to a strawberry grower, so the alternate cemetery site became known as Strawberry Field.[28]

After the city council approved the land swap, Larry Agran, along with local veteran groups that had advocated for the cemetery, gathered signatures to force a referendum on the proposal.[29] Following a bruising campaign, Irvine voters overturned the land swap in November 2018.[30]

For the next two years, the plan to build a cemetery at ARDA bubbled up and down on council agendas but could never achieve enough momentum to get close to genuine success.[31] In the meantime, some of the veterans who had joined the battle in 2018 began to look elsewhere for a credible cemetery site, fearing the Irvine site would never come to fruition.[32]

Sure enough, in late 2019, the council floated the idea of once again allowing commercial development on the ARDA site, citing concerns that cleaning up toxic waste and other debris there would overwhelm the park's diminishing resources.[33]

And once again, Agran turned to the ballot box, creating an initiative to rezone ARDA for a single purpose: veterans cemetery and memorial.[34] After gathering sufficient signatures to require an election, Agran brought the initiative to the city council. Faced with the prospect of a second divisive election in two years, the council capitulated and adopted the initiative.[35] Four months later, the COVID epidemic began its march of death and fear across the nation, bringing most activity in the Great Park to a halt.[36]

By late 2023 and early 2024, buoyed with growing revenues from the accelerating values of homes and other properties being built in Great Park Communities, the city announced an ambitious expansion plan for amenities in the park, including long-planned museums, a botanical garden, water elements, and a permanent amphitheater to replace the temporary one that Five Point had operated.[37] Funds were also to be expended to spruce up existing park facilities. There was even a proposal to build an elevated tram system connecting elements of the park.[38]

In late 2025, as the improvements to the park continue, the potential for a veterans' cemetery in Irvine has faded dramatically.[39] Frustrated with the city's continuous acrimony on the matter, former allies in the state legislature and veterans groups instead identified a site at Gypsum Canyon in Anaheim for the cemetery, acquiring support from officials charged with building state veterans cemeteries, as well as the County of Orange, which proposes to share the site for a new and much-needed county cemetery.[40] As of October 2025, a total of $55 million in state and county funds have been reserved for development of the cemetery.[41] Every city in the county is now on record in support of the Gypsum Canyon site.[42] Irvine is left with 133 acres of land that can only be used for a cemetery and veterans memorial unless voters approve a zone change.[43]

While 23 years after it was conceived the Great Park has made tremendous strides, as at the Tustin LTA base, there are still hundreds of acres of land left empty in the center of the huge property.[44]

***

A careful examination of the master plan for the Irvine Ranch created by William Pereira and Associates in 1961 shows a wide variety of land uses, with a notable exception: there is no designation for a hospital[45]

The plan anticipated that a medical school would be built as part of UC Irvine and a teaching hospital located on or near the campus. As described in Chapter 8, the campus hospital plan was spiked in favor of a community-based hospital to be located on Sand Canyon Avenue near its intersection with the San Diego Freeway.

From that single hospital, opened as Irvine Medical Center, a medical care complex with more than 600 beds, surrounded by medical offices, has emerged as one of the three concentrations of health care in the city Hoag Hospital Irvine, which includes an emergency room, orthopedic care and surgery center, two parking garages, birthing facilities, and comprehensive medical care, is scheduled to open an expanded hospital complex in 2026, fueled by more than $200 million in gifts from Audrey Steele Burnard and David and Diana Sun[46]

When complete, the Hoag complex will have 391 beds.[47] It will join the other major hospital at the Sand Canyon/San Diego Freeway location, Kaiser Hospital Irvine. Part of the giant health care system centered in California and the western United States, Kaiser built the 269-bed hospital in 2008. It has comprehensive medical services as well as a busy emergency room.[48]

Just over a mile south of the two general care hospitals on Sand Canyon Avenue, the Lennar Foundation City of Hope has been treating cancer patients since 2022, and in late 2025 is scheduled to open a six-story, 79-bed hospital built at a cost of $425 million.[49] City of Hope is a highly regarded cancer research and treatment center with headquarters in the San Gabriel Valley community of Duarte.[50]

Seeded by a donation from the Lennar Foundation (Lennar was the original developer of the community around the Great Park and has partial ownership in Five Point Communities), the hospital expects a substantial patient load from individuals in Orange County and northern San Diego County who would otherwise have been forced to make the long drive to Duarte for treatment.[51]

The third locus of medical care opened in early 2025 at the intersection of Campus Drive and Jamboree Road at the border of the Irvine Business Complex.

On a 13-acre site that was previously empty, adjacent to UCI maintenance shops and a small arboretum, a complex of hospital, medical offices, and parking structures has emerged with the name UCI Health Irvine. At the center is a 350,000-square-foot hospital with 144 beds, all of which come with private showers and sleeper sofas, as well as a 75-inch video screen. Ten operating rooms and four interventional surgical suites are located within the hospital[52]

Adjacent are two specialty buildings, the Joe C. Wen and Family Center for Advanced Care, with specialty care for children and adults and the UCI Center for Autism and Neurological Disorders; and the Chan Family Comprehensive Cancer Center and Ambulatory Care building.[53]

Total cost of the new center is $1.3 billion, according to UCI's promotional materials. Of special note is that the land where the new hospital complex is built is part of the original UCI Inclusion Area, land purchased around the campus and extending around the San Joaquin Marsh

***

Estimating the annual income and value of the Irvine Company is, in the words of former *Orange County Business Journal* editor Rick Reiff, "Orange County's favorite parlor game."[54]

The *Business Journal*, *Forbes*, other local and national publications, and real estate blogs all estimate a number of just over $19 billion in total value in 2025. Given the dramatic expansion of office space and apartments since 2006, when it was last reliably reported that Irvine Company income surpassed $3 billion, a current estimate of three or four times that amount seems credible.

What does that value and the income mean to Donald Bren as he approaches his 94th birthday?

Compared to other American and international billionaires, Bren's lifestyle and the worldly benefits of his wealth remain largely in the shadows, a reflection of his lifelong desire for privacy. But there are exceptions. He owns three homes in California, one in Holmby Hills, two others on Newport Bay. Because he chose to own his 6,800-acre ranch in Idaho as a business, as well as his 161-foot yacht (which in the past was available for charter), public records of the two are easily accessible on the internet. The same is true for his charter airplane company, headquartered in Van Nuys, California.

The homes are relatively modest, by billionaire standards. Bren has owned the Los Angeles residence since at least 1969, when it was listed as his address in a *Los Angeles Times* story about his helicopter crash off Laguna Beach.[55] Zillow estimates the value of the home at $18 million. *Forbes* wrote at length about the home on Harbor Island, which Bren built in the style of his favorite architect, Andrea Palladio.[56] And a story in the *Orange County Register* referenced Bren's home on Linda Island. The value of the Harbor Island home was estimated at $17 million by the Orange County Assessor in 2014.[57]

Bren has been collecting art for decades and was identified in *Art and Antique* magazine in 1989 as one of the 100 hundred most significant private art collectors in the nation. After Bren took majority control of the Irvine Company, some of his artworks were displayed in a hall by his office on the ninth floor of company headquarters.[58]

The cost of day-to-day operations of the Bren family (his wife and son) has never been officially revealed. The only credible reference was made during a legal action involving Bren, when it was alleged he was spending $3 to $4 million per month for living expenses, an amount his attorneys disputed.[59]

Whatever he spends on himself, Bren has amassed a record of philanthropy that routinely places him in the upper echelons of giving by individuals in the United States.

By the time he was 91, his total charitable giving was pegged at $1.63 billion, with higher education the principal recipient. Bren has given UC Irvine $60 million for both structures (the Bren Events Center) and academics. He gave UC Santa Barbara $15 million to establish the Donald Bren School of Environmental Science and Management, which funds both a building and ten endowed chairs for professors.[60]

In Irvine, in 2006, the Irvine Company announced the creation of the Excellence in Education and Enhancement Fund to hire teachers in the fields of science, music, and art. The company promised to fund the initiative for a total of $50 million at an annual contribution of $2 million for 20 years plus additional funds in later years to achieve the $50 million total.[61]

A private organization designed to help student achievement through after-school programs caught Bren's eye in 2008; he pledged $8.5 million to THINK TOGETHER, which at the time largely focused on low-income areas in Santa Ana and central Orange County.[62]

Bren's most recent and largest donation can legitimately be labeled as out of this world.

The gift was $100 million to the California Institute of Technology (Cal Tech) to fund the research and development of the Space Solar Power Project. The technology would convert solar energy collected in outer space and beam it back down to Earth as electricity. If it is able to be built and operated at a reasonable cost, it would literally change life on the planet, freeing mankind from fossil fuels forever.[63]

Bren reportedly became interested in the technology after reading an article about it in *Popular Science* as a young boy. After pondering its potential, he approached Cal Tech about a gift, which the prestigious institution accepted (Bren was a Cal Tech trustee for decades; his wife now has a seat on the board).[64]

Reflecting his service in the U.S. Marine Corps, Bren has given $3.46 million to the U.S. Marines Corps University. He has also given $150,000 to support scholarships for students at the university.[65]

Bren has typically made donations from three vehicles: the Irvine Company, his personal funds, and the Donald L. Bren Foundation, which, at the end of 2024, held assets valued at more than $333 million.[66]

***

Bren's reclusiveness worked against him in an unusual manner in February 2010, when a man walked into the Cerritos branch of East West Bank and opened four new accounts. He deposited a federal income tax refund check in the amount of $1.389 million, then strolled out. The bank's video camera showed a man in his early 40s with a round face, a small patch of hair on the top of his head, and fringes of hair on the sides.[67]

His name, according to the just-opened account, was Donald Bren.[68]

The man had presented the bank with a driver's license bearing Bren's name. It showed an address in the southern part of the City of Los Angeles. On bank documents, he listed his occupation as a smoke shop owner. Apparently, no one at the bank thought twice about any aspect of the transaction or the documentation. For the following four weeks, "Donald Bren" drained his account of $1.1 million in cash before authorities caught on to the scam.[69]

Federal officials broadcast the photo and got a hit when a guard at Orange County Jail recognized former prisoner Moundir Kamil, who had served time for . . . guess what . . . bank robbery. He was arrested, tried for bank fraud, and sentenced to jail in June 2011 after pleading guilty.[70] Precisely how he obtained the check, as well as Donald Bren's Social Security number, was never revealed in press reports. Also left to the imagination of many in the county was why one of the wealthiest men in the nation, if not the world, had a federal tax refund sent to a mailbox instead of directly to his bank account via electronic deposit.

***

In early October 2025, Donald Bren reluctantly emerged from self-imposed privacy to comment on an uncomfortable subject.

Bren's son, David, 32, was being sued by several people who claimed he had swindled them out of millions of dollars. A long, detailed story in the *Los Angeles Times* laid out the damning details of four lawsuits accusing David Bren of promising to create "the ultimate man cave," an exclusive club in Beverly Hills that would offer access to exotic, expensive automobiles as a benefit of membership.[71] The club was to be called The Bunker and had allegedly attracted a number of well-known celebrities and business figures as potential members.

David Bren was one of two children Bren had fathered with Jennifer McKay Gold. Bren and Gold never married. Bren acknowledged parenthood for David and his sister Christie and agreed to support them. Few people knew about David and Christie until 2003, when Gold sued Bren in Los Angeles Superior Court, seeking a higher level of support, arguing that Bren's wealth justified it.[72]

To the surprise of many people who knew him, Bren fought the suit and testified in court when it went to trial in 2010. The proceedings exposed a variety of elements of the Bren/Gold pairing, which lasted more than a decade, including private letters.[73]

Bren's statement on the accusations against David was terse, delivered from an Irvine Company senior executive: "We do not have a personal or business relationship with this individual," Bren said of his son.[74] Various news reports labeled the comment cold, a rebuke, and an act of disownment, while others pointed out that as David had allegedly used his father's name to attract investors, Bren was justified in distancing himself from any legal or monetary exposure.

As of November 2025, no criminal charges had been filed against David Bren, who had reportedly relocated to Florida. Donald Bren returned to his cocoon of privacy and has not spoken about the situation again.

***

A few days after the David Bren story faded into memory, veteran reporter Jonathan Lansner of the *Orange County Register* delivered a wrap-up story about a variety of projects underway on the Irvine Ranch and on other Irvine Company–owned properties, demonstrating that the company has a long development runway ahead of it.[75]

Lansner totaled up close to 12,000 apartment units under construction or in "various stages of planning" in Orange, San Diego, and Santa Clara Counties, which, when completed, would bring the total number of apartments owned by the company to more than 70,000. At that total, the Irvine Company would become the 12th largest apartment owner in the United States, according to Lansner's reference point, the National Multifamily Housing Council.[76]

One of the largest developments was in Santa Clara. The company had acquired a 100-acre commercial site in 2013 and, since then, had been tearing down old two-story buildings and creating a new mixed-use community called Santa Clara Square. There are 1,840 apartments, 28 retail stores, and 1.7 million square feet of office space in the development.[77] Irvine Company vice president of planning Rob Elliot said the new community was "highly walkable . . . the streets are narrow enough that it encourages pedestrian movement between the offices, retail, and apartments." **79** He added that the concept of locating housing within a business district "is just a different opportunity. There's quite a group, particularly younger people, looking for that kind of living environment."[78]

It was Donald Bren who had the last word in Lansner's story. As has been the case in recent years, Bren did not speak in person but sent a statement. Nonetheless, it provides a fitting coda to this book.

Bren said, "Fulfilling the vision of master planning has been a lifetime dedication and passion. I am hopeful our stewardship will live on in many new ways."[79]

Notes

1 Roger Bloom, "The City of Irvine to Acquire the All American Asphalt Plant and Incorporate the Site Into a 700-Acre 'Gateway Preserve'," *Irvine Community News and Views*, February 28, 2023, https://irvinecommunitynewsandviews.org.

2 City of Irvine, "All American Asphalt," *City of Irvine Official Website*, accessed 2023, https://www.cityofirvine.org/community-development/all-american-asphalt.

3 City of Irvine, "Staff Report: Gateway Preserve and Land Valuation," *Presented*, April 11, 2023, Irvine City Council Meeting. Available via City of Irvine Council Archives.
This report estimated the value of the 80-acre parcel at approximately $330 million, based on preliminary appraisals tied to the Irvine Company's land gift.

4 Dana Bartholomew, "Brookfield to Buy 70 Acres in Irvine for Housing Project," Orange County Register, April 2, 2024, https://therealdeal.com/la/2024/04/03/brookfield-to-buy-70-acres-in-irvine-for-housing-project/.

5 Christopher S. Elmendorf, *A Primer on California's "Builder's Remedy" for Housing-Element Noncompliance* (Davis, CA: UC Davis School of Law, 2022), https://law.ucdavis.edu/sites/g/files/dgvnsk10866/files/media/documents/Builder%27s%20Remedy%20Primer-1.pdf.

6 Tom Berg, "Reinvestment to Enhance The Market Place," *Irvine Standard*, July 3, 2024, https://www.irvinestandard.com/2024/reinvestment-to-enhance-the-market-place/.

7 Irvine Community Connection, "The Market Place Master Plan," *Irvine Community Connection*, accessed September 17, 2025, https://www.irvineconnection.com/tmp/.

8 Orange Heights Master Plan, Orange Heights Master Plan, Irvine Company, accessed September 17, 2025, https://orangeheightsmasterplan.com/.

9 Orange Heights Master Plan, Irvine Company.

10 Orange County Register, "Orange Heights Lawsuit," December 27, 2024.

11 Roger Bloom, "Irvine Company Proposes New Residential Village on Golf Course Site," *Irvine Community News & Views*, May 13, 2025, https://irvinecommunitynewsandviews.org/irvine-company-proposes-new-residential-village-on-golf-course-site/.

12 Bloom, "Irvine Company Proposes New Residential Village."

13 Bloom, "Irvine Company Proposes New Residential Village."

14 City of Irvine, "General Plan Amendment Application: Oak Park (Planning Area 52)," staff letter dated June 5, 2025, referencing City Council approval of Memorandum of Understanding (MOU) on May 13, 2025, https://legacy.cityofirvine.org/civica/filebank/blobdload.asp?BlobID=36100.

15 Bloom, "Irvine Company Proposes New Residential Village."

16 City of Irvine, "General Plan Amendment Application: Oak Park."

17 Woodbridge Village Association, "Proposed Oak Park Village," *Woodbridge Village Association*, accessed September 17, 2025, https://wva.org/page/21440~1120217/proposed-spectrum-district-village.

18 Thomas Martinez, "Irvine Turns to Trees to Help Fund Great Park," Orange County Register, June 22, 2012, https://www.ocregister.com/2012/06/22/irvine-turns-to-trees-to-help-fund-great-park/.

19 City of Irvine, "Open Space Voter Protection and Oak Creek Golf Course Development Ballot Measures," *Presentation to City Council*, July 22, 2025, https://irvine.granicus.com/MetaViewer.php?meta_id=170917.

20 Roger Bloom, "City Council Rushes to Approve a Plan that Could Add 57,656 Apartments & Housing Units," *Irvine Community News & Views*, August 13, 2025, https://irvinecommunitynewsandviews.org/city-council-rushes-to-approve-a-plan-that-could-add-57656-apartments-housing-units/.
21 Bloom, "City Council Rushes to Approve a Plan."
22 Irvine Company, "A New Vision for Oak Creek Golf Club." Press release, January 16, 2026, https://www.irvinecompany.com/press-release/a-new-vision-for-oak-creek-golf-club/.
23 Orange County Great Park renaming documented in city council records, 2020–2022.
24 California Legislative Information, "SB-1451 Veterans Cemetery: Orange County," https://leginfo.legislature.ca.gov.
25 City of Irvine, "Development Agreement Between the City of Irvine and Heritage Fields El Toro, LLC," July 2010.
26 Irvine Planning Commission Minutes, 2015–2017.
27 Hosam Elattar, "Asian American Residents Voice Concerns Over Cemetery Location," *Voice of OC*, May 2017, https://voiceofoc.org.
28 City of Irvine Council Agenda, June 6, 2017.
29 Orange County Registrar of Voters, "Measure B Initiative Filing," 2018.
30 Orange County Election Results Archive, November 2018.
31 Irvine City Council Agendas and Minutes, 2019–2020.
32 VALOR (Veterans Alliance of Orange County), "Position Statement on Cemetery Site," 2020.
33 City of Irvine Staff Report, "Environmental Assessment of ARDA Site," December 2019.
34 City of Irvine Ordinance 20-08, "Veterans Cemetery Land Use Initiative," May 12, 2020.
35 See City of Irvine Ordinance 20-08, "Veterans Cemetery Land Use Initiative," cited in note 35.
36 City of Irvine COVID Response Updates, March–August 2020.
37 City of Irvine, "Great Park Framework Plan," June 2023, https://cityofirvine.org/greatpark.
38 Noah Biesiada, "Irvine's Gondola Proposal Raises Eyebrows," *Voice of OC*, April 2025, https://voiceofoc.org.
39 City of Irvine Council Minutes, July 2025.
40 CalVet Press Release, "Gypsum Canyon Selected for Veterans Cemetery," February 2025; Orange County Board of Supervisors Resolution No. 25-017.
41 Orange County Budget Office, "Veterans Cemetery Funding Allocation," October 2025.
42 Orange County Council of Governments, "Joint Resolution Supporting Gypsum Canyon," August 2025.
43 City of Irvine Municipal Code, Section 9-51-ARDA.
44 Orange County Grand Jury, *Land Use and Development at Former Military Bases*, Final Report, June 2025, https://www.ocgrandjury.org/grand-jury-reports.
45 William L. Pereria, *The Irvine Ranch Master Plan (Revised Draft)* (Los Angeles, CA: Pereira and Associates, August 21, 1961).
46 Sun Family Foundation, "Sun Family Foundation Doubles Gift to Hoag Hospital Foundation," *Sun FamilyFoundation*, March 2024, https://www.sunfamilyfoundation.org/wp-content/uploads/2024/03/Sun-Family-Foundation-Doubles-Gift-to-Hoag-Hospital-Foundation.pdf
47 Hoag Memorial Hospital Presbyterian, "Hoag Hospital Irvine," *Hoag*, n.d., accessed January 15, 2026, https://www.hoag.org/locations/hoag-hospital-irvine
48 California Health and Human Services Agency, "Kaiser Foundation Hospital – Orange County – Irvine," *Cal Health Find*, n.d., accessed January 15, 2026, https://calhealthfind.california.gov/facility/acute-care-hospital/050000137
49 Orange Coast Magazine, "Lennar Foundation Cancer Center at City of Hope Orange County," *Orange Coast Magazine*, July 2022, https://www.orangecoast.com/features/lennar-foundation-cancer-center-at-city-of-hope-orange-county

50 City of Hope, "About City of Hope," *City of Hope*, n.d., accessed January 15, 2026, https://www.cityofhope.org/about
51 "About City of Hope," *City of Hope*, n.d.
52 UCI Health, "UCI Health Irvine Medical Center Opens to Patients," *UCI Health News*, March 2025, https://www.ucihealth.org/news/2025/03/uci-health-irvine-medical-center-opens
53 UCI Health, "UCI Health Irvine Medical Center Reaches Substantial Completion," *UCI Health News*, December 2024, https://www.ucihealth.org/news/2024/12/uci-health-irvine-substantial-completion
54 Rick Reiff, "Irvine Company: The Billion-Dollar Question," *Orange County Business Journal*, June 14, 1995.
55 *Los Angeles Times*, "Developer Injured in Helicopter Crash," July 31, 1969.
56 Mary Forgione, "A $13.9 Million Home in One of California's Most Exclusive Island Neighborhoods," *Forbes*, June 7, 2023, https://www.forbes.com/sites/forbes-global-properties/2023/06/07/a-139-million-home-in-one-of-southern-californias-most-exclusive-island-neighborhoods.
57 Jonathan Lansner, "Donald Bren: Most Influential 2014," Orange County Register, December 18, 2014.
58 "Donald Bren Child Support Lawsuit," *Los Angeles Times*, August 20, 2010.
59 *Art & Antiques Magazine*, "America's Top 100 Collectors," March 1989. Donald Bren was listed among the nation's most significant private art collectors. Details of the issue are cataloged by archival sellers such as eBid: https://www.ebid.net/us/for-sale/art-antiques-magazine-march-1989-top-100-collectors-221272175.htm.
60 Donald Bren Foundation, "Philanthropy," *DonaldBren.com*, accessed October 16, 2025, https://www.donaldbren.com/philanthropy/.
61 Irvine Company, "Philanthropy: In Support of Excellence," Irvine Company Newsroom, April 29, 2025, https://www.irvinecompany.com/press-release/philanthropy-in-support-of-excellence/.
62 Philanthropy News Digest, "Irvine Company Chairman Donates $8.5 Million to Expand Afterschool Programs," February 21, 2008, https://philanthropynewsdigest.org/news/irvine-company-chairman-donates-8.5-million-to-expand-afterschool-programs.
63 California Institute of Technology, "Caltech Announces Breakthrough $100 Million Gift," *Caltech News*, August 3, 2021, https://www.caltech.edu/about/news/caltech-announces-breakthrough-100-million-gift-to-fund-space-based-solar-power-project.
64 California Institute of Technology, "Powering Earth From Space," *Caltech Space Solar Power Project*, accessed October 16, 2025, https://www.spacesolar.caltech.edu/about.
65 Marine Corps University Foundation, "Donald Bren," *Donor Profiles*, accessed October 16, 2025, https://www.mcuf.org/donald-bren.
66 Donald Bren Foundation, "Foundation Overview," accessed October 16, 2025, https://www.donaldbren.com/donald-bren-foundation/.
67 *Los Angeles Times*, "Fake Donald Bren Cashes $1.3 Million Tax Refund," June 15, 2011.
68 *Los Angeles Times*, "Fake Donald Bren Cashes $1.3 Million."
69 *Los Angeles Times*, "Fake Donald Bren Cashes $1.3 Million."
70 *Los Angeles Times*, "Fake Donald Bren Cashes $1.3 Million."
71 James Gordon, "Property Billionaire's Dismissal of Heir," Daily Mail, October 6, 2025, https://www.dailymail.co.uk/news/article-15166319.
72 *Los Angeles Times*, "Jennifer McKay Gold v. Donald Bren," Coverage Summary, 2003–2010.
73 *Los Angeles Times*, Daily Mail, and The Real Deal, "Coverage Compilation," October 2025.
74 Gordon, "Property Billionaire's Dismissal."
75 Jonathan Lansner, "Irvine Company Has 12,000 Apartments in Pipeline," Orange County Register, October 15, 2025.

76 National Multifamily Housing Council, "2023 NMHC Top 50 Owners," https://www.nmhc.org/research-insight/the-nmhc-50/top-50-lists/2023-top-50-owners/.
77 Irvine Company, "Santa Clara Square," *Project Overview*, https://www.irvinecompanyoffice.com/locations/silicon-valley/santa-clara-square/.
78 Irvine Company, "Santa Clara Square."
79 Lansner, "Irvine Company Has 12,000 Apartments in Pipeline."

Epilogue

The 150th anniversary of the Irvine Ranch in single ownership will be celebrated in 2026, the year this book is published.

There are several cities and other government jurisdictions within the historic boundaries of the ranch. We believe it is the story of the City of Irvine that resonates with the greatest clarity and significance in this anniversary year.

Created by a handful of young planners and engineers as the nation was on the verge of tremendous social and political change in 1960, the New Town of Irvine would emerge in tandem with a fledgling campus of the University of California to fulfill the vision of self-described dreamer William Pereira.

Has the vision been fulfilled in Irvine? We believe Irvine in 2026 is a community that can best be described as: Safe, Diverse, Green, Progressive, Dynamic.

Safe

One of the first priorities for the newly elected Irvine City Council in 1972 was to organize and staff its own police department, which took place in 1975. It made Irvine unique among the master-planned communities of the era; the others contracted with the Orange County Sheriff for law enforcement. Since then, Irvine has consistently been rated by FBI crime statistics and numerous other sources as one of the safest cities in the nation and the state. Its officers are, by design, among the best educated and highly paid in Southern California. Reflecting the power of modern social media, the police department posts videos on Instagram and Facebook featuring Irvine's finest nabbing criminals using standard police procedures as well as drone surveillance, all with the tagline, *Don't Commit Crimes in Irvine.*

Diverse

An Asian diaspora that began three decades ago and accelerated in recent years has resulted in people of Chinese, Taiwanese, Korean, Vietnamese, and Filipino origin now accounting for just over 40 percent of the city's total population. Asian residents make up approximately one-third of the population in all parts of the city, except for Orchard Hills, where the number exceeds 68 percent. UC Irvine professor Yong Chen describes Irvine as "a third-generation Chinatown . . . the Chinese who moved to Irvine were not recent immigrants but those who had been educated and worked elsewhere in the United States."[1]

The Irvine school district proudly notes that its students speak 72 languages. The city has a dozen grocery stores where a shopper can as likely hear Mandarin as English, and it is not unusual in the gym, a retail outlet, or in line at Starbucks to catch phrases in Farsi, Russian, Hebrew, or related dialects. The city's previous mayor traced her lineage to Pakistan, and three of the current seven members of the city council are of Asian ancestry.

Green

The beneficiary of thousands of acres of land donated to the public by Donald Bren, the city traces its commitment to building parks to just three years after incorporation, with voters agreeing to tax themselves to build community parks and bicycle paths. National organizations regularly recognize Irvine. Most recently, the Trust for Public Land rated Irvine as having the best park system in California and the second best in the nation.[2] In addition to recreational and passive parks, the city owes it verdant nature to the Irvine Company's commitment to "making landscape architecture a critical role in making (our) communities attractive to residents . . . making a neighborhood or village more understandable, stronger and filled with character," according to company senior vice president Rob Elliot, who also estimates that since 1960, more than half a million trees have been planted as a result of development.[3]

Progressive

Like all of Orange County, Irvine has steadily evolved from red to blue. The most recent voter registration figures show just under 64,000 Democrats, 34,300 Republicans, and 43,200 No Party Preference in the city. There are also 1,400 Libertarians, 808 Green Party members, and a surprising 5,340 listed as American Independent (the authors speculate that many in Irvine who signed up for this deeply conservative party did so believing they would be listed as an "independent" voter. It is emblematic of the confusion that Governor Newsom's wife once mistakenly registered as American Independent.)

In the most recent special election for a city council seat, a female Hispanic public relations executive was heavily supported by labor unions, besting a Republican former councilman. In recent years, the council has embarked on several environmental initiatives, helping to form an electricity cooperative with other cities to promote energy generated from non-petroleum sources. The venture had a rocky start, resulting in the removal of its executive director and threats by the city to bail out of what it had initiated, but the organization now appears stable.

Mayor Larry Agran, channeling his earnest liberalism from early in his political career, has recently suggested that Irvine take the lead in organizing other Orange County cities to clean up nuclear waste from the closed San Onofre nuclear plant 26 miles south of the city. While achieving such a goal would clash with federal law, not to mention the current federal administration, the fact that the city is moving forward with staff resources to even consider such a long-shot proposal underscores the nature of the political climate in Irvine.

Dynamic

The intellectual locus of Irvine is UC Irvine. Ranked number 32 among all universities and number 9 of public universities by *US News and World Report*, UCI has recently finished its fundraising campaign with $1 billion in cash or commitments to its endowment. It continues to attract funding from Orange County's tech and other high wealth families, most recently demonstrated by the multimillion-dollar donations from the Joseph Wen family providing funding for construction and operation of the new hospital complex on the border of Newport Beach and Irvine.

Business start-ups are thriving in the Irvine Spectrum and University Business Park. Blizzard Entertainment, Obsidian Entertainment, and Ready at Dawn top the list of video game companies located in the city. Rivian, the electric vehicle company that recently received a cash infusion from Volkswagen, is headquartered in Irvine. Irvine Spectrum Entertainment Center attracts millions of visitors annually and shows no sign of slowing down.

The city continues to pour funding into the Great Park, most recently celebrating the groundbreaking for the first major museum at the park. The city recently purchased a vacant industrial building for a new indoor recreation center. And in early 2025, Irvine divested itself from the Orange County Library System to form and operate its own libraries, with the stated goal of building a new, modern, and sizable central library, probably in the Great Park.

***

In 2026, plans for what may be the last village of any size or consequence will be presented to the city by the Irvine Company. As noted in the previous chapter, what is proposed for construction on the Oak Creek Golf Course has generated significant controversy.

We believe that for the city and the Irvine Company, the new village presents a stunning opportunity in the year of the 150th anniversary of the Irvine Ranch. Everything that has been learned and implemented about community planning in the city can be factored into the creation of this new village. Outstanding elements can be highlighted, past mistakes corrected, and innovations and experiments implemented with the attention to detail and design that Donald Bren has insisted on throughout his entire lifetime as a builder of master-planned communities.

With modesty and sincerity, we suggest that the new village commemorate the anniversary of the Irvine Ranch and be named Anniversary Village.

H Pike Oliver
Scottsdale, AZ

C. Michael Stockstill
Irvine, CA

Notes

1 Excerpt from *Chinese Americans in Irvine*, published by South Coast Chinese Cultural Center.
2 Trust for Public Land. "ParkScore® Rankings." Accessed October 2025. https://www.tpl.org/parkscore.
3 Quoted from Irvine Company promotional materials.

Bibliography

ABILITY Magazine. "Joan Irvine Smith on Christopher Reeve and Spinal Injury Research." 1996.

Agran, Larry. *The Cancer Connection*. New York: Anchor Books, 1977.

Agran, Larry. "Opinion Piece." *Los Angeles Times*, February 1994.

Agran, Larry. *Ballot Language for Measure W*. City of Irvine, 2002.

Aldrich, Daniel G. Interview by Samuel C. McCulloch. May 24, 1989. UCI Oral History Project.

American Institute of Architects. "2008 Institute Honor Awards for Regional and Urban Design." *Architect Magazine*, January 30, 2008. https://www.architectmagazine.com/awards/aia-names-2008-institute-honor-awards-winners_o.

American Planning Association. "APA National Award of Excellence: Orange County Great Park Comprehensive Master Plan." *Planning Magazine*, April 2009. https://www.ocgp.org/2008/12/apa-national-award-of-excellence/.

American Society of Golf Course Architects. "No One Likes to Be Turned Down—Tom Fazio." June 1, 2008. Accessed September 13, 2025. https://asgca.org/no-one-likes-to-be-turned-down-tom-fazio/.

American Society of Landscape Architects. "2008 Professional Award of Honor for Analysis and Planning: Orange County Great Park Comprehensive Master Plan." *ASLA Professional Awards*, April 28, 2008. https://cdn-v2.asla.org/awards/2008/08winners/127.html.

American Youth Soccer Organization (AYSO), Region 213 Archives. Irvine, CA.

Anaheim Bulletin. "Rancher Irvine Negotiates with Navy." April 18, 1942.

Anderson, Rich. Interview with H. Pike Oliver and C. Michael Stockstill. February 23, 2024.

Argyros, George. "Argyros Profile." *Chapman Magazine*, Spring 2003.

Associated Press. "Joan Irvine Smith Sues Art Agent for $24 Million." *San Francisco Examiner*, September 14, 1993.

Babbitt, Bruce. *Cities in the Wilderness: A New Vision of Land Use in America*. Washington, DC: Island Press, 2005.

Baldassare, Mark. *Orange County Annual Survey*. University of California, Irvine, 1982.

Baldassare, Mark. *Trouble in Paradise: The Suburban Transformation in America*. New York: Columbia University Press, 1986.

Baldassare, Mark. *When Government Fails: The Orange County Bankruptcy*. Berkeley: University of California Press, 1998.

Bartholomew, Dana. "Brookfield to Buy 70 Acres in Irvine for Housing Project." *Orange County Register*, April 2, 2024. https://therealdeal.com/la/2024/04/03/brookfield-to-buy-70-acres-in-irvine-for-housing-project/.

Berman, Laura. "Taubman on the Irvine Sale." *Detroit News*, April 17, 1983.

Birmingham, Stephen. *California Rich*. New York: Simon & Schuster, 1980.

Black, Shirley Temple. *Oral History Interview*. Stanford University Libraries, 2025. Accessed July 2025. https://oralhistory.stanford.edu.

Bozeman Daily Chronicle. "Flying D Ranch Sold to Ted Turner." July 1989.
Brazil, Ben. "Orange County Great Park Auditor Surrenders Accounting License After Legal Battle with State Oversight Agency." *Daily Pilot*, January 29, 2020.
Brennen, Bonnie. "Malice in Wonderland: Hedda Hopper and Louella Parsons in Hollywood." Image of Journalism in Popular Culture (IJPC), Temple University. Accessed September 8, 2025. https://www.ijpc.org/uploads/files/Hedda%20Hopper%20and%20Louella%20Parsons%20-%20Gossip%20Columnists.pdf.
Broadcasting & Cable. "Times-Mirror Cable Acquires Community Cablevision." December 1992.
Bronte, William. Quoted in "School Sites Rejected Due to KC-130 Flight Paths." *Los Angeles Times*, October 5, 1981.
Brower, Martin A. *The Irvine Ranch: A Place for People*. Bloomington, IN: AuthorHouse, 2013.
California Association of Realtors. "Historical Home Price Data." Accessed September 13, 2025. https://www.car.org/marketdata/data/.
California Debt and Investment Advisory Commission. "Community Facilities Districts: A Primer." Accessed July 2025. https://www.treasurer.ca.gov/cdiac/publications/cfds.pdf.
California Department of Fish and Wildlife. "Natural Community Conservation Planning (NCCP)." Accessed November 2025. https://wildlife.ca.gov/Conservation/Planning/NCCP.
California Department of Parks and Recreation. *Crystal Cove State Park Acquisition Summary*, 1979.
California Department of Parks and Recreation. *Upper Newport Bay Nature Preserve Dedication*, 1989.
California Department of Transportation. "Pacific Coast Highway." 2025. Accessed September 11, 2025. https://dot.ca.gov/caltrans-near-me/district-12/district-12-projects/pch.
California Secretary of State. "Irvine Company Political Contributions, 1978–1985." Campaign Finance Archive. Accessed July 2025. https://cal-access.sos.ca.gov.
California Secretary of State. "Irvine Company Political Contributions, 2023–2024." Campaign Finance Archive. Accessed September 12, 2025. https://cal-access.sos.ca.gov.
CalMatters. "Dark Money and Political Giving in California." October 2024.
Campbell, William. "Closing Arguments." *Irvine Company vs Smith*, June 1989.
Chamberlain, Mark. *The Laguna Canyon Project: Refining Activism*. Laguna Beach: Laguna Wilderness Press, 2018.
Chen, Yong. *Chinese Americans in Irvine*. Irvine, CA: South Coast Chinese Cultural Center, n.d.
City of Irvine Council Minutes. "Park Bond Election Results." November 1974.
City of Irvine Council Minutes. November 1975.
City of Irvine Council Minutes. "Charles R. Bren Park Dedication." 1978.
City of Irvine Council Minutes. "Venta Spur Park Development Hearings." 1980–1982.
City of Irvine Council Minutes. "Memorandum of Understanding Implementing Initiative Resolution 88–1." 1988.
City of Irvine Council Minutes. "Competitive Aquatics Center Dedication: William Woollett Jr." 1995.
City of Irvine Council Minutes. "Measure W Full Text." *City of Irvine*, March 2000.
City of Irvine Council Minutes. Council Minutes. "Formation of the Great Park Corporation." June 24, 2003.
City of Irvine Council Minutes. Council Agenda. "Contract Approval for Ford & Mollrich." April 2006.
City of Irvine Council Minutes. Council Minutes. "Western Sector Plan Approval." February 9, 2010.
City of Irvine Council Minutes. "David Sills Park Naming Resolution." 2012.
City of Irvine Council Minutes. "Memorandum of Understanding with Irvine Company for Housing Development." Adopted March 14, 2023.
City of Irvine Council Minutes. "City Council Meeting Agenda." May 25, 2024.

City of Irvine Council Minutes. "Jeffrey Open Space Trail Extension Plan." 2024.

City of Irvine Council Minutes. "Special Election Proposal for Spectrum District Village." Drafted July 2025.

City of Irvine Council Minutes. "Irvine Business Complex Vision Plan." Accessed September 9, 2025. https://www.cityofirvine.org/irvine-business-complex/ibc-vision-plan/.

City of Irvine Council Minutes. "City Council and Mayor." Accessed October 2025. https://www.cityofirvine.org/city-council.

City of Irvine Council Minutes. Police Department. "Don't Commit Crimes in Irvine." Instagram Video Posts. Accessed October 2025. https://www.instagram.com/irvinepolice.

City of Newport Beach. "Airport Noise and Curfew Regulations." Accessed September 11, 2025. https://www.newportbeachca.gov/government/departments/public-works/airport-issues.

City of Tustin. "Tustin Legacy Redevelopment Plan Approved." 2001.

City of Tustin. Press release. "Veterans Sports Park Dedicated." October 2022.

Clarke, Athalie Irvine. *Testimony, Irvine Company v. Smith*. Case No. 1983-IRV-149, Bloomfield Hills, MI, 1987.

Cleland, Robert Glass. *The Irvine Ranch: A Time and Place Remembered*. Los Angeles: The Huntington Library, 1962.

Collins, Jeff. "Mothballing of Lennar Home Projects to Continue." *Orange County Register*, January 29, 2009. https://www.ocregister.com/2009/01/29/mothballing-of-lennar-home-projects-to-continue/.

Committee of 4000. "Company Response." Box 1, Folder 6. Committee of 4000 Records, 1979–1983. MS.R.160. Special Collections and Archives, The UC Irvine Libraries, Irvine, CA.

Committee of 4000. "Corona del Mar High School Meeting Materials." Box 1, Folder 4. Committee of 4000 Records, 1979–1983. MS.R.160. Special Collections and Archives, The UC Irvine Libraries, Irvine, CA.

Committee of 4000. "Financing Proposal." Box 1, Folder 6. Committee of 4000 Records, 1979–1983. MS.R.160. Special Collections and Archives, The UC Irvine Libraries, Irvine, CA.

Committee of 4000. "Leaseholder Bill of Rights Event Materials." Box 1, Folder 5. Committee of 4000 Records, 1979–1983. MS.R.160. Special Collections and Archives, The UC Irvine Libraries, Irvine, CA.

Committee of 4000. "Legal Proceedings and Court Orders." Box 1, Folder 7. Committee of 4000 Records, 1979–1983. MS.R.160. Special Collections and Archives, The UC Irvine Libraries, Irvine, CA.

Committee of 4000. "Nielsen Letter and Settlement Proposal." Box 1, Folder 8. Committee of 4000 Records, 1979–1983. MS.R.160. Special Collections and Archives, The UC Irvine Libraries, Irvine, CA.

Committee of 4000. "An Open Letter to the Irvine Company." Promotional advertisement, circa 1982. Box 1, Folder 3. Committee of 4000 Records, 1979–1983. MS.R.160. Special Collections and Archives, The UC Irvine Libraries, Irvine, CA.

Conway, Ann. "Oaks Classic Draws Elite Crowd." *Los Angeles Times*, June 6, 1986.

CRE Analyst. "Donald Bren: How America's Richest Real Estate Mogul Built His Empire." Accessed September 8, 2025. https://www.creanalyst.com/insights/donald-bren-how-americas-richest-real-estate-mogul-built-his-empire.

Daily Pilot. "Helicopter Crash Injures Two." September 1, 1969.

Daily Pilot. "Newport Coast Annexation Agreement Finalized." January 29, 2000.

Daily Pilot. "Jean Watt, Newport Environmental Icon, Dies at 95." December 2023.

Davis, Joe. Interview by H. Pike Oliver and C. Michael Stockstill, April 10, 2025.

Dennis, Bill. *California Dreaming: Family Foundations and a Marriage Lived Well*. Self-Published, 2019.

Detroit Free Press. "Taubman: A Forceful Leader." August 7, 1983.

Detroit News. "Taubman Celebrates Sale with Fisher and Ford." April 17, 1983.

Dollar Times. "Inflation Calculator." Accessed September 8, 2025. https://www.dollartimes.com/inflation/.

Eberling, Lanny. Interview by C. Michael Stockstill and H. Pike Oliver, March 2021.

Edison Miller v. Bruce Nestande. Orange County Superior Court Case No. 81-CV-1123. Affirmed on appeal, 1983.

Editor & Publisher. "Irvine World News Sold to *OC Register*." March 2000.

El Toro Reuse Planning Authority (ETRPA). "Millennium Plan Proposed." *ETRPA Archives*, 1999.

Elliot, Rob. Quoted in Irvine Company Promotional Materials. Irvine Company, n.d.

Ellis, David. Interview with C. Michael Stockstill, February 2024.

Ellzey, Michael. Deposition in HSNO Forensic Contract Performance Review. March 2015.

Environmental Protection Agency. *Environmental Baseline Survey for Former MCAS El Toro*. Prepared by Southwest Division, Naval Facilities Engineering Command. U.S. Navy, 2003.

"ETRPA Shifts Strategy." *Los Angeles Times* March 18, 2001. Quoting Tristan Krogius.

Federal Election Commission. "ICEPAC Formation Documents." 1980–1986.

Federal Reserve Board. "Historical Interest Rates." Accessed June 1990.

Ferguson, William. Interview by C. Michael Stockstill and H. Pike Oliver. February 12, 2024.

Five Point Holdings, LLC. "Annual Report (Form 10-K)." Filed December 2024.

Fix, Warren. Interview by H. Pike Oliver and C. Michael Stockstill. June 7, 2023.

Flagg, Michael. "Smith's Win Wasn't Big; Neither Was Bren's Loss." *Los Angeles Times*, June 27, 1990.

Florian, Monica. Interview by H. Pike Oliver and C. Michael Stockstill. May 28, 2024.

Freeman, Paul. Interview by C. Michael Stockstill. April 17, 2024.

Friedman, Howard. Closing Arguments, *Irvine Company v. Smith*, June 1989.

Friends of the Irvine Coast. "Settlement Agreement between Friends of the Irvine Coast and the Irvine Company." September 29, 1987. Friends of the Irvine Coast Papers, Special Collections, UC Irvine Libraries, Langson Library.

General Services Administration (GSA). *Post-Implementation Report on Base Realignment and Closure (BRAC)*, 2021. https://www.governmentattic.org/52docs/GSAbracPostImplementRpt2021.pdf

General Services Administration (GSA). "GSA Land Disposal Guidelines." https://www.gsa.gov/real-estate/real-estate-services/real-property-disposition/the-disposal-process.

Gfeller, Doug. Interview by H. Pike Oliver and C. Michael Stockstill. August 19, 2020.

Ghirado, Diane. "Design Jury Summary Report." *City of Irvine*, 2006.

Glen Haas, Jane. "Inside the Irvine Company Trial." *Orange County Register*, November 1987.

Hayes, Thomas C. "Developer Acquiring Control of Irvine Co." New York Times, April 16, 1983, Section 1, p. 29.

Hector, Gary. "Donald Bren: The Invisible Empire Builder." *Fortune*, April 27, 1990.

Hector, Gary. "America's Richest Land Baron." *Fortune*, August 27, 1990.

Hellis, Brad. *Oral History*. DeGraff Center for Oral and Public History at California State University, Fullerton, 1985.

HSNO. *Orange County Great Park Forensic Contract Performance Review*. March 2015.

Irvine Company. Internal Planning Documents and Maps, circa. 1970s–1980s.

Irvine Company. *Annual Report*. Newport Beach, CA: Irvine Company, 1976.

Irvine Company. *Legal Affairs Bulletin: "Gibson Dunn to Latham: Legal Counsel Transition."* August 1980.

Irvine Company. "Imperial Valley Holdings Overview." Internal Memo, October 1982.

Irvine Company. "Asset Inventory and Valuation Database." Compiled by Gary Weber, 1983–1986.

Irvine Company. "Internal Memoranda and Trial Graphics." 1983–1989.
Irvine Company. "Planning Department Reorganization Memo." Issued by Donald Bren, 1984.
Irvine Company. "Roots and Wings Exhibit." 1985 (No longer in existence).
Irvine Company. Executive Committee Meeting Minutes, 1986–1990. Ray Watson Papers. Langson Library, University of California, Irvine.
Irvine Company. "Gateway Preserve Land Gift Announcement." February 2023.
Irvine Company. "Spectrum District Village Proposal Briefing Materials." May 2024.
Irvine Ranch Water District. *San Joaquin Marsh Restoration Overview*. Internal Report, 1990.
Irvine Unified School District and City of Irvine. "Joint Use Parks Agreement." 1975.
John Wayne Airport. "Orange County Airport History." Accessed September 11, 2025. https://www.ocair.com/about/history/.
Jones Day. "Lennar Settles 10-Week Jury Trial After $200 Million Damage Claim Excluded." *Experience Highlights*, December 2010. https://www.jonesday.com/en/practices/experience/2010/12/lennar-settles-10-week-jury-trial-after-$200-million-damage-claim-excluded.
Jung, Carl. 1933. *Modern Man in Search of a Soul*. Translated by W.S. Dell and Cary F. Baynes. New York: Harcourt.
JZMK Partners. "Woodbury Masterplan." Accessed September 9, 2025. https://jzmkpartners.com/project/woodbury/.
Kenney, Bill. Interview by C. Michael Stockstill, 2025.
Kraft, Scott. "The Bren Legacy." *Los Angeles Times*, February 13, 2011, Business Section. Accessed September 8, 2025. https://www.donaldbren.com/2011/donald-brens-legacy.
Kranser, Len. Quoted in *Los Angeles Times*, "ETRPA Shifts Strategy." March 18, 2001.
Kremer, Peter. *Isn't That Nice? Reflections of a Blessed Life*. Self-Published, 2013.
Latham & Watkins. Internal memorandum, March 7, 1980. Alfred Taubman Papers, Bentley Historical Library, University of Michigan.
LeBlanc, Michael. "Public Comments at Irvine City Council Meeting." May 25, 2024.
Little League International. "Northwood Team Wins World Series." 1987.
Long Beach Independent. "Report on Proposed $5 Million Crystal Cove Hotel." June 26, 1956.
Loper, Mary Lou. "Oaks Classic Expands with $70,000 Purse." *Los Angeles Times*, June 7, 1987.
Los Angeles Citizen-News. "Claire Trevor on Child Rearing." October 29, 1949.
Los Angeles Times. "Country Club Estates Planned on Irvine Ranch." July 1927.
Los Angeles Times. "Profile of Milton Bren." March 5, 1939.
Los Angeles Times. "Irvine Company Plans Land-Lease Development." November 13, 1960.
Los Angeles Times. "Campus Drive Bridge Approved." December 21, 1972.
Los Angeles Times. "Virginia Slims Tournament Draws Stars to Mission Viejo." April 3, 1976.
Los Angeles Times. "Mission Viejo Mall Proposal." October 12, 1976.
Los Angeles Times. "Jet Noise and Development Restrictions." December 8, 1976.
Los Angeles Times. "TABI Partnership Wins Bid for Irvine Company." May 21, 1977.
Los Angeles Times. "Irvine Councilwoman Calls for Lawsuit over Jet Noise." June 9, 1979.
Los Angeles Times. "State Buys Irvine Coast." December 12, 1979.
Los Angeles Times. "Nestande Wins Bitter Supervisor Race." November 6, 1980.
Los Angeles Times. "El Toro Base Operations Continue." February 12, 1981.
Los Angeles Times. "Newport Center Development Limits Cut." March 19, 1981.
Los Angeles Times. "John Wayne Airport." February 5, 1982.
Los Angeles Times. "Leaseholders Bury Hatchet in Newport Celebration." April 11, 1983.
Los Angeles Times. "Joan Irvine Smith Threatens Lawsuit Over Bren Deal." April 21, 1983.
Los Angeles Times. "Bren Buys Out Irvine Company Partners." May 20, 1983.

Los Angeles Times. "Leaseholders File Suit Against Irvine Co." February 1983.
Los Angeles Times. "Taubman Defends Lease Terms Amid Newport Uproar." July 12, 1983.
Los Angeles Times. "Peter Kremer Sues Over Share Valuation." September 22, 1983.
Los Angeles Times. "Irvine Company Board Approves Bren Buyout." October 15, 1983.
Los Angeles Times. "Smith Sues to Block Irvine Company Merger." October 19, 1983.
Los Angeles Times. "Irvine Company Ends Cattle Operations." June 25, 1985.
Los Angeles Times. "Irvine Co., UCI Plan Research Park." October 22, 1985.
Los Angeles Times. "Joan Irvine Smith Finds New Passion in Equestrian Center." January 12, 1986.
Los Angeles Times. "Irvine Co. Restructures Management." June 11, 1986.
Los Angeles Times. "Catalano Appointed to Irvine Council." June 23, 1986.
Los Angeles Times. "Irvine Co. Lays Off 100 in Restructuring Move." July 9, 1986.
Los Angeles Times. "Irvine Company Files $300 Million Expansion Plan." August 17, 1986.
Los Angeles Times. "Newport Center Referendum Defeated." October 1986.
Los Angeles Times. "Fashion Island Remodel Cost $60 Million." November 23, 1986.
Los Angeles Times. "Kremer Settlement Revealed." May 6, 1987.
Los Angeles Times. "Testimony by Ray Watson on Coastal Development Profit Estimates." August 12, 1987.
Los Angeles Times. "Taubman Testifies in Irvine Valuation Trial." October 8, 1987.
Los Angeles Times. "Leisure World Developer Pays Marines $8 Million." January 27, 1988.
Los Angeles Times. "Joan Irvine Smith: Horsewoman and Businesswoman." February 13, 1988.
Los Angeles Times. "Laguna Canyon Conservancy Offers $10 Million for Land." April 26, 1988.
Los Angeles Times. "Oaks Classic Weekend Blends Sport and Society." June 3, 1988.
Los Angeles Times. "Open Space Initiative Passes." June 8, 1988.
Los Angeles Times. "Irvine Mayor's Race Turns Bitter." November 6, 1988.
Los Angeles Times. "Architect Ernie Vasquez Builds Corte Bella." February 5, 1989.
Los Angeles Times. "Thousands March to Save Laguna Canyon." November 12, 1989.
Los Angeles Times. "Joan Irvine Smith Hosts Gala at the Oaks." January 22, 1990.
Los Angeles Times. "County Buys Land from Marines for Road Projects." May 8, 1990.
Los Angeles Times. "Native American Burial Sites on Newport Coast." September 23, 1990.
Los Angeles Times. "Laguna Canyon Deal Reached." October 9, 1990.
Los Angeles Times. "Bren Begins Talks with Nature Conservancy." October 12, 1990.
Los Angeles Times. "Marketplace Repainting Ceremony Led by Councilwoman." April 7, 1991.
Los Angeles Times. "Bramelea Homes Debut on Newport Coast." November 26, 1991.
Los Angeles Times. "Toxic Spill Concerns Halt Apartment Project." January 11, 1992.
Los Angeles Times. "Joan Irvine Smith's Art Spree Preserves California's Scenic Past." January 15, 1993.
Los Angeles Times. "Joan Irvine Smith Plans New Riding Park." March 6, 1995.
Los Angeles Times. "Spectrum Entertainment Center Opens." November 22, 1995.
Los Angeles Times. "NCCP Agreement Signed in Shady Canyon." July 18, 1996.
Los Angeles Times. "Measure S Defeated." November 6, 1996.
Los Angeles Times. "Trainer's Lawsuit Reveals Rift with Joan Irvine Smith." February 2, 1997.
Los Angeles Times. "Irvine Spectrum Emerges as Tech Hub." May 18, 1997.
Los Angeles Times. "Giannulli Lists Newport Coast Lots." February 24, 1998.
Los Angeles Times. "El Toro Closure Timeline." July 3, 1999.
Los Angeles Times. "Ferris Wheel and Ice Rink Additions." December 10, 1999.
Los Angeles Times. "The Spiritual Keeper of the Flame for the Company's Planners." December 30, 1999.
Los Angeles Times. "ETRPA Shifts Strategy." March 18, 2001.
Los Angeles Times. "Measure F Ruled Unconstitutional." August 12, 2001.

Los Angeles Times. "Eighty Percent of Orange County's Jews Within Fifteen to Twenty Minutes of Campus." January 26, 2003.
Los Angeles Times. "Pelican Hill Resort Opens with Palladian Design." June 21, 2003.
Los Angeles Times. "Lennar Analyzes El Toro Site." February 17, 2005.
Los Angeles Times. "Crystal Cove Trailer Residents Evicted." February 27, 2006.
Los Angeles Times. "Balloon Rises Over Great Park." October 2007.
Los Angeles Times. "Santa Ana Unified Settles Base Dispute." August 22, 2008.
Los Angeles Times. "Centex Withdraws from Tustin Legacy Partnership." October 12, 2009.
Los Angeles Times. "Irvine Council Removes Great Park Board Members." January 9, 2013.
Los Angeles Times. "Audit of Great Park Spending Approved." January 28, 2013.
Los Angeles Times. "Great Park Audit Faults Oversight." January 28, 2014.
Los Angeles Times. "Marian Bergeson, Former State Senator and Education Leader, Dies at 90." August 17, 2016.
Los Angeles Times. "Banning Ranch Saved by $50 Million Gift." March 2022.
Los Angeles Times. "Historic Tustin Hangar Destroyed by Fire." November 8, 2023.
Los Angeles Times. "Irvine to Buy Asphalt Plant for $284 Million, Shut It Down." February 28, 2023.
Marshall, Roberta. Interview by H. Pike Oliver and C. Michael Stockstill, April 14, 2024.
Martin, John. *Inspiration From the Mediterranean—Doubling Land Values with an Attached Motor Court Project: The Corte Bella Story*. Martin and Associates LLC, 2020.
McDermott, Carol. Interview by C. Michael Stockstill and H. Pike Oliver. March 20, 2024.
McFarland, William. Interview by H. Pike Oliver and C. Michael Stockstill, January 8, 2024.
Melching, Jeff. Legal Opinion on 1988 Open Space Initiative. May 2024.
Mello, Henry, and Michael Roos. *Mello-Roos Act Legislative History*. California State Assembly Archives, 1982.
Michigan Court System. *Irvine Company v. Smith*, Case No. 1983-IRV-149, Bloomfield Hills, MI, 1983–1990.
Mudgett, David. Interview by H. Pike Oliver and C. Michael Stockstill, February 3, 2025.
National Trust for Historic Preservation. "Preservation Challenges for Tustin Hangar." Accessed September 13, 2025. https://savingplaces.org/stories/tustin-hangar-preservation.
New Worlds Magazine. "Industrial Neighborhoods and Walkable Design." May 1976.
New York Times. "Pete Wilson Wins Senate Seat." November 3, 1982.
New York Times. "Seattle Mariners Sold." August 1, 1989.
New York Times. "Taubman Convicted in Price-Fixing Case." April 23, 2002.
Newport Beach Independent. "Allan Beek, Ferry Heir and Growth Foe, Dies at 96." February 2024.
Novaquatics. *Heritage Park Aquatics Center History*. Internal Publication, 2000.
OC Metro Magazine. "Donald Bren Interview." March 6, 2003.
OC Metropolitan. "Ray Watson Reflects on Coastal Planning." December 1, 1992.
OCThen.com. "Top Gun Filming Locations." Accessed September 13, 2025. http://www.octhen.com/2006/11/top-gun-filming-locations.htm.
OC Weekly. "Village of Hope Nears Capacity." March 10, 2010.
OC Weekly. "The Wizard of Agran." November 2010.
Olson, Brad. Interview by H. Pike Oliver and C. Michael Stockstill, March 25, 2024.
Orange Coast Magazine. "Donald Bren's Alhambra Inspiration." April 16, 1998.
Orange County Business Journal. "Irvine Ranch Water District Authorizes $1 Billion in Bonds." May 1978.
Orange County Business Journal. "Spectrum Infrastructure Investment Tops $227 Million." March 12, 1996.
Orange County Business Journal. "Retail Struggles After Financial Crisis." February 2, 2009.
Orange County Business Journal. "FLIGHT Office Campus Opens." October 1, 2019.
Orange County. "Orange County Central-Coastal NCCP/HCP White Paper." Undated.

Orange County Parks Commission Archives. Reference to Florence Koch, 1920.
Orange County Parks Department. "William Mason Regional Park Dedication." 1973.
Orange County Register. "Donald Bren Profile." April 12, 1987.
Orange County Register. "Measure a Passes." November 9, 1994.
Orange County Register. "Irvine Company Builds 20-Story Tower." 1995.
Orange County Register. "Measure F Approved." March 7, 2000.
Orange County Register. "Gehry Building Demolition Criticized." February 11, 2007.
Orange County Register. "The District Opens in Tustin." August 15, 2007.
Orange County Register. "Haddad Announces Revised Funding." January 2011.
Orange County Register. "Council Approves $172M Great Park Deal." November 27, 2013.
Orange County Register. "Forde and Mollrich Fired from Great Park Project." January 15, 2013.
Orange County Register. "Standard Pacific Buys 78 Acres in Tustin." May 14, 2014.
Orange County Register. "Agran Audit Campaign Targets Council Race." October 15, 2014.
Orange County Register. "Celebrity Residents of Newport Coast." June 18, 2020.
Orange County Register. "Spectrum Draws 17 Million Visitors Annually." July 7, 2023.
Orange County Register. "Navy to Spend $100 Million on Hangar Cleanup." November 15, 2023.
Orange County Register. "Lawsuit Filed over Orange Hills Development." December 2024.
Orange County Registrar of Voters. "Measure A Election Results." November 1984.
Orange County Registrar of Voters. "Voter Registration Statistics by City." Accessed October 2025. https://www.ocvote.gov/data/voter-statistics.
Orange County Superior Court. Smith v. Irvine Company, Case No. 83-OC-1427. Ruling issued November 1983.
Orange Juice Blog. "The Yellow VW Deathtrap." August 2025.
Peltason, Jack W. Interview by Ann Lage. Oral History Office, University of California, Berkeley, 2001.
Perlman, Jeffrey. "Slow-Growth Initiative Stirs County Debate." *Los Angeles Times*, June 21, 1987. Petersen, Larry. "Joan Irvine Smith: Heiress in an Armed Camp." *Orange Coast Magazine*, May 1984.
Pinsky, Mark I. "Irvine Hospital Dispute Ends in Compromise." *Los Angeles Times*, October 13, 1985.
"Proposition 13: How It Changed California." Sacramento Bee. June 6, 2018.
RBF Consulting. *The Tustin Hangars: Titans of History*. Irvine, CA, 2008
Reese, Richard. Interview by C. Michael Stockstill and H. Pike Oliver. August 8, 2020.
Rivian. "Volkswagen Group and Rivian Announce Joint Venture." Press release, June 25, 2025. https://rivian.com/newsroom.
Roan, Sharon, and John Westcott. *UCI: Bright Past, Brilliant Future*. Regents of the University of California, 2015.
Roulac, Stephen. *Strategic Real Estate: The New Economy of Property*. New York: McGraw-Hill, 1988.
Sanders, Greg. Interview by H. Pike Oliver and C. Michael Stockstill, 2024.
Santa Ana Register. "Santa Ana Chamber Confirms Navy Lease." March 20, 1942.
Santana, Norberto. Interview with C. Michael Stockstill, March 10, 2025.
Schatz, Thomas. *The Genius of the System: Hollywood Filmmaking in the Studio Era*. New York: Henry Holt, 1988.
Schout, Robert. Interview by C. Michael Stockstill and H. Pike Oliver, December 12, 2022.
Schout, Robert. Interview by C. Michael Stockstill and H. Pike Oliver, March 2023.
Schumacher, E. F. 1973. *Small Is Beautiful: Economics as if People Mattered*. New York: Harper & Row.
Segerstrom, Henry. "Henry Segerstrom Biography." *HenrySegerstrom.com*. Accessed September 12, 2025. https://www.henrysegerstrom.com.

Shea, Christina. "Public Statement at Irvine City Council Meeting." *Save Irvine Open Space*, June 14, 2025. https://saveirvineopenspace.com/public-comments-june-14-2025.
Shute, Mihaly, and Weinberger. Legal Representation for Friends of the Irvine Coast, referenced in Settlement Agreement, 1987.
Sim, Richard. Interview by H. Pike Oliver and C. Michael Stockstill, January 18, 2024.
Sim, Richard. Interview by H. Pike Oliver and C. Michael Stockstill, February 29, 2024.
Smith, Joan Irvine. *A California Woman's Story*. Irvine, CA: The Irvine Museum, 2006.
Solar Decathlon Comes to Irvine. 2025. "U.S. Department of Energy." Accessed September 13, 2025. https://www.energy.gov/eere/solar/solar-decathlon.
Southwest Division, Naval Facilities Engineering Command. *Environmental Baseline Survey for Former MCAS El Toro*. U.S. Navy, 2003.
Stars and Stripes. "Edison Miller Dies at 84." March 14, 2022.
State of California, Department of Finance. "Redevelopment Agency Dissolution." 2011. https://dof.ca.gov/Programs/Redevelopment/.
Stern, Robert A.M. Quoted in *Orange County Register*, February 11, 2007.
Sukut Construction. "Project Summaries (Irvine Coast)." Accessed September 13, 2025. https://www.sukut.com/completed-projects/.
SWA Group. "Jeffrey Open Space Park." Accessed January 14, 2026. https://www.swagroup.com/projects/jeffrey-open-space-park/.
SWA Group. "Shady Canyon Narrative." Accessed September 9, 2025. https://www.swagroup.com/projects/shady-canyon/.
Taylor, John. 1987. *Storming the Magic Kingdom: Wall Street, the Raiders, and the Battle for Disney*. New York: Knopf.
Taubman, Alfred. Letter to Irvine Company Ownership Group, September 12, 1980. Alfred Taubman Papers, Bentley Historical Library, University of Michigan.
Transportation Corridor Agencies. "Foothill/Eastern Transportation Corridor." Accessed September 11, 2025. https://thetollroads.com/about/projects/foothill-eastern/.
Transportation Corridor Agencies. "San Joaquin Hills Transportation Corridor." Accessed September 11, 2025. https://thetollroads.com/about/projects/san-joaquin-hills/.
Trust for Public Land. "ParkScore® Rankings." Accessed October 2025. https://www.tpl.org/parkscore.
Tustin Unified School District News. "Legacy Magnet Academy Opens." August 2021.
United Press International. "Peter Kremer, President of the Irvine Co., Announces Resignation." UPI Archives, October 12, 1982.
United Press International. "Irvine Co. and Its Directors Have Been Sued for $5 Billion by Dissident Minority Shareholder Joan Irvine Smith." October 19, 1983.
University of California, Irvine. Daniel Aldrich Biography. Chancellor's Office Archives.
University of California, Irvine. Langson Library, Special Collections and Archives. Friends of the Irvine Coast Papers.
University of California, Irvine. Langson Library, Special Collections and Archives. Raymond L. Watson Papers, MS-R001.
University of California, Irvine. San Joaquin Marsh Reserve Acquisition Summary. 1969.
University of California, Irvine. "University Research Park Lease Agreement Announcement." July 1998. In Office of the Chancellor Records: Ralph J. Cicerone (1998–2005). Special Collections and Archives, UC Irvine Libraries. https://oac.cdlib.org/findaid/ark:/13030/c828069d/.
University of California, Irvine. UCI Long Range Development Plan. 2007.
University of California, Irvine. "Brilliant Future Campaign Surpasses $1 Billion Goal." *News Release*, September 2025. https://news.uci.edu.
University of California, Irvine. Press Release. "UCI Acquires Orange County Museum of Art." October 2025.
University of California, Irvine. "Thesaurus Linguae Graecae." Accessed September 9, 2025. www.stephanus.tlg.uci.edu.

University of California, Irvine. "University Hills." *Irvine Campus Housing Authority*. Accessed September 9, 2025. www.icha.uci.edu.

University of California, Irvine Foundation. "Establishment of the National Water Research Institute and Clarke Prize." 1993.

University of Michigan Record. "Generous U-M Supporter A. Alfred Taubman Dies at 91." April 18, 2015. https://record.umich.edu/articles/generous-u-m-supporter-alfred-taubman-dies-age-91.

U.S. Census Bureau. "Housing Density Trends in Southern California." 2023.

U.S. Congress. House. Committee on Ways and Means. *Subject of Tax Reform*, 473–504. Tax Reform Act of 1969. Pub. L. No. 91-172, 83 Stat. 487.

U.S. Department of Defense. "Congressional Medal of Honor Citation: William Barber." Korean War Records.

U.S. News & World Report. "University of California, Irvine Rankings." Accessed October 2025. https://www.usnews.com/best-colleges/uci-1314.

Vardoulis, Basil. Interview by C. Michael Stockstill and H. Pike Oliver. March 2021.

Variety. 2002. "Regal Acquires Edwards Theaters." August 15, 2002.

Voice of OC. "Irvine's Open Space Controversy Reignites." June 2025.

Wall Street Journal. "AirCal Sold to American Airlines." July 15, 1986.

Wall Street Journal. "Full-Page Ad: Irvine Company's Legacy." February 2020.

Washington Post. "Taubman Buys Sotheby's." December 4, 2001.

Watson, Raymond L. Memo on Leasehold Policy. Raymond L. Watson Papers, MS-R001. Special Collections and Archives, University of California, Irvine Libraries.

Watson, Raymond L. Oral history interview by Ann Lage. *Planning and Developing the New Town of Irvine, California, 1960–2003*. Berkeley: Bancroft Library, University of California, 2005. Accessed July 2025. https://digicoll.lib.berkeley.edu/nanna/record/218181/files/watson_ray.pdf.

Watt, Terry. Interview by H. Pike Oliver and C. Michael Stockstill, March 26, 2024.

Weber, Gary. Interview by authors, March 3, 2025.

Wikipedia. "Pete Wilson." Accessed September 6, 2025. https://en.wikipedia.org/wiki/Pete_Wilson.

Wikipedia. "Milton H. Bren." Accessed September 8, 2025. https://en.wikipedia.org/wiki/Milton_H._Bren.

Wilson, Pete. *Oral History Interview by Lisa Rubens*. California State Archives State Government Oral History Program, 2001. Accessed July 2025. https://archives.cdn.ca.gov/oralhistory/wilson_pete.pdf.

Witt, Norm. Interview by H. Pike Oliver and C. Michael Stockstill, January 16, 2025.

Woodbridge Community Association. "Mike Ward Memorial Statement." 2013.

Worden, Jon. Interview by C. Michael Stockstill, 2024.

Yankelovich, Daniel. *New Rules: Searching for Self-Fulfillment in a World Turned Upside Down*. New York: Random House, 1981.

In memoriam

Hundreds of individuals played a role in transforming the Irvine Ranch. Many shared their memories with us as we researched and wrote this book and the one that preceded it. We remember here those who have left us since publication of the first book.

John Burton, Member, Irvine City Council
David Booher, Irvine Company state government advocate
Lanny Eberling, Irvine Company executive
William Ferguson, Irvine Company federal government advocate
Mary Ann Gaido, Member, Irvine City Council
Jeanne Keevil, *Irvine World News* editor
Peter Kremer, Irvine Company president
Wesley Marx, Irvine planning commissioner
Robert Nelson, Public affairs consultant
Jerry Patterson, Member of Congress
Gabrielle Pryor, Member, Irvine City Council
Richard Reese, Irvine Company executive
Al Treviño, Irvine Company executive
Basil "Bill" Vardoulis, Member, Irvine City Council

Index

For Product Safety Concerns and Information please contact our
EU representative GPSR@taylorandfrancis.com Taylor & Francis
Verlag GmbH, Kaufingerstraße 24, 80331 München, Germany